Lecture Notes in Economics and Mathematical Systems 589

Michael Schwind

Dynamic Pricing and Automated Resource Allocation for Complex Information Services

Reinforcement Learning
and Combinatorial Auctions

With 83 Figures and 53 Tables

 Springer

Dr. Michael Schwind
Faculty of Economics and Business Administration
Johann Wolfgang Goethe University
Campus Bockenheim
Mertonstraße 17
60325 Frankfurt am Main
Germany
mail@michael-schwind.de

Library of Congress Control Number: 2007920183

ISSN 0075-8442

ISBN 978-3-540-68002-4 Springer Berlin Heidelberg New York

Springer is part of Springer Science+Business Media

springer.com

© Springer-Verlag Berlin Heidelberg 2007

Production: LE-TeX Jelonek, Schmidt & Vöckler GbR, Leipzig
Cover-design: WMX Design GmbH, Heidelberg

SPIN 11941217   88/3100YL - 5 4 3 2 1 0   Printed on acid-free paper

Preface

Writing a PhD thesis is like climbing a high mountain in unknown terrain. There are many ups and downs on the way to the summit, and sometimes you think you will never reach it, especially if the summit is temporarily out of sight because you have somehow lost the orientation in the vast amount of academic literature and related problems. The best method to reach the summit was not to look at it permanently, but only to watch your steps on the way to the next peak, like a conference paper or a journal article. In the end, if you have to put things together and to prove that your initial idea did not lead into no man's land, but did help you to discover some new route in the "terra icognita" of the scientific landscape, the air will be very thin. The last meters are in particular so exhausting that you hardly can enjoy the view from above when you have finally reached the summit.

I would like to thank all the people that helped me to reach the summit. First of all I have to thank my academic adviser Prof. Dr. Wolfgang König at the Institute for Information Systems (IWI) at Johann Wolfgang Goethe-University for giving me the opportunity to venture on my academic adventure. He always provided me with institutional support that was necessary on my long way up and encouraged me to give my best. The second important person on my journey was Prof. Dr. Oliver Wendt who not only showed me how to make the first expeditions into academic terrain, but also was a very good friend in the IWI base camp, where he never stopped teaching me how complex the scientific landscape could be. The entire expedition would have been impossible without my fellow mountaineer at the other end of the rope, Dr. Tim Stockheim who had the necessary courage and motivation to keep us on track, when I struggled during the climb.

I am also very glad for the good times I shared with the other members of the IWI base camp, Prof. Dr. Tim Weitzel, Dr. Roman Beck, Dr. Norman Hoppen, Dr. Sven Grolik, Dr. Rainer Fladung, Tobias Keim, Jochen Franke, and Oleg Gujo. They were not only good fellows in having fun, but always supported me with a helping mind when necessary, like proof reading the final version of this work. I hope that we all can continue our friendship and research partnership in the future.

I am also grateful to the German National Science Foundation and the German Federal Ministry of Education and Research for providing the resources of my scientific projects 'Pricing of Distributed Information Services' and 'Dynamic Bundle Pricing'.

I would also like to thank my mother Eva-Maria who has always backed my efforts and given me the motivation to try the PhD adventure by showing me the fascinating face of nature and science when I was a little boy.

Finally, I proudly can say that the most important person in the last three years of my expedition and in my life, has been Katrin who has given me the confidence and strength to reach the summit. Hopefully this rope will never be torn apart.

Michael Schwind   Frankfurt, January 2007

Foreword Prof. Dr. König

Information services are often used to distribute valuable information products to customers via modern information and communication systems, e.g. the departure time of a train given on the Internet. The provision of such services requires specific resources that are normally only available with a limited capacity, like e.g. the server capacity for communication processes that is necessary to bring the time-table information to the web users.

In many cases these capacities are offered by different institutions at zero marginal cost, e.g. the German railway corporation provides the server capacity for free and the customers naturally have Internet access via their personal computers, such that a single request does not generate additional costs.

The developed economies that are undergoing a massive change from industrial production to the service society are especially confronted with an increasing 'digitalization' of production. This means that, for example, an increased coordination effort - realized by more intensive data exchange measured in 'megabits per second' - leads to a reduction of logistics expense measured in 'tons kilometers'. This is one of the reasons that a further substantial increase in the production and distribution of digital products and services has to be expected in the future and with this the questions arises:

If, as shown in the previous example, the consumption of simple information products is not necessarily linked to an increase in expenses for the users for the services, why should service providers then further offer more complex information services for free at a satisfying quality level? Will service providers for these reasons offer more differentiated services above a basic level at all?

Indeed, an increasing number of 'complex information products' is necessary to further develop a modern economy. These complex products consist of elements of other information products and so forth. Let me explain this interdependence and nested structure by using the example of a video conference: The conference users need the availability of n video devices at different places jointly with communication capacity between these devices which require more than one phone line. All these resources are required to be available at the same time. If they are not, the conference can not take place and the value of the disjoint resources that are reserved in advance will be zero. This type of interdependence is called complementarity. The traditional way to handle such a resource allocation problem from an economic viewpoint is to use m disjoint auctions ($m \leq n$) for each resource type. However, this does not resolve the complementarity problem: You might receive e.g. three of six required resources for an acceptable price, the other three resources, however, might be very expensive or even unavailable, and therefore devalue the user utility of the entire resource package in the video conference case. On way to handle this problem is to use combinatorial auctions, that are able to deal with this problem. And the cost of performing the combinatorial auction may be included in the calculation of the 'business case'. Another way is to have a mediator learn the yield optimal allocation of the task bundles to the system resources. Resource allocation tasks that include the complementarity problem are fairly widespread - the practical relevance of a solution for this kind of problems is very high.

Michael Schwind decides to solve this demanding problem with the objective of achieving the efficient allocation and dynamic pricing of the resources for complex information services in the domain of distributed systems. For this purpose he uses the method of agent-based computational economics, a young discipline that promises to provide insights into economic processes by formulating the behavior of interacting subjects as software agents in a simulation environment and by analyzing the evolving complex system properties in various application scenarios that should be close to real world situations.

Altogether, the work of Michael Schwind not only presents, a broad spectrum of simulation methods, algorithms, and experiments especially in the domain of agent-based computational economics, but also provides unique and groundbreaking insights by applying and advancing recent methods of artificial intelligence, like neural networks or reinforcement learning, and nature oriented optimization methods, like genetic algorithms and simulated annealing, to the problem of combinatorial resource allocation under complementarities.

Wolfgang König   Frankfurt, January 2007

Foreword Prof. Dr. Wendt

Since the beginning of economic research, the efficient allocation of resources to production and consumption processes has always been the central focus of the discipline. During the last century modern micro-economic theory has made tremendous progress along this path by proving that Pareto-efficient allocations can be achieved by a market mechanism, adapting a global price vector for all goods until an equilibrium is reached and no consumer or producer wants to buy or sell any quantity of any goods. Furthermore, the proof that this equilibrium can even be reached by a fully decentralized "peer to peer" mechanism not requiring any central control (like an auctioneer or commodity exchange) seemed to provide the theoretical underpinning for the market economies' superiority compared to the eastern economies' obsolete concept of a benevolent global planner.

However, most of the theory relies on strong assumptions: The preferences of producers and consumers must be representable by concave utility functions (prohibiting any complementarities) and all goods must be divisible. We do not even have to resort to the "digital economy" of information production to realize how far fetched these assumptions are: A dish can only be prepared when all ingredients are available, a nice vacation requires a flight, a hotel and a rental car for the same period of time and no canceled flight could ever be substituted by an increased number of hotel rooms or rental cars. The smaller the number of suppliers in each of these markets, the higher the risk of possibly getting stuck with one of the three resources not being available when bidding in three separate auctions.

In his thesis Michael Schwind addresses these issues of complementarity by pricing an allocation of resource bundles from the perspective of (automated) information production. Here, most processes typically require a combination of network bandwidth, CPU usage, main memory and mass storage. Interestingly enough, even though peer to peer file sharing protocols or initiatives like UC Berkeley's SETI@home project motivate thousands of users to (temporarily) share these resources for free, none of the major operating systems offer a "pricing protocol" to implement a market mechanism. Most of today's operating systems still much rather rely on simple pre-defined process priorities and "fairness criteria" to allocate their resources.

Michael Schwind shows that with the advent of GRID systems these deficiencies have become obvious and resorting to economic mechanisms of resource allocation has recently been promoted by many researchers in the computer science domain. His thesis gives a comprehensive and thorough review of the different fields of economic theory and computer science which contribute to the theoretical foundations of what would be needed to construct an incentive compatible GRID operating system, allowing for a Pareto-efficient allocation of all its resources by a dynamic pricing mechanism.

He classifies the existing research into two major strands, namely the (general) game theoretic approaches and the market-based allocation approaches. The former frame the allocation problem as a problem of mechanism design for a cooperative n-agent game and frequently but not necessarily use prices to solve this problem. The latter explicitly picture the information system network as a global market on which buyers compete for scarce processing and storage capacity provided by the owners of these hardware resources. To this strand of research the use of prices and money to solve the allocation problem appears to be natural and is not questioned.

However, the focus of the decision processes analyzed varies: While combinatorial yield management as well as combinatorial auctions both allow reservation of future resource bundle usage (in contrast to "spot markets" only permitting ad-hoc requests for immediate task processing), yield management assumes that each of the competing requests has to be decided upon at request time (rendering the pricing decision be a stochastic combinatorial optimization problem), whereas combinatorial auctions postpone the winner determination and pricing decision to a specific point in time pre-selected by the auctioneer in most cases (rendering the dynamic pricing decision an iterated deterministic problem). For both cases - which lead to different implications for the bidders' strategic behavior - Michael Schwind not only reviews and contrasts recent literature but also develops innovative dynamic pricing approaches to solve them, based on reinforcement learning, local search heuristics and relaxations of the set packing problem posed by the combinatorial auctions.

As an empirical study which is part of this work reveals, what prevents most companies engaged in e-commerce from using dynamic pricing mechanisms is not the fear of losing customer goodwill by imposing a pricing model on them, which might be perceived as unfair and obscure but much rather the companies' fear that the costs of implementing such a mechanism outweigh the potential profits it can generate. Although this thesis does in fact confirm that dynamic pricing and allocation of resource bundles being are a complex issue it also illustrates that the potential gains are tremendous and the road towards an automated and cost-efficient decision making is about to be paved, rendering the above-mentioned perception of the cost-benefit ratio a misperception.

Oliver Wendt   Kaiserslautern, January 2007

Contents

1

Introduction

1.1 Pricing of Distributed Information Services

The optimal allocation of resources is traditionally one of the most fundamental topics in economics as well as in the design and implementation of technological production processes. The scarcity of resources dominating the optimization of these processes is normally expressed by prices, which can act as a control variable in economic processes. The work presented here deals first of all with *price controlled resource allocation* for the provision of *information services and information products* in the context of a research project called the *"pricing of distributed information services" (PRISE)*[1]. The research was continued within the project *"Preis und Erlösmodelle im Internet, Umsetzung und Marktchancen" (PREMIUM)*[2] focusing on the subtopic "dynamic bundle pricing".

1.2 Motivation

The use of the computing power provided by distributed computer infrastructure, like the Grid, is of increasing interest for *information technology* infrastructure *(IT)* and service providers in growing *business-to-business (B2B)* and *business-to-consumer (B2C)* markets (Foster et al. 2001). The simultaneous use of network resources and computing capacity to enable *web-based video conference* and *peer-to-peer (P2P) telecommunication services* between corporations is an example of a *business-to-business* related application of *information service and information production* tasks, which utilizes such IT

---

[1] The project was funded by the German National Science Foundation, the "Deutsche Forschungsgemeinschaft" (DFG).

[2] The project was funded by the German Federal Ministry of Education and Research, the "Bundesministerium für Bildung und Forschung" (BMBF).

infrastructures.[3]. In B2C markets the provision and accounting of *video-on-demand services* has similar application properties. An exemplary scenario for the allocation of IT services to distributed IT resources is depicted in figure 1.1. Requests for the resources that are necessary to provide services like a video conference or database processing job have to be allocated to various networked PC systems including different types of resources like central processing units, volatile memory, or disk capacity. The challenge of this work is now to find suitable economically inspired mechanisms for this allocation task. In doing this the preferences of the users and their willingness-to-pay play an important role: whereas the data processing service might not be very time critical for the user the video conference is. Accordingly to this, the video conference users should be willing to pay more for the immediate availability of the resources than the user of the database service. The transfer of a considerable amount of data that is collected during daily business activities, in times of low infrastructure resource load, could be seen as a further instance of an IT task that has no high time criticality and where a price controlled allocation of resources is valuable for the users and the service providers.

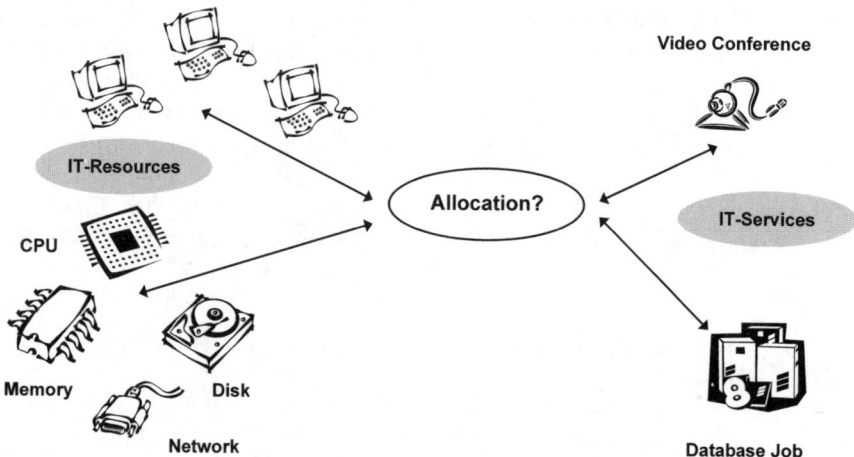

Fig. 1.1. Scenario for the allocation of resources in a distributed IT infrastructure

The main goal of this thesis is to find and construct appropriate methods and applications for the efficient allocation of information services and information production processes in distributed IT infrastructures. The previous examples illustrate that this distributed character is very common in modern IT applications and therefore motivates concentration on such systems. Normal IT infrastructure, - like *Internet-based systems for Web-services* -, provides the technical resources needed to execute diverse information services,

---

[3] An example is the P2P-based telecommunication system Skype (www.skype.com).

by making computing power, network connection bandwidth, disk storage capacity, and volatile memory capacity available to the hosted software processes. Theses IT structures are usually requested *simultaneously* in particular combinations for the execution of the tasks. Therefore it seems to be reasonable to gather and book the required capacities in advance before starting a particular *information services and information production* process.

An intelligent way to manage such a challenging resource allocation process for the provision of information services is the use of *economic inspired mechanisms*. For this reason economic principles have been increasingly used for such computerized allocation processes in the last twenty years (Clearwater 1996) and they have turned out to provide a superior performance. Recent applications of economic resource allocation mechanisms have been adapted to work in Grid systems (Buyya & Vazhkudai 2001). The use of economic resource allocation methods for allocation, e.g. in Grid systems, implies the readiness of the users to accept prices for information services and the products that are dynamically changed depending on the load situation of the IT system and its different resource types respectively. For this reason, the thesis illustrates economic resource allocation methods from the perspective of *dynamic pricing* mechanisms. Used this way, the consumers' *willingness-to-pay (W2P)* can be transformed into a *market price* for the resources and acts as a *control variable* for the allocation process.

The thesis provides an overview of such allocation systems, evaluates empirically the prospects of dynamic pricing in industry sectors related to the information services business, and elaborates two sophisticated methods for price controlled resource allocation in distributed computer systems. Both approaches, after having been tested and assessed in different scenarios, are finally compared with each other. Consequently recommendations for the application of the methods in real world scenarios are given. The order and structure in which this is done within this work will be outlined in the next section.

1.3 Structure of the Thesis

The argumentation of this work starts with a definition of the thesis' main objects of investigation, - *information services and information production* and *price controlled resource allocation* -, in connection with a brief description of the methodology that will be applied to gain the insights striven for. Then a detailed specification of the resource allocation scenario that will be used for all experiments carried out within this work is given. The scenario and the definitions are then put into the context of a *framework of philosophical thought* by making a short foray into some classical and contemporary economic ideas and concepts that touch the concern of the chosen research approach. This underpins the bilateral and interdisciplinary character of this thesis, operating on the area of pure technical optimization on the one hand and employing principles of economic modeling on the other hand. Encouraged by this philosophical background, the thesis tries to find an appropriate definition and taxonomy for *dynamic pricing* and *automated resource allocation* in the second chapter. The subject of automated resource allocation itself is analyzed and categorized by identifying resources properties that are relevant in the context of the thesis' objects of investigation. Subsequently, the argumentation is intensified by providing a taxonomy of economically-oriented resource allocation mechanisms that have been developed for use in the context of *distributed computer systems*. In order to deal with the characteristics of such distributed computer systems, *multi-agent technology* will be used in the experimental settings. Therefore two subsequent sections are devoted to illustrating this technology and the associated communication standards. The second chapter ends with a short view of the interdependency of dynamic pricing and automated resource allocation within the scenario investigated. The third chapter is concerned with an empirical study that evaluates industry preferences for the application of dynamic pricing methods. After analyzing the attitude of companies towards different methods of dynamic pricing according to various user groups, and especially with respect to information services and information production, several practical implications are drawn from the study which ultimately impact on the assessment of the resource allocation methods investigated. The fourth chapter turns to the application of *reinforcement learning* techniques for dynamic pricing in the context of a *yield maximizing* allocation strategy. After explaining the basics of reinforcement learning and yield management, two adaptive allocation models are presented and evaluated. The first uses reinforcement learning to adapt an optimal acceptance policy for task allocation in a single resource system, the second is concerned with the simultaneous allocation of tasks to multiple resources. *Combinatorial auctions* are introduced into the framework as an alternative to reinforcement learning in yield maximizing systems in the fifth chapter of the thesis. The first section of this chapter is concerned with the basics of combinatorial auctions like the formalization of the winner determination problem, bidding languages, and incentive compatibility issues. A comprehensive analysis of winner determination algorithms in combinatorial auctions follows. The fifth part of the thesis ends with the development of a decision support and evaluation system for the design of combinatorial auctions. By integrating the insights gained from the design issues in the previous chapter, a dynamic pricing and resource allocation system employing combinatorial auctions is developed and evaluated in chapter six. Two main methods of dynamic resource pricing are provided within this part of the thesis while solving a non-trivial valuation problem. This chapter also investigates various bidding mechanisms in the context of system stability and allocation efficiency. The thesis ends with a comparison of yield maximizing reinforcement learning and combinatorial auctions with respect to dynamic pricing and automated resource allocation. Finally advice for the practical application of the proposed pricing and allocation methods is derived from the experimental and empirical results acquired in the thesis.

1.4 Methodology, Definitions and Scenario

1.4.1 Methodology

The prevalent research methodology of this work is based on simulations. Following the paradigm of *agent-based computational economics* that will be extensively discussed in section 1.5.5, an economic system is constructed by using multiple interacting agents. This step is taken following a diligent study of similar models and a careful and extensive review of the algorithmic methods that are required to operate these models. After the engineering phase is finished and the system implemented, various simulation runs give answers to the research questions posed in the beginning. Benchmarks and test problems are applied to find objective information about the capabilities of the solution developed. The process of construction and test can be iterated to improve the solution in the desired direction. Having reached a satisfying stage of maturity the system is confronted with other 'real world' situations and its reaction is measured (see e.g. section 6.3). Thus, by using a very engineering-related approach, new insights about economic behavior in realistic environments can be drawn from the experiments. As well as using this mode of proceeding, classical socio-scientific research techniques will be applied in this work, including, e.g., the statistical analysis of a survey conducted in several industries (see chapter 3). Together with the theoretical deduction and aggregation of results from literature leading to decision theoretical models (see section 5.3), the statistical insights thus acquired experimentally and empirically will be subsumed in recommendations for the implementation and operation of dynamic pricing and resource allocation systems at the end of this work (see section 7.2).

1.4.2 Definition of Information Services and Information Production

The term *information service* has not yet been defined precisely within the *information science* community.[4] In the context of this thesis the term information service comprises: "... a computer-based information system which is used by the service provider to perform value creating communication actions in a social interaction context. In this interaction the consumer takes part by using the system to acquire knowledge, and / or performing communication actions through the system, for the purpose of achieving a particular result."[5] The notion of information services is strongly linked to the term *information product* (or *information production*) in the framework used in this work. The definition of information production is derived from an expression coined in

---

[4] This statement corresponds to the result of a conversation with Prof. John L. King on June 10th 2005.

[5] The definition presented here is based on a notion provided by Hultgren & Eriksson (2003).

the German language area where it is defined as "Informationserarbeitung" and describes the "mapping of information resources (e.g. online databases) onto information of relevance by employing methods of 'information work'. Information production uses information systems to access data."[6] The definition of information production, as it is used in this thesis, also comprises information refinement procedures. The work presented here shares the view of Kuhlen (1995) for a definition: "Information refinement describes the mapping of a set of relevance information on the set of processed information by means of 'information work'. Formal methods of information refinement include all forms of media refinement. Pragmatic value-adding services are the adaptation of informations to diverse user requirements."[7]

Both terms, - *information services* and *information products* -, are subsumed under the expression *information services and information production (ISIP)* for further use.

Examples of the ISIP processes investigated here are mainly instances of remote information services provided over the Internet, such as those practiced in the context of Web services:

- the customized retrieval, replication or generation of customized stock chart data,
- the search and replication of shared files in a P2P system,
- the risk evaluation of a portfolio by a bank customer with defined defaults,
- the broadcast of a sporting event to paying viewers via TCP/IP protocol,
- the calculation of an optimal transportation route for a logistics provider.

All these ISIP processes are based on IT resources allocated to the tasks for their execution. Market prices for the resources serve as control variables for the allocation processes considered in the thesis' simulation models. This should guarantee efficient, balanced, and stable system behavior. For this reason the second crucial definition that has to be presented in this section is a paraphrase of the term 'price controlled resource allocation'.

---

[6] "Unter Informationserarbeitung versteht man die Abbildung von Informationsressourcen (z.B. Online-Datenbanken) auf Relevanzinformation durch Anwendung von Methoden der Informationsarbeit (z.B. Recherchen). Durch Formen der Informationserarbeitung wird auf Informationssysteme zugegriffen." (Kuhlen 1995, p. 82 et seq.)

[7] "Unter Informationsaufbereitung versteht man die Abbildung der Menge der Relevanzinformation auf die Menge der aufbereiteten Information durch Methoden der Informationsarbeit. Zu den formalen Verfahren gehren alle Formen der medialen Aufbereitung. Zu den pragmatischen Mehrwertleistungen der Aufbereitung gehören Verfahren zur Anpassung von Informationen an unterschiedliche Benutzerbedürfnisse, unterschiedliches Informationsverhalten oder unterschiedliche Ziele." (Kuhlen 1995, p. 82 et seq.)

1.4.3 Definition of Price Controlled Resource Allocation

As already mentioned above, economic mechanisms, especially market mechanisms employing prices, have turned out to be a suitable tool for efficient resource allocation in computer systems.[8] For this reason, the work presented in this work deals mainly with *price controlled resource allocation (PCRA)* for the provision of *information services and information products*. PCRA denotes the use of a *market price*, or at least the W2P given by the consumer, for the efficient allocation of resources in a distributed computer system. This price serves as a *control variable* in the *allocation process* performed by a single or multiple *market mediators* in analogy to economic mechanisms in microeconomic approaches.

The PCRA consists of two basic optimization elements:

- *dynamic pricing* that allows the adaptation of the resources prices in the resource allocation system according to the demand function and the current resource load,
- and *automated resource allocation* enables resource allocation in dependency on the W2P of the resource users.

Both optimization elements stand in an antagonistic relationship, that will be elaborated in section 2.5 of the following chapter in connection with a detailed definition of *dynamic pricing* in section 2.1 and *automated resource allocation* in section 2.2.

1.4.4 Scenario for Price Controlled Resource Allocation for Information Services and Information Production

In order to provide a clearer view we should look briefly at the *resource allocation scenario* for ISIP provision used in the course of this work. Multiple types of resources, underlying the ISIP processes, can be conceived as beeing allocated in the PCRA process from the technical point of view. For the sake of simplicity, the scenario of this thesis concentrates on four resource types:

- *Central processing units (CPU)* that are mainly responsible for the processing of the data in the ISIP processes.
- *Volatile memory capacity (MEM)* which is necessary to store short term processing data for the central processing units.
- *Non-volatile storage capacity (DSK)* which is necessary to hold mass data on databases and preserve program code for the ISIP processes.
- *Network bandwidth (NET)* that is required for the data interchange between different computer units.

The general scenario used within this thesis is constructed as follows:

---

[8] This topic will be exhaustively discussed in section 2.3.

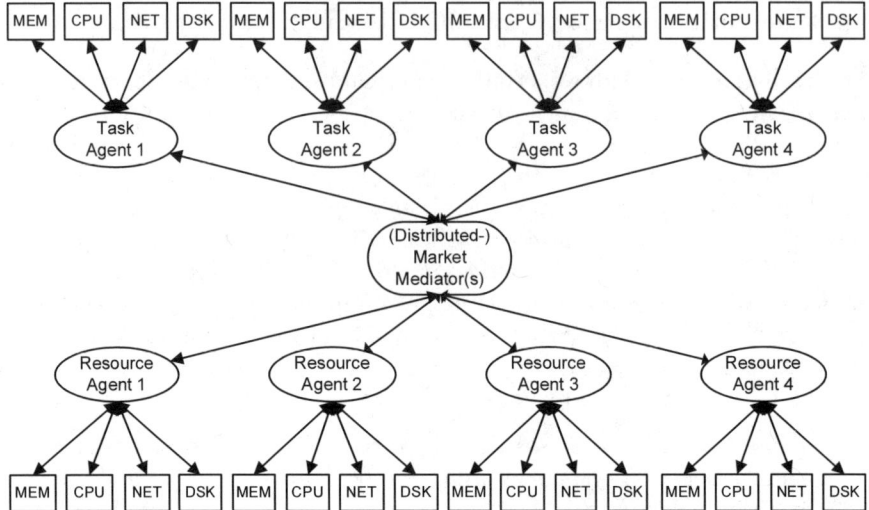

Fig. 1.2. Scenario for the (distributed) allocation of ISIP resources

- *Task agents* are engaged in acquiring the resources needed to process the ISIP task within the computer system on behalf of real world clients. They do this by bidding for the required resource combination via mediating agents.
- *Mediating agents* receive the resource bids and calculate an allocation profile for the available resources managed by the *resource agents* according to the allocation mechanism chosen. After a successful allocation process bidders are informed about the acceptance of their bids.
- *Resource agents* collect information about available resources on their particular host IT systems within a network of distributed computers and provide this information to (distributed) market mediator(s). The resource agents offer the available capacities to the task agents via the mediating agents. In the case of a bid acceptance via the mediating agent(s) the resources needed for the task accomplishment are reserved in advance for the corresponding winning agent.

Figure 1.2 depicts the ISIP allocation scenario. Resource agents administrate available MEM, CPU, NET and DSK capacities on their particular host computers systems on the supply side. On the demand side task agents collect the required resource combinations including MEM, CPU, NET and DSK capacity needed to accomplish their production tasks. In between of resource and tasks agents stand(s), the market mediator(s) employing different allocation mechanisms, that are the subject of the work following in this thesis. The necessity of the existence of a centralized or distributed market mediator(s) will be discussed in section 1.5 from the econo-philosophical point of view. For most scenarios employed in this thesis the mediator will have a centralized character to facilitate the use of allocation methods based on mathematical economics like the combinatorial auction.[9] The author of this work, however, is aware of the fact that this assumption can be called into question, if one is relying on the simple daily observation of market processes. We will therefore return to the postulation of a distributed mediator in the context of the yield maximizing learning system in chapter 4.

A crucial element for the efficient[10] allocation of resources in a PCRA system is the users' ability to express their preferences exactly. These preferences often give rise to non-linearities in the consumers' service valuations. This may be due to individual preferences, although in the case of our PCRA scenario these non-linearities are mostly derived from technological dependencies in ISIP provision.

Three main types of non-linearities in consumers' resource valuations are of interest in the domain of ISIP provision presented here:[11]

- *Quantity preferences*: The valuation of a good or service does not increase linearly with the requested quantity, e.g. quantity discounts are often given for higher consumption rates.
- *Temporal preferences*: This means consumers have different valuations for the same good or service at different times, e.g. an earlier fulfillment of an ISIP task leads to a higher W2P than a tardier ISIP task completion.
- *Bundle preferences*: Different types of goods or services are requested as a bundle, so that the valuation depends on the combination of the goods or services, e.g. in production environment the availability of an input factor has to be synchronized with a production service at specific points of time.

The implications of quantity and bundle preferences, - *subadditivity* and *superadditivity* in the goods and services valuation - will be discussed in section 5.1. In connection with the PCRA scenario for ISIP provision, the question of an appropriate formal representation of the agents preferences and the resulting valuations comes into perspective. One way to formulate such time dependent valuation functions for connected resources with discrete execution points in time is to use two-dimensional matrix representations. One dimension describes time $t \in \{1, \ldots, T\}$ at which the resource is required within the request period T.[12] The other dimension $o \in \{1, \ldots, O\}$ denotes the resource types[13] required for the ISIP provision process. The allocation (request) for a quantity of a individual resource o at time t is denoted by a matrix element $q(o, t)$.

---

[9] Inspiring discussions with Prof. Uwe Walz yielded the awareness that the term 'combinatorial auction', as it will be used in this book, is not always compatible with the auction definition used by 'traditional economists'.

[10] The definition of *'efficiency'* will be elaborated in detail in section 2.2.1 of the next chapter together with the properties of allocation mechanisms.

[11] Further impacts of consumer valuations are elaborated in section 5.3.

[12] Time period T is divided into equidistant intervals (*time slots*) within this model.

[13] Usually MEM,CPU,NET,DSK within the thesis' model.

Table 1.1. Example of aggregation of a bid matrix BM into an allocation matrix AM resulting in a new allocation matrix AM' controlled by constraint matrix CM

Initial Allocation Matrix: AM

Resource	1	2	3	4	5	6	7	8	9	10	11	12	13	14	15	16	17	18	19	20	21	22	23	24
o_1	4		2		3		5			2		2	1	1		4		3	2			2		5
o_2	3			1	2	3		1		3	4		2	3		2	3		2	1	5			2
o_3			5		1		1				2			1		5	2	1		2	2	1		
o_4	1				1					1	3	1		5	4	2			5	2	3			3

Bid Matrix: BM — Time Slot t

Resource	6	7	8	9	10	11
o_1	2				3	3
o_2	1		3	1		2
o_3			4		1	
o_4		2			1	

Constraint Matrix: CM

Resource	1	2	3	4	5	6	7	8	9	10	11	12	13	14	15	16	17	18	19	20	21	22	23	24
o_1	5	5	5	5	5	5	5	5	5	5	5	5	5	5	5	5	5	5	5	5	5	5	5	5
o_2	5	5	5	5	5	5	5	5	5	5	5	5	5	5	5	5	5	5	5	5	5	5	5	5
o_3	5	5	5	5	5	5	5	5	5	5	5	5	5	5	5	5	5	5	5	5	5	5	5	5
o_4	5	5	5	5	5	5	5	5	5	5	5	5	5	5	5	5	5	5	5	5	5	5	5	5

Resulting Allocation Matrix: AM'

Resource	1	2	3	4	5	6	7	8	9	10	11	12	13	14	15	16	17	18	19	20	21	22	23	24
o_1	4		2		3	2	5			5	3	2	1	1		4		3	2			2		5
o_2	3			1	2	4		4	1	3	5		2	3		2	3		2	1	5			2
o_3			5		1		1	4		1	2			1		5	2	1		2	2	1		
o_4	1				1		2			2	3	1		5	4	2			5	2	3			3

Three types of matrices play a crucial role in the framework of this thesis:

- *Bid matrix* (BM) denotes a request for a resource bundle submitted by a task agent. The whole bid matrix is assigned a price p according to the agents' W2P.
- *Allocation matrix* (AM) describes the current allocation for the resources o and time slots t that is awarded by the mediating agent(s) within the time period T.
- *Constraint matrix* (CM) expresses the maximum quantity q_{max} of resource o that can be awarded to the task agents by the mediating agent(s) at a time point t. For simplicity q_{max} is a constant for all resources and time slots t in the framework of this thesis.

Resource allocation is done via market mediator(s) while selecting the requests submitted by the task agents according to criteria defined in the mediators' optimization function. The chosen bids are integrated into the AM by aggregating the resource requests into the current AM in the form of the BMs. This process is depicted in Table 1.1. The requested resource load of the BM is simply added to the AM at the desired point in time resulting in AM'. This process is trivial as long as there is no violation of a *resource load constraint*. This can simply be controlled by the market mediator while comparing the resulting AM' with the CM comprising the upper resource bounds. However, if the mediator has the choice between several different AMs associated with various bid prices p, the allocation problem turns into a complex *combinatorial optimization problem* as described in section 5.1.6.

Time dependent valuation functions for the provision of a particular ISIP task can be formulated by repeated bidding of the same BM with a varying W2P in following time periods T. Table 1.3 depicts such a time dependent valuation function with a *declining W2P function*. This bidding behavior will be of special interest in the context of yield management applications because bidders are normally prepared to pay a higher amount for a task allocation with an earlier acceptance notification of a required service, due to risk aversion and preference for earlier production time. At a later notification point in time the bidders' W2P for the same task will be lowered due to the shortened planning horizon, which remains for an alternative resource procurement.

The opposite behavior on the part of the bidding agents in terms of their valuation of the tasks can occur in the competitive environment of an auction. The bidders will increase their indicated W2P associated with the bid for the resource bundle if they recognize that their estimated resource price lies below the real resource price. This should lead to an *inclining W2P function*.

It is important to mention in the context of the PCRA scenario used here, that task agents can resubmit their bids only after a total time period T has elapsed. This time period T, however, contains T time slots that can be booked separately within a BM for ISIP provision purposes.

1.5 Classic Economists and Paradigms in Pricing and Resource Allocation

The optimal allocation and coordination of scarce resources by a price controlled mechanism is one of the central themes in the economic thought of the last two centuries. In this section a short introduction to economic approaches relevant for the pricing and allocation issue in the information services and information production domain, as addressed by this thesis, will be given. This is done by depicting in historical order selected classical approaches of economic thought, that seem to be useful to provide broad philosophical and socio-economic foundations for the technical argumentation presented in the

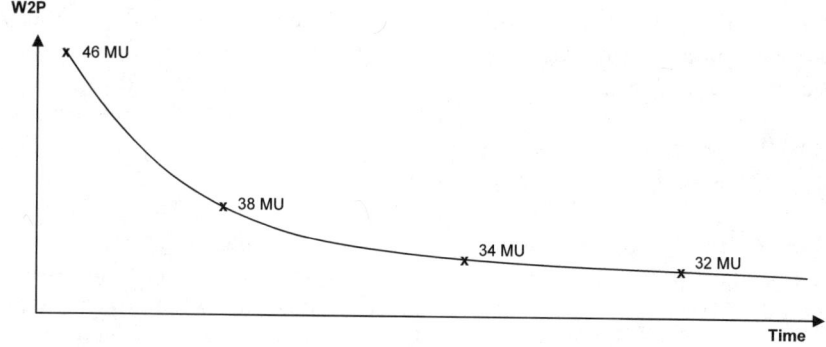

Bid Matrix: p = 46MU

	Time Slot t					
Resource	1	2	3	4	5	6
o_1	2				3	3
o_2	1		3	1		2
o_3			4		1	
o_4		2			1	

Bid Matrix: p = 38MU

	Time Slot t					
Resource	1	2	3	4	5	6
o_1	2				3	3
o_2	1		3	1		2
o_3			4		1	
o_4		2			1	

Bid Matrix: p = 34MU

	Time Slot t					
Resource	1	2	3	4	5	6
o_1	2				3	3
o_2	1		3	1		2
o_3			4		1	
o_4		2			1	

Bid Matrix: p = 32MU

	Time Slot t					
Resource	1	2	3	4	5	6
o_1	2				3	3
o_2	1		3	1		2
o_3			4		1	
o_4		2			1	

Fig. 1.3. Time dependent valuation functions with a declining W2P

following parts of this thesis. The relationship of these socio-economic foundations to the topic of pricing in computational environments is then outlined in connection with examples of their application in a literature survey of economic resource allocation in distributed computer systems in chapter 2.[14]

1.5.1 Léon Walras and the "Equilibrium Tâtonnement Process"

Léon Walras (1834-1910) was the first economist to put the neoclassical theory in a formal *general equilibrium* setting. Walras tried to demonstrate how an economy with many goods fits together and reaches equilibrium. For this purpose Walras (1896) built a system of simultaneous linear equations to describe an economic equilibrium model. Walras showed that the whole system is solvable because the number of equations equals the number of unknowns. Their solution provides the goods' *equilibrium prices* and quantities in allocation. The demonstration that price and quantity are uniquely determined for each commodity in such an economy seems to be his greatest contribution to economic science. Since the emergence of such an equilibrium is not clear in the real world environment, Walras simulated an artificial market process to calculate, whether such an approach would bring the system to equilibrium. For this reason he postulated a *'tâtonnement' process* in Walras (1896). Tâtonnement can be regarded as a trial-and-error process in which a price is called out and market participants subsequently reveal how much they are willing to demand or supply at that price. If there is an excess of supply over demand, then the price will be lowered so that less would be supplied and more would be demanded. Thus the prices should converge towards the equilibrium. To keep the equilibrium constant until the commodity prices are adjusted, Walras assumed a lack of exchange until an equilibrium is reached. This assumption is however very unrealistic and limits the usefulness of the process as an aid to better understanding how real markets work.

In his 1906 published work "Manuale di Economia Politica" Vilfredo Pareto (1848-1923), a disciple of Léon Walras, reformulates the Walras equations, while focusing on formulating equilibrium in terms of solutions to individual problems of objectives and constraints (Pareto 1906). This way Pareto became on of the founders of modern microeconomics by introducing the notion of Pareto-optimality, the idea that a society is enjoying maximum *'ophelimity'*, when no one can be made better off without making someone else worse off.

This thesis will take up such a tâtonnement process in connection with dynamic pricing in markets for computational resources and will also introduce the question of stable utility-maximizing equilibria in such pricing systems.

---

[14] Some inspiration for the following argumentation is extracted from the websites:
   http://cepa.newschool.edu/het/home.htm
      http://www.economyprofessor.com
      http://www.politicsprofessor.com

1.5.2 Schumpeter's Theory of Economic Development and "Evolutionary Economics"

Joseph A. Schumpeter (1883-1950), who was an enthusiast of Léon Walras, early adapted the idea that the interconnection of allocation and pricing must be the crucial element of an economic theory based on an equilibrium concept. He claimed that the equilibrium price extracted by using a maximization calculus is only valid for a fixed state in a system of various possible allocations.[15] In accordance with the general equilibrium theory of Walras, Schumpeter postulated the concept of a disappearing excess demand at the equilibrium point (see mathematical formulation of the general equilibrium in section 1.5.4) He characterized this state as the disappearance of all barter activities and the drying up of all commodity flows in the economic system.[16] However, like Walras, Schumpeter was not able to formalize this process exactly, let alone provide a mathematical solution of the equilibrium problem. As well as his adherence to Walras' ideas, Schumpeter's emphasis in price and allocation equilibrium was on its dynamic behavior properties: "For this reason it is very important not to approach these problems having the apparatus of statics in view. Despite of what has been so far said, we do not reject all cooperation between statics and dynamics; it was suggested while discussing the variation method, that their power extends a specific distance into the domain of dynamics...".[17]

The greatest contribution of Schumpeters' work to recent economic models, however, seems to be his famous theory of entrepreneurship, published first in the book "Theorie der wirtschaftlichen Entwicklung" (1911). Within this theory Schumpeter argues that risk-friendly entrepreneurs create technical and financial innovations in the face of competition and falling profits.

---

[15] "Um aus den unendlich vielen möglichen den uns interessierenden Gleichgewichtspreis herausgreifen zu können, müssen wir eben zum Ausdrucke bringen, dass er und die unter seiner Herrschaft sich ergebende Güterverteilung sich zu erhalten streben, d.h., dass die weitere Veränderungen zum Ausdrucke bringenden Symbole zu Null werden. Unter welchen Bedingungen das letztere geschieht und ob nur einer oder mehrere solcher Gleichgewichtspreise in einem gegebenen Zeitpunkte und für ein angegebenes Gut bestehen können, sind dann weitere Fragen, die uns hier nicht berühren." (Schumpeter 1908, p.199)

[16] "Den Glcichgewichtszustand der ökonomischen Quantitäten zu beschreiben - und gewisse Variationen desselben - ist das Problem der Ökonomie. Alle Tauschakte tendieren danach ihn zu realisieren, d.h. einen Zustand zu realisieren, in dem keine Veränderung der Quantitäten erfolgt, der sich daher zu erhalten strebt und deshalb besonders interessant ist." (Schumpeter 1908, p. 198)

[17] "Aus diesem Grunde ist es auch sehr wichtig nicht mit dem Apparate der Statik vor unseren Augen an diese Probleme zu gehen. Wir leugnen nämlich trotz allem Gesagten nicht jede Zusammenarbeit zwischen Statik und Dynamik; es wurde schon bei der Erörterung der Variationsmethode angedeutet, dass ihre Macht eine gewisse Strecke in das Gebiet der Dynamik hineinreicht..." (Schumpeter 1908, p. 622)

The intrinsic evolutionary process of the birth of new and dying of established enterprises spurred by the survival of only the fittest companies, leads to a first version of *'evolutionary economics'* - and is responsible for irregular economic growth generated by this selection process. For this reason Schumpeter is broadly considered as the father of evolutionary economics (Andersen 1991). Evolutionary economics focuses mainly on two crucial aspects, the selection effect and the innovation effect: "The selection effect covers what economic evolution has in common with biological evolution. This effect has a negative part – destruction – that reflects the fact that firms and other economic entities with subnormal characteristics shrink and disappear. It also has a positive part – creation – that reflects the fact that firms and other economic entities with supernormal characteristics are promoted. ... More specifically, the innovation effect includes innovation in the narrow sense as well as imitation and learning."(Andersen 2004)

Schumpeter provides a condensed characterization of his notion of evolutionary economics in the description of the capitalistic process: "The essential point to grasp is that in dealing with capitalism we are dealing with an evolutionary process ... [It is a process] that incessantly revolutionizes the economic structure *from within*, incessantly destroying the old one, incessantly creating a new one. This process of *'creative destruction'* is the essential fact about capitalism." (citation taken from Andersen (2004)).[18] Despite the fact that Schumpeter never explicitly addressed learning as a driving factor for a continuous innovation process, a branch of evolutionary economics and innovation literature that stands in the tradition of his economic approach, refers to such techniques to explain the development of economic equilibria over time (Fagerberg 2003). The thesis will refer to this tradition with two questions:

- How can *learning* be introduced into the pricing process for ISIP provision systems to guarantee a *yield maximizing* behavior that converges to a stable equilibrium, while increasing the survival probability of the participating companies in a tough e-market?

- Can the use of *dynamic pricing* be regarded as a crucial *innovation* in e-markets, which also contributes to a competitive advantage for the participating companies?

---

[18] Der Kapitalismus ist also von Natur aus eine Form oder Methode der ökonomischen Veränderung und ist nicht nur nie stationär, sondern kann es auch nicht einfach sein... Der fundamentale Antrieb, der die kapitalistische Maschine in Bewegung setzt und hält, kommt von den Konsumgütern, den neuen Produktions- und Transportmethoden, den neuen Märkten, den neuen Formen der industriellen Organisation, welche die kapitalistische Unternehmung schafft. ... den gleichen Prozess einer industriellem Mutation - wenn ich diesen biologischen Ausdruck verwenden darf - , der unaufhörlich die Wirtschaftsstruktur *von innen heraus* revolutioniert, unaufhörlich die alte Struktur zerstört und unaufhörlich die neue schafft. Dieser Prozess der 'schöpferischen Zerstörung' ist für den Kapitalismus wesentliche Faktum. (Schumpeter 1950, p. 136-138)

1.5.3 Hayek's Economic Competition and the "Theory of Spontaneous Order"

One of the greatest admirers of Walras was the Austrian economist and Nobel laureate Friedrich August von Hayek (1889-1992). He is considered to be one of the most important twentieth century scholars in economics, epistemology, philosophy, and psychology. Some of his concepts are transferable to software agent technology and the design of distributed information systems, especially the spontaneous emergence of order in an uncoordinated organization. Hayek regarded the idea that rational economic order could be constructed by solving a simple mathematical optimization problem based on a central planner's knowledge about individual preferences and the available resources as a major flaw in a concept which he called 'constructivist rationalism'. Due to the obvious non-existence of a central instance where the necessary optimization information is accessible to a central planner in real world economic systems, he proposed that the economic coordination problem can be seen as a problem of knowledge diffusion and utilization in a system of distributed individuals. "The basic economic problem of society is therefore not only a question of the allocation of given resources... It is rather a problem of ensuring the best utilization of the resources, which are known to some members of society, for purposes whose relevance is only known to some individuals. Briefly this is a problem of the use of knowledge that is not given to any individual in its entirety."[19] Hayek proposes leaving the economic decisions to the affected individuals in a decentralized way because the "knowledge of the special circumstances of time and place" only exists in a scattered way in people's minds. This should guarantee a fast adaptation to economic reality (Hayek 1976a, p. 107 et seq.).

Hayek interprets competition as an evolutionary process, which contributes to the discovery of new knowledge that would have been left undiscovered otherwise. It is not correspondence but competition that spurs the efficiency of the economically acting individuals (Hayek 1996a, p.119). However, the results of this competition are not predictable, which leads to one of Hayek's core propositions, the 'constitutional ignorance'[20] of economic acting individuals.

---

[19] "Das ökonomische Grundproblem der Gesellschaft ist daher nicht nur eine Frage der Allokation gegebener Ressourcen ... Es ist vielmehr ein Problem der Sicherung der besten Verwendung gegebener Ressourcen, die manchen Mitgliedern der Gesellschaft bekannt sind für Zwecke, deren relative Bedeutung nur einzelnen Individuen bekannt sind. Um es kurz zu sagen, es ist ein Problem der Nutzung von Wissen, das in seiner Gesamtheit keinem Individuum gegeben ist." (Hayek 1976a, p. 104)

[20] "... the impossibility for anyone of knowing all the particular facts on which the overall order of the activities in a Great Society is based. It is one of the curiosities of intellectual history that, in the discussions of rules of conduct, this crucial fact has been so little considered although it alone makes the the significance of these rules intelligible. Rules are a device for coping with our constitutional ignorance." (Hayek 1976b, p. 108)

The market has the function of encouraging out an adaptation process on the part of the individuals to make use of the novel facts (Hayek 1996a, p. 119).

Hayek uses the expression '*catallaxy*' derived from the Greek idiom 'katallattein', which denotes exchange and reconciliation, to describe the market economy.[21] The coordination mechanism in Hayek's interpretation of a market economy is based on a mechanism employing individuals' signaling of subjective value creation. In detail Hayek refers to patterns of human activities which can also be considered as methods of knowledge acquisition that occur under certain circumstances (Hoppmann 1993, p. 16 et seq.):

- Individuals have market access and acquire income
- Individuals weigh up alternatives and rank them according to their preferences
- Individuals pursue their own interests which are not necessarily selfish
- Constitutional ignorance prevents individuals from being able to anticipate the actions of other individuals and avoids the correct assessment of future market states as well

The catallaxy principle requires adaptive behavior on the part of the individuals, especially by using learning methods and coordination with other agents to deal with the constitutional ignorance. Behavioral rules caused and defined by societal traditions are also responsible for the emergence of catallaxy (Hoppmann 1993, p.19 et seq.).

The exchange of information through catallaxy leads to the formation of prices, representing a subjective exchange value for each agent based on his individual information, while this valuation is not visible for all agents. Enduring market relationships are the epitome of a grown non-purpose-specific spontaneous order, denoting an unpredictable pattern that depends only on the impact of the agents' behavioral rules (Hoppmann 1993, p.21 et seq.).[22]

The concept of a catallactic market with its decentralized allocation and price finding mechanism is an important alternative to the centralized model of a virtual auctioneer in the Walras equilibrium economy. For this, Hayek's catallaxy plays an prominent role in the process of selecting an appropriate resource allocation mechanism in distributed computer systems. However,

---

[21] "Since the name 'catallactics' has long ago been suggested for the science which deals with the market order and has more recently been revived, it would be appropriate to adopt a corresponding term for the market order itself. From this we can form an English term catallaxy which we shall use to describe the order brought about by the mutual adjustment of many individual economies in the market. A catallaxy is thus the special kind of spontaneous order produced by the market through people acting within the rules of the law of property, tort, and contract." (Hayek 1976b, p. 108)

[22] Adam Smith had already postulated the existence of a mechanism responsible for the emerging economic order driven by the individuals pursuing their own utility without being able to identify particular development patterns. He therefore called this mechanism the 'invisible hand'. (Hayek 1996b, p.10)

Hayek never formalized such a catallactic process, so that we have to resort to various methods to map such a socio-economic system onto a self-organizing technical system. The introduction of decentralized auction processes, as they are used in the context of this thesis, might be one feasible approach to resolve the contradiction between optimization effectiveness in allocation processes and the supply of required information from the decentralized individuals. Learning in distributed systems, also in the sense of evolutionary economics, could be another way to deal with the problems raised by Hayek.

1.5.4 Arrow's General Equilibrium Theory and "Neo-Walrasian Economics"

Kenneth J. Arrow (1921-) made some fundamental contributions to extending the *equilibrium theory* of Léon Walras by founding the 'Neo-Walrasian' school. Arrow addressed the topic of *'stability'* topic in the context of a competitive equilibrium with multiple markets. The Walrasian model with a separate market for each good assumes gross substitutability between goods, implying the absence of complementarities (Fan et al. 2001). Together with Debreu, Arrow formulated economic Walras' model and showed that there are always prices to equilibrate all markets (Arrow & Debreu 1954). The resulting allocation of the goods is 'efficient' in the sense that there is no other allocation which increases the welfare of every agent in the economy (Kubler 2005, p. 1).[23] The next section introduces the formulation of the *general equilibrium theory* according to the notation used by Sandholm (1995).

Formulation of the General Equilibrium Theory

The *general equilibrium economy (GEE)* depicted in this section consists of n goods g and m agents a that are divided into consumer agents a_i and producer agents a_j. The market also has prices $\mathbf{p} = [p_1, \ldots, p_n]$, where $p_g \in \mathbb{R}$ is the price of good g. The two classes of agents characterize the equilibrium economy:

- *Consumers:*
  A fraction f of all agents in the GEE is assigned to be consumers a_i. They can sell, buy, and consume bundles of goods according to their preferences. Each consumer i has an utility function $u_i : \mathbb{R}_+^k \to \mathbb{R}$ that denotes these bundles and ranks them into an order. The initial allocation $e_{i,j}$ of good g to the consumer agent i is denoted as his endowment. The amount of good g that agent i finally consumes is described by $x_{i,g}$. The consumer optimizes his utility by choosing a feasible bundle $\mathbf{x}_i = [x_{i,1}, \ldots, x_{i,n}]^{-1}$.

---

[23] This proposition is known as the first *welfare theorem*. The second welfare theorem states that every efficient allocation can be implemented as a Walrasian equilibrium with suitable redistribution of income and endowments.

A consumed bundle is feasible if its cost does not exceed the agent's endowment employing the current price vector $\mathbf{p}$. Each consumer i has an initial endowment $\mathbf{e}_i = [e_{i,1}, \ldots, e_{i,n}]^{-1}$ at the beginning of the equilibrium tâtonnement process.

- *Producers:*
The remaining $(m - f)$ agents in the GEE system are assigned to be producers a_j. They can transform types of goods into other sorts of goods by employing a *technology*. A technology is defined as a combination of inputs and outputs. The consumption of a good for the purpose of production is denoted by a negative number. A *production function* describes a technology for producer agent a_j with output $y_{j,g}$ of good g. A technology is denoted by a production vector $\mathbf{y}_j = [y_{j,1}, \ldots, y_{j,n}]^{-1}$. The set of all feasible technologies available to producer a_j is expressed as Y_j. The profit of a producer agent a_j is $\mathbf{p} \cdot \mathbf{y_j}$, where $\mathbf{y_j} \in Y_j$. All consumers a_i in this model are assumed to be in the possession of producer shares. Subsequently the profits earned by the producers are distributed among the consumers according to the ratio of shares they hold on the producer agents. The fraction that consumer a_i owns at producer a_j is expressed by $\theta_{i,j}$.

A general (Walrasian) equilibrium with the vectors $(\mathbf{p}^*, \mathbf{x}^*, \mathbf{y}^*)$ exists if:

- *Markets clear:*

$$\sum_i \mathbf{x}_i^* = \sum_i \mathbf{e}_i \sum_j \mathbf{y}_j^* \tag{1.1}$$

- *Each consumer a_i maximizes its preferences given the prices:*

$$\mathbf{x}_i^* = \arg \max_{\mathbf{x}_i \in \mathbb{R}_+^n \mid \mathbf{p}^* \cdot \mathbf{x}_i \leq \mathbf{p}^* \cdot \mathbf{e}_i + \sum_j \theta_{i,j} \mathbf{p}^* \cdot \mathbf{y}_j} u_i(\mathbf{x}_i) \tag{1.2}$$

- *Each producer a_j maximizes its profits given the prices:*

$$\mathbf{y}_i^* = \arg \max_{\mathbf{y}_j \in \mathbf{y}_j} \mathbf{p}^* \cdot \mathbf{y}_j \tag{1.3}$$

In the GEE agents are assumed to behave *competitively* ignoring the impact of their bidding on the resulting price. This imposes some restrictions on the model. A competitive equilibrium is given for a *price* and *quantity vector* stated by each agent only if

(i) for each agent a, there is a solution for the optimization problem, that either results in a balanced quantity vector $\mathbf{x}_i^*$ or an equilibrated price vector $\mathbf{p}^*$,

(ii) the amount of *produced* and *consumed* goods $\mathbf{x}_i^*$ is *balanced* and equals the initial endowment $\mathbf{e}_i$.

The existence of competitive equilibria requires the assumption of *continuity*, *monotonicity*, and *concavity* of the utility and production function (Mas-Colell et al. 1995). If these conditions apply strictly, these equilibria are unique. Competitive equilibria have two desirable properties deriving from the fundamental welfare theorems of economy:

- *All* competitive equilibria are *Pareto optimal*.
- *Any* feasible Pareto optimum is a *competitive equilibrium* for some initial allocation of endowments.

Computing Competitive Equilibria

A major problem with the GEE is that, although it can be proved that price-allocation equilibria exist under certain conditions (see previous section) by using *Uzawa's equivalence theorem* (Uzawa 1962) and *Brouwer's fixed-point theorem* (Brouwer 1912), or *Kakutani's fixed-point theorem* (Kakutani 1941), the algorithmic way of reaching such an equilibrium is not clear. This derives from the fact that these proofs are *not constructive*.[24], meaning that these proofs assume the use of alternative decisions, where it can be left open which alternative to use for the purpose of the proof. For this reason a calculation algorithm for the equilibrium cannot be derived directly from such a proof.[25] Additionally the complexity of a general GEE calculation scheme seems not to be clear. The lower bound for the worst-case complexity of Brouwer's fixed-point calculation is stated to be exponential in the quantity of goods and services of the GEE by Hirsch & Papadimitriou (1989). The upper bound of the complexity of a GEE solution, which is strongly related to the question of the complexity of the Nash equilibrium in a n-player game, is not yet clear either: "Is there a polynomial algorithm for computing a (mixed) Nash equilibrium in such a game? ... Because of the guaranteed existence of a solution, the problem is unlikely to be $\mathcal{NP}$-hard; in fact, it belongs to a class of problems "between" $\mathcal{P}$ and $\mathcal{NP}$, characterized by reliance on the parity argument[26] of the existence proof." (Papadimitriou 2001, p. 2). Costa & Doria (2005), however, have a different view of the computability of general equilibrium theory. In accordance with a question by Vela Vellupillai, derived

---

[24] "Constructivism can broadly described as an 'intolerance of methods that had to affirm the existence of thing's sort without to find one'. Constructivism does not allow indefiniteness in proofs." (Bry 1989, p. 36)

[25] This phenomenon is defined as "Unreasonable In-effectiveness of Mathematics in Economics" by Velupillai (2004b), who ventures the optimistic conjecture that "every result in economic theory that depends on topological - non-constructive and uncomputable - fixed-point theorems can be derived, with imperceptible change in content, with recursion-theoretic fixed-point theorems that are also constructive." (Velupillai 2004a, p. 12)

[26] The impact of the parity argument in this argumentation is explained in Papadimitriou (1994).

from the domain of classic recursion and complexity theory: "Is it possible to construct a Turing machine equivalent of such a dynamic system such that it halts at a particular configuration?", they put the question: "Given some mathematical model for an arbitrary market economy, can we algorithmically decide whether it has reached some equilibrium set prices?" For this purpose they use "the recursion-theoretic version of Rice's theorem for a fragment of real analysis in a way that makes it directly applicable to deriving undecidability results for continuous vector fields"[27] (Velupillai 2004a, p. 7). Costa & Doria (2005) finally show that for non-cooperative games within a strong theory T (like Zermelo-Fraenkel set theory plus axiom of choice), there exists a non-cooperative game Γ where each strategy set s is finite, but such that for some "s^* a Nash equilibrium for Γ" is undecidable. This means that for some constellations of the GEE, finding a equilibrium solution is $\mathcal{NP}$-hard.[28] Leaving behind the discussion about the complexity of the GEE, a couple of approximation algorithms have been developed to solve the price-allocation problem.

One of the first algorithms proposed is a modified simplex method by Scarf & Hansen (1973). The simplex operates on g goods, while normalizing the prices so that their sum is equal to one. The set of feasible price vectors forms an n-dimensional simplex in $\mathbb{R}^g$ that is subdivided into a large number of of sub-simplices. Scarf proves the existence of such a sub-simplex, that encloses the searched equilibrium price, and shows how to optimally subdivide the simplex to provide an excess demand within a pre-specified tolerance. Scarf's algorithm, which has a computational complexity of the fourth power to the basis of the number of commodities involved, is subject to various improvement proposals, outlined e.g. in Scarf (1982) and Judd (2002).

The second prominent approach to calculating the fixed-points of a GEE is presented by Smale (1975). The algorithm claims to calculate a trajectory for the prices p_g. To find an equilibrium price vector $\mathbf{p}^*$ with $\mathbf{p} = (p_1 \ldots p_n)$ where the excess demand function $z(\mathbf{p}^*) = 0$, Smale defines a price matrix $(n \times n)$ with one of the commodities being a numeraire and formulates the global Newton differential equation $D_z(\mathbf{p})\frac{dp}{dt} = -\lambda z(\mathbf{p})$ where $D_z(\mathbf{p})$ is the matrix of the first partial derivatives of z. λ switches the sign of $z(\mathbf{p})$ according the sign of D_z and causes the tâtonnement process to take the interdependencies of the demand for the various commodities and the price into account[29]. The approximation algorithm converges from almost any initial price system on the boundary of the price simplex (Smale 1976, p. 294).

In an alternative approach to the centralized algorithmic solutions for the calculation of GEE solutions, Axtell (1999) proposes the use of decentralized

---

[27] That describe the excess demand function in the GEE (remark by the author).

[28] Further aspects of this discussion about the computability of economic equilibria can be found in Schwind (2005b).

[29] In contrary to Smale's approach, the Walras tâtonnement equation is simply $\frac{\partial p}{\partial t} = z(\mathbf{p})$.

exchange process allowing for bilateral trading between single agents as well as groups of agents. In his model Axtell makes assumptions very close to the classic Arrow-Debreu model. By proposing a *'first welfare theorem of decentralized exchange'*, he shows that k-lateral exchange between agents leads to a globally stable equilibrium reaching a Pareto-optimal allocation (*second welfare theorem of decentralized exchange*). In contrast to classical tâtonnement processes, the application of k-lateral trading leads to altered wealth for the participating agents.[30] Axtell finds the complexity of decentralized exchange processes to be in $\mathcal{P}$ when employing Cobb-Douglas-like preference functions for the agents. As a result of Axtell's approach, it can be seen that decentralized multilateral exchange processes are more realistic than classic GEE solution approaches, coming nearer to an economic equilibrium process like that described by Hayek.

Competitive Equilibria with Complementarities and Indivisibilities

Shortly after the formulation of the Arrow-Debreu GEE it became clear that in the presence of indivisibility of the allocated goods and services, the existence of competitive equilibria with unique prices is not guaranteed (Papadimitriou 1981). Additionally, the assumption of decreasing or at least constant utility functions came into question: "Both linear programming and the Walrasian model of equilibrium make the fundamental assumption that the production possibility set displays constant or decreasing returns to scale; that there are no economies associated with production at a high scale. I find this an absurd assumption, contradicted by the most casual of observations... If the technology giving rise to a large firm is based on indivisibilities, then this technology can be described by, say, an activity analysis model in which the activity levels referring to indivisibility goods are required to assume integral values, like 0,1,2,..., only." (Scarf 1994, p. 114-115). One way to address this indivisibility problem is the use of *combinatorial allocation mechanisms* that allow participants to express their individual valuations as different combinations of integer units of goods and services. These mechanisms allow the consideration of substitutionality and complementarity effects to avoid the inefficiencies of simple Walrasian markets. Bikhchandani & Mamer (1997) propose a model for an exchange economy that is related to the assignment market by Koopmans & Beckmann (1957). In their model agents try to acquire a bundle of objects by bidding based on their nonlinear valuations over the bundles, while the particular objects only exist once. The resulting optimization problem is $\mathcal{NP}$-hard due to the *indivisibility* of the goods (Deng et al. 2002). Bikhchandani & Mamer (1997) show that market clearing prices exist in such an assignment market if any optimal solution to *integer programming (IP)* is an optimal solution to a relaxed IP problem. In such a

---

[30] This condition is called the *Walras law*, which is violated in the case of k-lateral exchange processes due to agents' trading at non-equilibrium prices.

case the solution of the IP and the relaxed IP is identical and the dual linear program yields prices that decentralize efficient allocations in assignment markets (Bikhchandani & Mamer 1997, p. 397) (see section 5.2.1 for discussion of IP and relaxed IP in connection with duality). In other cases, the existence of a price equilibrium is problematic. However, the equivalence between the linear programming solution approach and the calculation of the pricing equilibrium persists (Bikhchandani & Ostroy 2002, p. 402). Bikhchandani & Ostroy (2005) develop an incentive compatible *combinatorial auction* with an ascending price mechanism from the package assignment model, which algorithmically implies a $\mathcal{NP}$-hard *combinatorial allocation problem* (Vries & Vohra 2001), being aware that there is a trade-off between simplicity and efficiency of multi-object auctions. Combinatorial auctions provide the opportunity to map almost all variants of technical and economic allocation problems as an extension of the classic Arrow-Debreu GEE. Combinatorial auctions mark the *borderline of recent economic allocation model research*, however, producing a considerable game-theoretical and computational complexity.

1.5.5 Tesfatsion's "Agent-based Computational Economics"

As a reaction to the "Unreasonable In-effectiveness of Mathematics in Economics" claimed by Velupillai (2004b) due to the increasingly complicated analytical models for the description of economic processes, like the extension of the Arrow-Debreu (1954) economy into a combinatorial allocation economy described by Bikhchandani & Ostroy (2005), a new branch of economics called *'agent-based computational economics'(ACE)* appeared:

"Agent-based computational economics is the computational study of economies modeled as evolving systems of autonomous interacting agents" (Tesfatsion 2001). ACE is a specialization of complex adaptive systems *(CAS)* in the economic domain. A CAS can be defined as follows "A CAS behaves /evolves according to three key principles: order is emergent as opposed to predetermined, the system's history is irreversible, and the system's future is often unpredictable. The basic building blocks of the CAS are agents. Agents are semi-autonomous units that seek to maximize some measure of goodness, or fitness, by evolving over time. Agents scan their environment and develop schemata representing interpretive and action rules. These schemata are often evolved from smaller, more basic schema. These schemata are rational bounded: they are potentially indeterminate because of incomplete and/or biased information; they are observer dependent because it is often difficult to separate a phenomenon from its context, thereby identifying contingencies; and they can be contradictory. Schemata exist in multitudes and compete for survival." (Dooley 2005)

Using ACE for constructing and testing a socio-economic model normally requires two basic steps (Tesfatsion 2006, p. 8):

- The ACE modeler starts by computationally constructing an economic world comprising multiple interacting agents.

- The modeler steps back to observe the development of the world over time.

According to Tesfatsion (2006, p. 8 et seq.) the following properties can be assigned to the agents in an ACE model:

"The agents in an ACE model can include economic entities as well as social, biological, and physical entities ... Each agent is an encapsulated piece of software that includes data together with behavioral methods that act on these data. Some of these data and methods are designated as publicly accessible to all other agents, some are designated as private and hence not accessible by any other agents, and some are designated as protected from access by all but a specified subset of other agents. Agents can communicate with each other through their public and protected methods."

Tesfatsion (2006, p. 8) identifies four basic objectives of insight in ACE research:

- One objective is to achieve better *empirical understanding*: why have particular global regularities evolved and persisted, despite the absence of centralized planning and control? ACE seeks explanations for this phenomenon grounded in the repeated interactions of agents operating in realistically rendered worlds. Agents should have the same flexibility in these worlds as their corresponding entities in the real world do. This implies e.g cognitive and learning capabilities.
- A second objective of ACE is *normative understanding*: how can agent-based models be helpful for the discovery of good economic designs? ACE researchers are interested in evaluating designs for economic policies, institutions, and processes such that they provide socially desirable system performance. The ACE world is often populated by self-interested agents with learning capabilities that allow them to develop over time. The outcome of such artificial worlds should be efficient, fair, and orderly, despite the agent's self-oriented strategic behavior.
- The third objective is *qualitative insight* and *theory generation*: how can economic systems be better understood through a systematic examination of dynamic behaviors? An example of this is the questions of economists such as Smith, Schumpeter, and Hayek that have still not been answered: what are the self-organizing capabilities of decentralized market economies? A typical approach to this question is to construct an agent-based world that captures key aspects of decentralized market economies (circular flow, limited information, strategic pricing,...), introduce self-interested traders with learning capabilities, and to observe the development over time.
- A fourth objective is *methodological advancement*: which tools are needed to undertake the study of economic systems through computational experiments? ACE researchers have to model the structural, institutional, and behavioral characteristics of economic systems for this purpose and subsequently to evaluate the logical validity of the propositions made together with the theoretical foundation of experimental design. Finally,

the experimentally-generated theories have to be tested against real-world data.

Due to the properties and goals described above, ACE has some crucial advantages over classic approaches to economic modeling:

- *Emergence of complex properties from self-organization:*
  The phenomenon of spontaneous order was firstly observed in CAS related systems by John Holland. He called this phenomenon *'emergence'*, in the context of networked technical systems, constructed to artificially mimic principles of nature: "Emergence occurs in systems that are generated. The systems are composed of copies of a relatively small number of components that obey simple laws. Typically these copies are interconnected to form an array that may change over time under control of the transition function... The whole is more than the sum of the parts in these generated systems. Interactions between the parts generate a composed behavior that cannot be obtained by summing the behaviors of the isolated components. Said another way, there are regularities in system behavior that are not revealed." (Holland 1998, p. 225) The emergence principle can be seen as a direct connection between CAS (or ACE) and Smith's *'invisible hand'*, Schumpeter's *'evolutionary economy'* as well as Hayek's *'catallaxy'*:[31] "Many economic systems can be classified as complex systems. Such a system is complex in a special sense: It consists of a network of interacting agents (processes, elements); it exhibits a dynamic, aggregate behavior that emerges from the individual activities of the agents and its aggregate behavior of the individual agents. (Holland & Miller 1991, p. 365)

- *Computability and constructiveness:*
  By definition, ACE models are inherently capable of satisfying the claim of *computability and constructiveness*. ACE models can not be modeled and evaluated beyond recent computational possibility, unlike their pure mathematical counterparts. The ACE property of emergence seems to allow the formulation of simple behavioral rules for the agents, while expecting complex system behavior.[32] Holland & Miller (1991, p. 365) argue at this point: "Current theoretical constructs, based on optimization principles, often require technically dramatically demanding derivations. It is an obvious criticism of these constructs that real agents lack the behavioral sophistication necessary to derive the proposed solutions. This dilemma is resolved if it is postulated that adaptive mechanisms, driven by market forces, lead the agents to act as if they were optimizing." Especially in

---

[31] See the work of Vriend (2002) for a further discussion of the relationship between Hayek and ACE.

[32] An often quoted argument at this point is that the $\mathcal{NP}$-hardness of the GEE-problem with indivisibilities and complementarities does not prevent real markets from developing of a (stable?) equilibrium. This *complexity reduction* effect may be exploitable in ACE formulations.

the field of evolutionary economics and evolutionary game theory substantial advances substantial advances have been made using ACE methods (Axelrod & Hamilton 1981, Axelrod 1987).

By now there are a large number of ACE models for the simulation of Walrasian (Tesfatsion 2006, Wellman 1993, Cheng & Wellman 1996), game theoretically oriented (Vriend 2004) as well as combinatorial markets (Hudson & Sandholm 2002, Boutilier & Goldszmidt 1999, Sandholm & Lesser 1997, Parkes & Ungar 2000) providing reasonable results.[33] ACE seems to be at least an innovative element in the exploration of modern economics, possibly replacing traditional manners of thinking about economic analysis.

The previous sections demonstrated the complexity of pricing and allocation mechanisms in a brief discussion of the historical development of economic theory that affects the object of investigation of this thesis.

Two points in the discussion seem to be of crucial importance and should therefore be pointed out in our context:

- The complexity of modern equilibrium theory is prohibitive with respect to the computability of real world allocation and pricing problems. If solution algorithms have been found at all, they are extremely time consuming especially for the most interesting and challenging case of combinatorial resource allocation.
- One way to tackle the inherent problems of 'realistic' resource allocation processes is to employ distributed systems that operate in virtual economic worlds, as the new branch of ACE research proposes. This could be partially done based on paradigms deriving from classics of economic thought, like those of e.g. Walras, Schumpeter, or Hayek, but has also to be enriched with novel techniques that are made possible by the recent computational possibilities, like the introduction of learning capabilities and evolutionary principles.

Equipped with these mathematical, historical, and philosophical prerequisites from classical economics, the next chapter can now address in detail the specifics of dynamic pricing and automated resource allocation that are of interest especially in the context of information production and information services provided by distributed computer systems.

---

[33] The use of ACE simulations is especially popular in the auction and negotiation context, as well as in networked economies and stock-market micro simulations.(Schwind, Stockheim & Rothlauf 2003, Stockheim et al. 2005, Stockheim, Schwind & Weiss 2006, Hein et al. 2005, Hein et al. 2006)

Dynamic Pricing and Automated Resource Allocation

The main concern of this thesis is to evaluate the application of economically organized resource allocation methods for the provision of information services in distributed computer systems. A *price variable* is used to control this allocation process by dynamically reacting to supply and demand fluctuations. On this account the first part of this chapter deals with the appropriate definition of dynamic pricing accompanied by a taxonomy of dynamic pricing methods. The second part of this chapter deals with the application of such dynamic pricing mechanisms especially for efficient resource allocation in distributed computer systems. This is done by providing a further categorization of these systems together with an analysis of the resource properties and allocation mechanism characteristics typical of the ISIP provision tasks treated in this thesis. Subsequently the applicability of *multi-agent systems* for the simulations aimed at, is discussed while including the topic of ontologies for agent communication purposes. The chapter ends with a short introduction to the two dynamic pricing mechanisms that will be investigated in more detail in the rest of the thesis.

Returning to the pricing issue, it can be stated that the price variable can fulfill two functions in the context of our PCRA scenario:

- Firstly the consumers can express their *willingness-to-pay* by using the price variable associated with their demand for the resources to enable the market mediator(s) to control the automated allocation mechanism.
- Secondly the *reservation price* claimed for the use of a resource can be adjusted by the service provider according to the particular load situation to reduce demand and increase the yield and efficiency of the system.

Applied in this way, the resource acts as an *information entity* that enables a decentralization of resource allocation control. The pricing of resources provides a number of benefits in distributed computer systems, like the Internet or Grid applications (Gupta et al. 1999): decision making, especially the pricing process, becomes decentralized for the participants, pricing changes the

user behavior and makes the allocation process more transparent. In extreme
load situations dynamic pricing avoids the devaluation of the entire resource
network by preventing it from congestion due to exhaustive use. This topic
is traditionally addressed by congestion pricing mechanisms (MacKie-Mason
& Varian 1995). Due to the *volatility* of computational load in computer networks the application of *dynamic prices* is being increasingly discussed as a
way to replace *fixed pricing* methods for load balancing and resource allocation purposes in computational environments (Gupta et al. 1997, Ganesh
et al. 2001).

2.1 Dynamic Pricing

The concept of *dynamic pricing* has not yet been defined precisely in the
literature. For this reason Figure 2.1 tries to give a short taxonomy of the *dynamic pricing methods* referred to in the context of this thesis. Pricing can be
mainly divided into *fixed pricing*, denoting *prices posted by the sellers* that are
changed only in the long term following market fluctuations, like traditional
retail prices, and dynamic prices that result either from an interactive price
discovery process *(interactive pricing)* or a short term price setting process
carried out by the sellers depending on consumer behavior *(dynamic price
posting)* (Elmaghraby & Keskinocak 2003, p. 1288)(Kauffman & Wang 2001,
p. 7034). With dynamic price posting, goods are sold under *take-it-or-leave-it
prices*, where the seller dynamically changes prices over time *(intertemporal
prices)* based on factors such as time of sale, demand information, and supply availability. Dynamic price posting includes *dynamic price discrimination*
and *yield management* methods, which are both explained in the following
section.

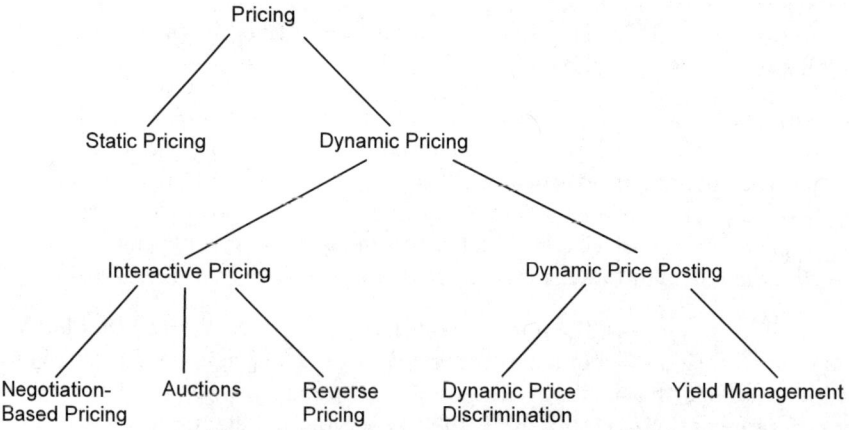

Fig. 2.1. Taxonomy of pricing mechanisms

Dynamic pricing touches on almost all classic economic price setting mechanisms, like *auctions* or *negotiation-based pricing*, and in addition innovative pricing mechanisms like *reverse pricing*. Dynamic pricing has been of increasing importance in recent years. Elmaghraby & Keskinocak (2003, p. 1287) give three reasons for this phenomenon:

(i) the increased availability of demand data
(ii) the ease of changing prices due to new technologies
(iii) the availability of decision support tools for analyzing demand data for dynamic pricing

In the context of this thesis dynamic pricing denotes 'a flexible price setting process with respect to market dynamics (supply and demand fluctuations), sectoral price discrimination according to the W2P of the individual customer (customer group) regarding specific product properties like e.g. quality, intertemporal price discrimination as well as the capacity and inventory conditions of the sellers/producers facilities'.

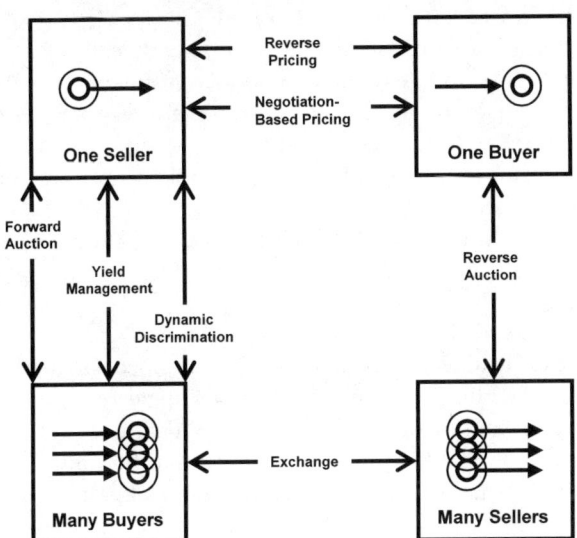

Fig. 2.2. Buyer seller constellations in relation to suitable dynamic pricing mechanisms

The following section gives a brief overview of the dynamic pricing mechanisms mentioned (see Table 2.1) according to the given criteria. These criteria involve seller-buyer interaction cardinality (number of sellers and buyers involved, see also Figure 2.2), ability to respond to market dynamics, pricing processing time (time period from pricing request to price setting), costs for the application of the pricing method, as well as allocation and pricing efficiency (with respect to a welfare or return maximization objective) and are based on a discussion of Internet pricing principles given in (MacKie-Mason et al. 1996).

2.1.1 Negotiation-based Pricing

Negotiation-based pricing (NBP) in the traditional way, is conducted as a bilateral face-to-face, mail or fax process (buyer-seller cardinality is one-to-one)[1]. For this reason traditional price negotiations are difficult to manage, have a high response time, and are prone to misunderstanding (Bichler et al. 2003, p. 313). It is the time consuming bilateral negotiation process itself that makes NBP costly. The supply-demand reaction dynamics of negotiation-based pricing depends strongly on the frequency of the (re-)negotiation process. Due to the high cost of traditional negotiation processes the dynamics to market tends to be in the mid-range compared with the other DP methods. The efficiency and fairness of traditional NBP depends strongly the negotiating talent of the agents. To avoid this drawback, *electronic negotiation tables*, *decision* and *negotiation support systems* as well as artificial *negotiation software agents (NSA)* have been introduced. Electronic (price) negotiations promise higher level of process efficiency and effectiveness, and most importantly, a higher quality and faster emergence of agreements (Bichler et al. 2003, p. 314). This effect relies on an efficient design negotiation protocol design and a high repetition rate of the negotiation process itself. Examples of electronic negotiation platforms (besides multiple other implementations) are SARDINE (Morris et al. 2000), Kasbah (Chavez & Maes 1996), RETSINA (Sycara et al. 2003), Tête-à-Tête (Maes et al. 1999) and COALA (Tsvetovat et al. 2000).

2.1.2 Auctions

Auctions, "in their traditional form, are resource allocation mechanisms based on a competitive bidding process over a single issue (e.g. price) of a single, well defined object, and involve" (Bichler et al. 2003, p. 317) "a set of auction rules that specify how the winner is determined and how much he has to pay" (Wolfstetter 1996). Auctions can be seen as market institutions "with an explicit set of rules determining resource allocation and prices on the basis of bids from the market agents" (McAfee & McMillan 1987) while the bids indicate the bidders' W2P for an object (Milgrom 1989). Classical non-electronic auctions are costly and time-consuming to perform. For this reason traditional auctions are normally implemented for valuable goods (fine art, real estate) or large quantities of identical goods (stock exchange, flower auction). Electronic auctions allow one to reduce cost-per-item drastically, which is a partial reason for the success of web-based auctions.[2] The repetition frequency

---

[1] Sometimes negotiation-based pricing is defined for all bi- and multilateral bargaining processes including auctions (Bichler et al. 2003, p. 315). In this thesis auctions etc. are not treated as a subset of NBP.

[2] E.g. the auction portal: www.amazon.com/auctions

of electronic auction can be increased up to millisecond time intervals (electronic exchanges)[3] and thus provides low response times to supply-demand changes.

According to the seller-buyer cardinality, auctions can be divided into three classes:

- **Forward Auctions**: one seller and multiple buyers (classical auction format e.g. the English art-auction)
- **Reverse Auctions**: one buyer and multiple sellers (procurement auction increasingly popular in Electronic-markets)
- **Exchanges**: multiple sellers and multiple buyers (exchange format)

Forward auctions are normally designed for revenue maximization, whereas reverse auctions should minimize purchase costs. The effectiveness and fairness of auction mechanisms (Pareto optimality, welfare maximization etc.) are the subject of a great deal of auction and mechanism design literature (see (Milgrom 2004, Klemperer 2004)). Besides multiple professional B2C and B2B auction platforms,[4] a number of more sophisticated e-auction platforms enabling combinatorial and multi-attribute auctions have been developed for research purposes in the recent years. Examples of such platforms are e.g. eMediator (Sandholm 2002b), Michigan Internet AuctionBot (Wurman et al. 1998) and iBundler (Giovanucci et al. 2004).

2.1.3 Reverse Pricing

Reverse pricing (RP) enables both buyer and seller to influence the final price of a product or service. Transactions take place for a price denoted by a buyer's bid if this bid equals or surpasses a secret threshold price initially set by a seller. If a buyer's bid does not surpass the seller's threshold price, the option to place additional bids depends on characteristics of the mechanism design as defined by the seller or a third party (Bernhardt & Hinz 2005). The response of RP to the market dynamic depends on the adjustment of the threshold price to the consumers' price reaction function. Reverse pricing is primary enabled by e-commerce *automated negotiation system (ANS)* systems and can be customized by various design variables depicted in detail by Bernhardt (2004). The IT infrastructure of RP systems and the simplicity of the pricing mechanism lead to short response times and low pricing costs. RP has a one-to-one buyer-seller cardinality and exploits the individual W2P of the customers and leads therefore to an increased revenue compared to other pricing methods (Chernev 2003, Spann et al. 2004, Il-Horn & Terwiesch 2003). The most prominent implementation of reverse pricing allows consumers to select from a choice of discrete prices.[5]

---

[3] Providers of high frequency data: www.tickdata.com , www.olsonresearch.ch

[4] Most popular examples are: www.ebay.com , www.covisint.com

[5] E.g. reverse pricing for flight sales: www.priceline.com

2.1.4 Dynamic Price Discrimination

Dynamic price discrimination (DPD) estimates the consumer's W2P based on statistical data and consequently adjusts a product or service price to increase seller return (Bichler et al. 2002, p. 287). DPD can be executed as *first-degree* price discrimination which is considered to be "perfect" price differentiation (Varian 1996), *second-degree* price discrimination (producer sells different units of output for different prices e.g., quantity discounts and premiums), and *third degree* price discrimination (customers are priced based on group classifications e.g., senior citizen discounts), of which the latter two can be considered the most common dynamic pricing methods. First degree DPD in particular is made possible by the increase in electronic *customer relationship management (CRM)* systems. Customers are charged individual prices that are based on their estimated W2P when buying in online shops. Electronic CRM systems are used to collect the individual data necessary to calculate the estimated W2P of the customer for the application of first degree DPD. Because of the automated W2P calculation in electronic first degree DPD, the pricing response time is in the short range. DPD's reaction time to market behavior is slow due to the long term evaluation of the consumers' reaction function and the resulting slow change of pricing schemes. The cost of DPD is situated in a medium to upper level compared to other dynamic pricing methods. DPD should be Pareto efficient due to its first degree price differentiation property (Varian 1996). At the moment, only a couple of e-commerce-based DPD models exist (Ghose et al. 2002, Ulph & Vulkan 2000). Real world applications of first degree DPD are rare because consumers do not accept the emerging "price inequality", especially if they are price takers (Bergen et al. 2003).[6]

2.1.5 Yield Management

Yield management (YM) is designed to find the optimal revenue management for perishable assets through price segmentation (Weatherford & Bodily 1992, p. 833). YM is frequently found in industries with perishable inventories and fixed or least only gradually changeable capacities like the airline, hotel, computer or telecommunication industry (Belobaba 1989, Bitran & Gilbert 1996, Nair & Bapna 2001, Humair 2001). YM normally includes a dynamic pricing process due to the inter-temporal value decline of the goods and services and an additional product differentiation component to make the pricing policy less transparent. YM is a seller-centric approach, - meaning that the seller sets the price -, addressing multiple buyers. The pricing response time is in the mid range compared with other dynamic pricing mechanisms, because price calculation in YM usually employs precalculated pricing schemes using historical pricing data. YM methods are designed for sellers' revenue optimization, while customers' self selection of the pricing class in connection with product differentiation leads to a high level of customer satisfaction.

---

[6] See also chapter 3.

Table 2.1. Outline of different dynamic pricing methods with respect to buyer-seller interaction, response to market dynamics, processing time and pricing costs, and allocation efficiency.

	Seller-buyer interaction	Description of the pricing mechanism	Response to market dynamics
Negotiation-based Pricing	$1:1$ $(1:n)$ $(n:1)$	Bilateral negotiations lead to exchange contracts based on individually bargained prices. Multiple parties can be involved by repeating negotiations with alternating partners.	Dynamics of NBP depend on the repetition frequency of the negotiation process and the possibility of renegotiation.
Auctions	$1:n$ $1:n$ $n:1$	Interactively determine the price and allocation of goods and services based on the bids of the participants according to predefined rules.	Depends on the repetition frequency of the auction. Response times can be very low e.g. for electronically managed continuous double auctions.
Reverse Pricing	$1:1$	Customer places a bid (typically below his W2P). Seller accepts bid if threshold price is outbid. Threshold price is unknown to buyers.	RP's response dynamic to the market depends on setting the price threshold according to the consumers' price reaction.
Dynamic Price Discrimination	$1:1$	Differentiates prices according to an individual customer classification following the estimated W2P of the buyers.	Dynamics depends on the sellers' learning rate for the consumers' reaction function. Low pricing time is possible if ANS are used.
Yield Management	$1:n$	YM price differentiation follows a fixed pricing scheme. YM uses consumers self-selection related to service level and timing conditions.	Dynamics strongly relate to the learning rate for the consumers' W2P and the quality of the YM model.

	Processing time and pricing costs	Allocation and pricing efficiency
Negotiation-based Pricing	NBP usually requires long processing time and has high pricing costs per negotiation process. B2B pricing applications strive for ANS to reduce costs.	The efficiency of NBP allocation corresponds to the negotiating talent and bargaining position of the parties. ANS can help to achieve balanced pricing and allocation.
Auctions	Processing time and pricing cost of auctions strongly depends on the auction type and implementation. Recent e-commerce applications of auctions lead to low per item pricing costs.	Allocation and pricing efficiency of auctions varies heavily with the chosen mechanism. Mechanism design literature discusses this subject in detail.
Reverse Pricing	RP systems have low processing time for requests as a result of the simplicity of the pricing mechanism.	RP methods aim to exploit the individual W2P of the customer. Due to the invisibility of the threshold price, pricing in-equality is not perceived.
Dynamic Price Discrimination	If DPD is used in connection with e-commerce applications low response time and low level transaction costs can be achieved.	DPD methods aim to exploit individual W2P. This leads to a higher return for the seller. Perceived inequality of DPD can cause customer disaffection.
Yield Management	YM systems allow for short processing time of pricing requests for goods and services. Pricing costs lie in the mid-range due to well-developed YM systems and fair experience from airline industries.	YM pricing targets high revenue from sales and achieves a satisfying allocation if pricing classes and contingents are appropriate. Customer satisfaction is high due to self-selection of the pricing class.

2.2 Automated Resource Allocation

The dynamic and efficient allocation of resources has an outstanding importance in the economic use of distributed computer systems. The complexity of this task is determined by the size (number of systems and number of users), and heterogeneity in applications (online transaction processing, multimedia, intelligent information search) and resources (central processing unit capacity, memory, bandwidth, services etc.) is further increased by the following factors (Ferguson et al. 1996, p. 157):

- Resources are owned by multiple organizations.
- Continuous change of systems, set of users and applications.
- Dependency of systems performance on simultaneous allocation of multiple resources.

In the following, the types and properties of resources and allocations will be characterized shortly.

2.2.1 Properties of Resources

Resources in a distributed system can be regarded as dynamic system elements which are subject to fluctuations in the course of time. A central impact factor for the design of resource allocation systems is the nature of the resources themselves. Chevaleyre et al. (2005, p. 16 seq.) discern the following types of resources according to their properties in the context of resource allocation systems:

- **Continuous vs. Discrete:**
  A resource may be either *continuous* (e.g. bandwidth) or *discrete* (e.g. number of CPUs). Continuous resources are regarded as infinitely *divisible*, whereas discrete resources are only divisible into units of an *in-divisible* entity. Divisible resources, however, can be regarded as discrete resources with very small in-divisible units. This physical property has a strong impact on the selection of allocation methods. Indivisibilities have to be treated by allocation methods that satisfy the integrity conditions, like integer programming. These methods are computationally complex.
- **Divisible or Indivisible:**
  As shown in the previous point *(in)divisibility* and *continuousness* (discreteness) are tightly coupled. In this thesis all resources are assumed to be indivisible. This enables the users of the allocation methods presented here to form bundles of discrete resource units and allows them to optimize the allocation by using combinatorial assignment methods.
- **Shareable or Exclusive:**
  A *shareable* resource can be simultaneously used by a number of agents. *Non-shareable* resources can only be exclusively used by one agent at a time. In the context of this thesis resource agents provide discrete units of

their production capacity which can be acquired in different quantities by task agents for exclusive use according to their bids.

- **Static or Consumable:**
  A resource may be *consumable* in the sense that the agent holding the resource may use up the resource while performing a given action (e.g. providing a specific web service). Furthermore resources may be *perishable* in the sense of vanishing or losing its value over time (e.g. CPU, NET capacity that is not used). In this work, resources are considered to be *static* as well as to be perishable depending on the application context.

- **Single-unit vs. Multi-unit:**
  Multi-unit settings support the agents' bidding for a number of discrete units of the same resource type. *Single-unit* settings trade each unit of a resource as a different resource type by assigning a specific signature to each resource unit. The distinction between single-unit and multi-unit settings is a matter of representation. Multi-unit representations can be converted into a single unit notation by assigning a unique identifier to each resource unit, even if the particular resource unit has the same type as other resource units in the bundle. Multi-unit representations are more compact than single-unit formulations. In this thesis multi-unit representations are used, while the single-unit problem can be seen as a degenerate multi-unit problem.

- **Resources vs. Tasks:**
  A *task allocation* problem can be regarded as a *resource allocation* problem at a sufficiently high level of abstraction, see also section 2.2.3. From the viewpoint of a producing agent, tasks may be seen as resources with negative payoff in economic models, because the agents have to pay for the use of these resource. This consideration, however, is too simple if constraints and complementarities are taken into account while defining specific resource allocation problems.

2.2.2 Properties of Allocation Mechanisms

Allocation mechanisms can be evaluated by using different quality criteria. The selection of these criteria, however, depends on the individual objectives of the allocation system designer. In the context of this thesis, the evaluation criteria for negotiation and coordination processes proposed by Sandholm (1995, p. 12 et seq.) are of interest:

- **Social Welfare:**
  Social welfare is the sum of all agents' payoffs or utilities in a given allocation solution. It measures the global degree of satisfaction of the agents in system. Social welfare can be used as a criterion for comparing alternative mechanisms by evaluating the solutions that the mechanisms lead to. If social welfare is used to rate allocation mechanisms, the definition of utility must be made comparable for all agents. All welfare maximizing solutions are a subset of Pareto efficient solutions.

- **Pareto Efficiency:**
  A solution x is Pareto efficient (Pareto optimal) if no other solution x' exists such that at least one agent is better off in x' than in x and no agent is worse off in x' than in x. Pareto efficiency is a measure of global allocation quality. Once the sum of the payoffs is Pareto optimal, an agent's payoff can increase only if another agent's payoff decreases. The Pareto efficiency measure does not require an inter-agent utility comparison.

- **Individual Rationality:**
  The participation of individuals in a allocation process must lead to a positive individual payoff compared with the payoff yielded by a solution in which the agent does not participate. An allocation mechanism is individually rational if participation is individually rational for all agents (see section 6.3.5).

- **Stability:**
  It is individually rational for self-interested agents to manipulate allocation mechanisms such that they provide a higher payoff, reducing the welfare of the remaining agents and avoiding Pareto efficient solutions. Therefore allocation mechanisms should be non-manipulable. Sometimes it is possible to design mechanisms with dominant strategies, which means that an agent is best off by using a specific strategy no matter what strategies the other agents use (see section 5.1.8). In many cases the agents' best strategies depend on the strategy of the other agents in the allocation mechanism. In such settings other stability criteria are needed, because dominant strategies do not exist. A solution for this is the use of mechanisms that provide *Nash equilibria*. In a Nash equilibrium, each agent chooses a strategy that is the best response to the other agents' strategies. However, two main problems arise with the use of the Nash equilibrium. First, in some games a Nash equilibrium does not exist. Second, some games provide multiple Nash equilibria. In such cases it is not clear which strategy the agents should play. Even if there is a Nash equilibrium and it is unique, in some cases it is not clear what it guarantees regarding the outcome for the agents. Additionally, conflicts between efficiency and stability objectives arise while employing Nash strategies.

- **Symmetry:**
  A criterion which is often subsumed under the stability criteria is the symmetry criterion. It denotes that no agent should be given preferential treatment, nor should any participant in the allocation process be treated unequally.

- **Computational Efficiency:**
  Advanced allocation mechanisms are computationally complex. This often leads to system congestion resulting from a high computational overhead. In many cases the optimization objectives of resource allocation mechanisms have to be put into perspective. An explicit trade-off between computational cost and allocation quality has to be made e.g. by using heuristics (see section 6.3.1).

- **Distribution and Communication Efficiency:**
  To enhance redundancy in allocation mechanisms e.g. robustness with respect to a single point failure or a performance bottleneck, distributed mechanisms should be preferred. Computational and communication efficiency often stand in a trade-off relationship. For example, negotiation mechanisms in an economic allocation system that uses simple bilateral bargaining process require relatively low computational power compared to centralized allocation mechanisms like combinatorial auctions. Communication complexity, however, behaves the other way round in this example: with multiple bilateral negotiations the communication effort is high, whereas auctions have low communication requirements. Computational efficiency, distribution and communication efficiency therefore stand in a trade-off relationship (see section 2.3.).

For the analysis and design of ISIP allocation mechanisms it seems to be helpful to look further at additional coordination system classification criteria.

2.2.3 Classification of Resource Allocation and Scheduling Problems

Scheduling can be regarded as a special case of the resource allocation problem (Casavant & Kuhl 1988, p. 1). In Table 2.2 a survey of allocation and scheduling problems in the context of general resource allocation is provided. Scheduling decisions are generally described as allocation decisions with *time constraints* and *interdependencies* and *utility complementarities* between different tasks on the distributed resources. Additional precedence constraints lead to *job-shop scheduling* related allocation problems which are $\mathcal{NP}$-hard to solve (Garey et al. 1976). Allocation problems in time without any precedence interdependencies are often referred to as *load balancing* problems (Mehra & Wah 1993, Mehra & Wah 1997).

Scheduling problems do not usually have any utility or price-dimension-like economic resource allocation problems. However, they can be enhanced with an economic dimension by assigning economic weights (utilities, prices) to the jobs. A class of these problems, called *economic job-shop scheduling* problems was recently introduced by (Elendner 2003, Conen 2003, Stockheim et al. 2005). The expression 'allocation' perse mostly has an economic connotation, denoting that the utility which exists for the resource requesting agents, is taken into consideration in the allocation decision. To avoid semantic conflicts, this type of allocation is called *economic allocation*. As an extension of this allocation notation *combinatorial allocation* problems cover all the criteria mentioned in Table 2.2. Substitutionalities as well as complementarities (utility interdependence) are included into combinatorial allocation decisions.

Combinatorial allocation methods are also capable of dealing with most scheduling problems because time constraints and interdependences can be expressed in the form of strong complementarities (Collins et al. 2002, Elendner 2003, Conen 2003, Stockheim et al. 2005).

Table 2.2. Classification of resource allocation and scheduling problems

Assignment Problems	Time Constraints	Time Inter-dependence	Precedence Constraints	Utility Constraints	Utility Inter-dependence
Load Balancing	✔				
Simple Scheduling	✔	✔			
Job-Shop Scheduling	✔	✔	✔		
Economic Job-Shop Scheduling	✔	✔	✔	✔	
Economic Allocation				✔	
Combinatorial Allocation	✔	✔	✔	✔	✔

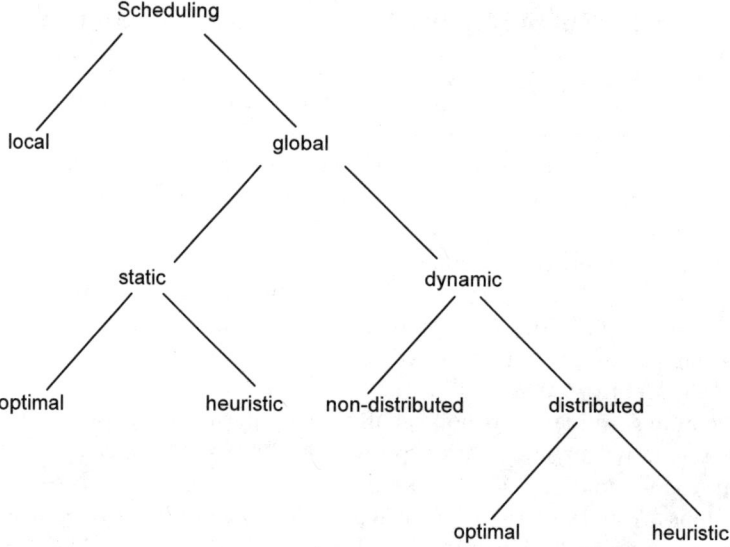

Fig. 2.3. Taxonomy for allocation and scheduling problems used in computer science

Casavant & Kuhl (1988, p. 5) provide a hierarchical taxonomy for allocation and scheduling problems which is widely used in computer science. Figure 2.3 depicts a slightly modified version of this taxonomy. The following five characteristics used in the classification graph could also help to categorize allocation systems for ISIP provision in distributed computer systems:

- **Local versus global:**

Local scheduling denotes the assignment of processes to the time slots of a single resource. The allocation decision in local scheduling environments is reached without regarding any interdependencies between production tasks on other resources. Global scheduling decisions are concerned with the allocation of tasks in resource networks like the job-shop scheduling problem.

- **Static versus dynamic:**
  The choice between static and dynamic scheduling indicates the time at which the scheduling or assignment decisions are made. In the case of static scheduling the allocation is calculated before the execution of the production plan starts. This is especially useful if the complete planning information is available before the plan's execution starts. Dynamic scheduling, however, can deal with new information arriving during execution time by updating the current schedule.

- **Optimal mechanisms versus heuristics:**
  An allocation mechanism can be considered as optimal if efficient resource allocation, according to the criteria below, is achieved. For this purpose complete information about resource requirements and system load must be provided to the scheduler before the optimization starts. Sub-optimal allocation methods are distinguished into heuristics and approximation solutions. The latter terminate the optimization process if the solution quality is near a predefined range. Heuristics use a simplified optimization model to deal with high problem complexity ($\mathcal{NP}$-hardness).

- **Centralized versus decentralized:**
  If scheduling is performed by a centralized instance, the entire allocation mechanism can be called "centralized" allocation. A decentralized approach allows the collaboration of multiple actors in the allocation process.

- **Cooperative versus non-cooperative:**
  Scheduling and allocation mechanisms are called "cooperative" if schedulers allow the use of the resources of one profit center for other profit centers without asking for immediate compensation. Non-cooperative mechanisms are often employed in the context of autonomous agent modeling and imply that the agents try to keep the resources for their exclusive use.

Additionally, a further pair of criteria for scheduling system classification are useful in the context of this thesis:

- **Learning versus non-learning:**
  While using dynamic scheduling, it is helpful for the decision system to learn the arrival patterns of tasks in order to anticipate resource congestion situations. Learning systems, unlike non-learning systems, are able to adapt to system properties and can help to increase the allocation quality. Learning from other actors' behavior also ameliorates performance in decentralized scheduling systems (Boyan & Moore 1998, Zhang 1996).

2.3 Economic Resource Allocation in Distributed Computer Systems

As shown in the introduction, (1.5) the dichotomy of centralized and decentralized approaches gave rise to the discussion of optimal coordination and resource allocation mechanisms in the classical microeconomic literature of the last eighty years. This discussion was transferred to the domain of automated resource allocation in *computational economies* with the networking capabilities of information technology that arose in the early eighties. The development leads from the first economic resource allocation approaches for distributed systems published in "Ecology of Computation" (Miller & Drexler 1988*a*, Miller & Drexler 1988*b*) to more refined allocation and scheduling mechanisms in computer environments offered by "Market-based Control" (Clearwater 1996) in the mid nineties. A large number of publications on economic resource allocation mechanisms has been produced since, bringing the topic into the Grid computing domain recently (Buyya et al. 2000, Buyya & Vazhkudai 2001, Buyya et al. 2001). In the following section a brief taxonomy (Figure 2.4) of economic resource allocation mechanisms is presented and illustrated with prominent allocation system approaches.

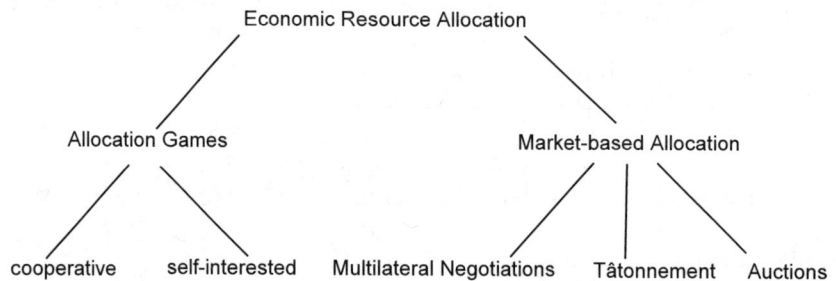

Fig. 2.4. Taxonomy of economic resource allocation mechanisms in distributed computer systems

Economic allocation mechanisms employ resource use measures in connection with explicit utility scales. Normally, monetary units are used to express the utility of the resource consumption. Prices (e.g. equilibrium prices) often serve as control variable in the context of market-based allocations. A feasible discrimination criterion for the class of economic allocation mechanisms in distributed systems is therefore the existence of global equilibrium price variables for each of the services and goods (cf. to Ferguson et al. (1996, p. 162 seq.)):

- If system participants only interact in the form of an *allocation game* with different payoffs as a result of specific actions employing various strategies, a *game theoretical* setting can be assumed. Pricing is not necessary in allocation games; however *local prices* are often used as an information entity in recent approaches.

- If agents interact as buyers and sellers aiming for a *global equilibrium price* by means of *negotiation or exchange procedures*, the mechanism is regarded as *market-based allocation*.

Both approaches, however, are not totally separable. Markets can, for example, be seen as coalition games with possible utility transfer (Osborne & Rubinstein 1994, p. 268). Besides such overlapping, the taxonomy offered by Ferguson et al. (1996, p. 162 seq.) is valuable for a main categorization.

2.3.1 Allocation Games

The allocation of tasks in a network of freely accessible shared resources often leads to congestion of single resources and in the worst case to a devaluation of the entire resource network utility (MacKie-Mason & Varian 1995). From the game theoretical point of view this can be compared to the 'tragedy of the commons' metaphor given by Hardin (1968). Two main solution approaches suggest themselves in this context:

- To avoid excessive use of the resources, one can introduce *self-interestedly* acting economic agents that manage and own individual resources in the network (Wellman 1995, p. 88).
- To achieve a load-balancing effect the *unselfish* distribution of tasks between the resources can be targeted by the agents in the sense of *cooperative problem solving* strategies (Durfee 1999).

Table 2.3 depicts an overview of several categories of allocation game approaches together with the related literature that will be discussed in the following sections.

Table 2.3. Categories of allocation games and related approaches from literature

Allocation Games	
cooperative	self-interested
Nash Bargaining (Network Games) Dziong & Mason (1996), Chang & Subramanian (2000), Yaiche et al. (2000), Cao et al. (2002)	**Nash Bargaining** Feldman et al. (2004), Kwok et al. (2005), Maheswaran & Basar (2001) **Minority Games** Galstyan et al. (2003), Savit et al. (2003) **Iterative Prisoner Dilemma** Feldman et al. (2005), Bredin et al. (2000)
Worth / State / Task Oriented Domains Zlotkin & Rosenschein (1992), Rosenschein & Zlotkin (1994), Zlotkin & Rosenschein (1996)	

In a well-known game theoretical approach Rosenschein & Zlotkin (1994) address the problem of truth-revelation and cooperation in a system of insincere self-interested agents that have to solve tasks according to their own goals. The setting unexceptional regards the interaction between two agents that have symmetric capabilities and equal costs for the handling of their tasks. Their model satisfies the assumptions of individual rationality, stability, symmetry, simplicity and Pareto efficiency. Rosenschein & Zlotkin (1994, p. 19 seq.) present a domain classification into *worth oriented domains (WOD)*, *state oriented domains (SOD)* and *task oriented domains (TOD)*:

- *Worth oriented domains:*
  In these domains agents assign a worth to each potential state. Resources are scarce in WODs. Agents in a WOD only can reach their goals by making compromises (Zlotkin & Rosenschein 1993).
- *State oriented domains:*
  In SODs, as in WODs, agents have to resolve conflicts over limited resources. SODs, however, unlike WODs, capture binary worth functions. This means that agents try to reach some explicit goal states under the constraint that no states that partially satisfy the agents exist. For this reason, SODs represent a subclass of WODs (Zlotkin & Rosenschein 1996).
- *Task oriented domains:*
  TODs, in turn, are a subset of SODs. In TODs the tasks can be carried out without regarding the interaction with other agents. All resources needed to accomplish a task are available to an agent. It is, however, possible that other agents can reach the goals with lower costs. For this reason, within TODs agents try to redistribute their tasks, so that they can be dealt with less expensively and mutually beneficially. TODs are inherently cooperative domains (Zlotkin & Rosenschein 1992).

In the Rosenschein & Zlotkin framework, three alternative types of deals are analyzed. In the first category, the *pure deals*, agents allocate deterministically exhaustive, disjoint task sets. The second category comprises *mixed deals* that define a probability distribution over such disjoint task sets. With *all-or-nothing deals*, the third category describes mixed deals where agents can choose from deals that include partitions with one agent handling the tasks of all other agents.

To test the strategic implications of the particular domains, Rosenschein & Zlotkin identify three types of lying in task revelation. The variants of strategic lying with respect to the existence of a task are:

(i) *Hidden task:* a task that is existent, but unrevealed.
(ii) *Phantom task:* a task that does not exist, but can be generated.
(iii) *Decoy task:* a task that does not exist and can not be generated.

As a result of these analyse it can be seen that in more general forms of TODs (in terms of utility function shapes), many different lying methods are profitable here, and in consequence also for the subclass of WODs.

The work of Rosenschein & Zlotkin laid the theoretical foundations for a significant number of negotiation systems operating in distributed computer systems and networks. While the Rosenschein & Zlotkin framework considers cooperative as well as self-interested agents in resource allocation networks, practically-oriented applications normally employ only one of the two agent behavior categories:

- **Cooperative agents:**
  The use of cooperative agents in resource allocation games for distributed computational resources is not widely spread. Recent game theoretically-based resource allocation schemes are mainly used in wireless network environments. Often these approaches are subsumed under the decription *'networking games'*. Cooperative games for resource sharing often employ the *'Nash bargaining'* approach, where the bargainers negotiate for a fair *'contract point'* from among the set of all feasible solutions (Nash 1950). The outcome is chosen based on a-priori defined *'fairness criteria'*. These criteria, formulated as axioms in so-called *'axiomatic bargaining theory'* are: *symmetry, Pareto optimality,* and *invariance with respect to utility transformations* (scalability). A number of resource sharing algorithms for telecommunication and wide area networks based on variants of the Nash bargaining solution exist, e.g. Dziong & Mason (1996), Chang & Subramanian (2000), Yaiche et al. (2000), Cao et al. (2002). See Altman et al. (2004) for an extensive discussion of allocation games in networks. The latest approaches in cooperative resource allocation games for distributed computer systems concentrate on self organization in ad-hoc networks (Fang & Bensaou 2004).

- **Self-interested agents:**
  The concept of using *self-interested* agents to formulate allocation mechanisms in a game theoretical setting is closer to the classical market concept than solutions employing cooperative strategies.
  Non-cooperative game theoretical approaches to resource allocation in distributed computer systems are very common in the domain of electronic networks. Network resource management is often carried out by using *congestion-based pricing* (MacKie-Mason & Varian 1995, Korilis et al. 1995, Korilis et al. 1997) and *routing capacity allocation,* e.g in the TCP/IP protocol (Roughgarden 2002, Tardos & Roughgarden 2002, Cole et al. 2003, Correa et al. 2004). These applications are not discussed in this thesis: for a further overview on selfish routing and congestion pricing in networks cf. e.g. Altman et al. (2004).
  Most non-cooperative allocation and scheduling strategies in distributed computer systems first postulate (*i*) *utility functions* for the system participants. In the second step (*ii*) the formulation of *best response strategies* follows. For these mostly simple decision rules the *existence of a Nash equilibrium* is proved in a system of multiple agents as the third step (*iii*). Within two further steps the (*iv*) *efficiency* is measured (compared to achievable welfare) and the (*v*) *fairness* (symmetry) of the simulation results is checked. Feldman et al. (2005) use this methodology to implement and test a *finite parallelism model* where bidders asynchronously

bid for resources and receive the ratio of the resources requested within the bid to the sum of all bids submitted for the resource.[7] Their model uses a price anticipating mechanism for the machines, where the resource price is estimated by the total bids placed on a machine. The calculation of the Nash equilibrium in the finite parallelism setting is done by employing a *best response strategy* and a *greedy strategy*. The algorithms are examined in terms of convergence, efficiency, and fairness. The greedy algorithm has problems in converging to a stable equilibrium for a small number of agents; however, the convergence speed for an increased number of agents is high. The greedy allocation scheme performs better than the best response strategy in terms of efficiency, but behaves worse in terms of fairness. Bredin et al. (2000) construct a game theoretical allocation mechanism similar to the Feldman et al. synchronous bidding model. Mobile Internet agents bid for CPU resources located on specific host servers (differentiated services) via network connections. In various simulations they show that the resource allocation mechanism effectively prioritizes agents according to their endowments.

In another Nash equilibrium-oriented allocation system Kwok et al. (2005) propose a hierarchically organized game in a Grid. They integrate an agent-reputation index into the job-acceptance decision mechanism to prioritize the resource attribution instead of using a virtual budget of monetary units. A similar allocation model proposed by Maheswaran & Basar (2001) also relies on the construction of a stable Nash equilibrium; however, it tends to be a mixture between a pure game theoretical and a market approach.

One way to close the gap between cooperative and non-cooperative solutions in allocation games is the use of *evolutionary game theory*. Arthur (1994) presents an ACE model for the presence of visitors in an overcrowded location ('El Farol' bar). In his ACE setting he demonstrates that agents quickly find the optimal visiting strategy by learning from past data, regardless of which optimization heuristics they use to make their decisions. In 1997 Challet & Zhang expanded the concept of Arthur's attendance problem to an evolutionary game, the so called 'minority game'. The primarily non-cooperative game is played by an odd number of players, each with a finite number of strategies. At each step, the players have to choose to be in side A or side B. The fixed payoff of the game is attributed proportionally to the players that are in the minority side after their decision has been met. Played in such a way, it is an advantage for the bidders to be on the minority side together with as few other game participants as possible. As in the 'El Farol' problem, repeated iterations of playing the game and participants' learning leads to the convergence of the resource load to a mean value, preset as the threshold in the minor-

---

[7] This allocation model is related to auctions, however, more simple, responsive, and scalable.

ity game. Approaches proposed by Galstyan et al. (2003) and Savit et al. (2003) employ this effect for efficient resource allocation in distributed computer systems.

Feldman et al. (2004) also use the dynamics of evolutionary game theory to provide robust incentives for cooperation in a P2P network. In a generalized version of Axelrod's (Axelrod & Hamilton 1981) *iterative prisoner dilemma* that allows an asymmetric payoff matrix, one player is the client and the other player has the role of a server. The payoff matrix of the *generalized prisoner dilemma (GPD)* has the property that mutual cooperation leads to higher payoffs than mutual defection, mutual cooperation leads to higher payoffs than one player exploiting the other and defection weakly dominates cooperation at the individual level. The GPD payoff matrix models the reward of sharing files for the P2P participants. In the Feldman et al. model, agents take independent actions like: alter the current strategy, learn about the strategies' performance or stay with the same strategy. To learn the player collects information about the performance of its own strategy and the strategy of its neighbors. The introduction of a *reciprocal decision function* (trust function) sets the basis for the cooperation incentives. It maps from the history of the players's actions to the decision whether to cooperate with the player or to defect. In the case of the Feldman et al. approach, the agents simply take their own willingness to share files in the system as an estimator for the willingness to share of the peers. The probability of cooperating with a peer depends on this estimate. As an outcome of simulations the incentive model significantly increases the level of cooperation while the agents learn suitable response strategies for their peers' behavior. This effect can be observed for a medium number of agents (60); for a higher number of agents (120), however, cooperation fails in this setting.

An increasing number of game theoretical allocation models uses reputation and trust to replace virtual or real currencies in systems for the exchange of computational resources. This topic will not addressed in this thesis: see e.g. Mui et al. (2002), Huberman & Wu (2002), Stockheim, Schwind & König (2003a) and Xiong & Liu (2003) for a further discussion.

2.3.2 Market-based Allocation

As stated above, *market-based allocation* is characterized by the use of *equilibrium prices* as a central *information and coordination instrument* for the market participants. The emergence of such equilibrium prices can be explained in different manners. A first approach is strongly related to the interpretation of market forces by early economists like Adam Smith and F.A. Hayek (see section 1.5.3). Market pricing relies on the efficiency of *multilateral negotiations*. Through information diffusion, - communication of the temporary prices for particular goods and services -, these multilateral negotiations lead to an equilibrium market state driven by the strive for individual utility optimization. These negotiation process can be *standardized* and *effectivized* leading to institutionalized market mechanisms like *auctions*. A third approach that lies between these two concepts is the use of *tâtonnement procedures* that equilibrate *supply* and *demand* in markets by employing a *virtual auctioneer*. This more theory-related concept of market understanding, derived from mathematical economics, however, has its place in economically founded computational resource allocation systems. For this reason the next sections discuss the three methods from the perspective of resource allocation in distributed computer systems.

Table 2.4 gives an overview of the market-based mechanisms associated with the literature that will be discussed subsequently.

Table 2.4. Categories of market-based allocation and related approaches from literature

Market-based Allocation		
Multilateral Negotiations	Tâtonnement	Auctions
Contract Net Protocol ENTERPRISE: Malone et al. (1988) MARIPOSA: Stonebraker et al. (1994) NIMROD-G: Buyya & Vazhkudai (2001), Abramson et al. (2002) **Catallactic Self-organization** CATNET: Eymann et al. (2003), Eymann (2001), Reinicke et al. (2003)	**General Equilibrium** WALRAS' ALGORITHM: Cheng & Wellman (1996) Ygge & Akkermans (1999), Ygge (1998) SMALE'S ALGORITHM: Wolski, Plank, Bryan & Brevik (2001)	**Forward, Reverse, Double** SPAWN: Waldspurger et al. (1991) Ferguson (1989), Stoica et al. (1995), Miller et al. (1996) POPCORN: Regev & Nisan (1998) MAJIC: Levy et al. (2001) TYCOON: Lai, Huberman & Fine (2004) **Combinatorial Auctions** MIRAGE: Chun et al. (2004) SHARE: Ng et al. (2003b) AuYoung et al. (2004)

- **Multilateral Negotiations**

  Presumably the first protocol using multilateral negotiations in distributed computer systems is the *contract net protocol (CNP)* (Davis & Smith 1983). The CNP is an interaction protocol for the cooperative solution of tasks in a system of autonomous agents. In addition, the CNP can be seen as a paradigm for the structuring of large, complex problems in a distributed computing environment. Two types of agents are present in the CNP:

  - *Manager agents* that manage and control the execution of distributed tasks in the agent network.
  - *Contractor agents* that have special capabilities (resources) to execute tasks.

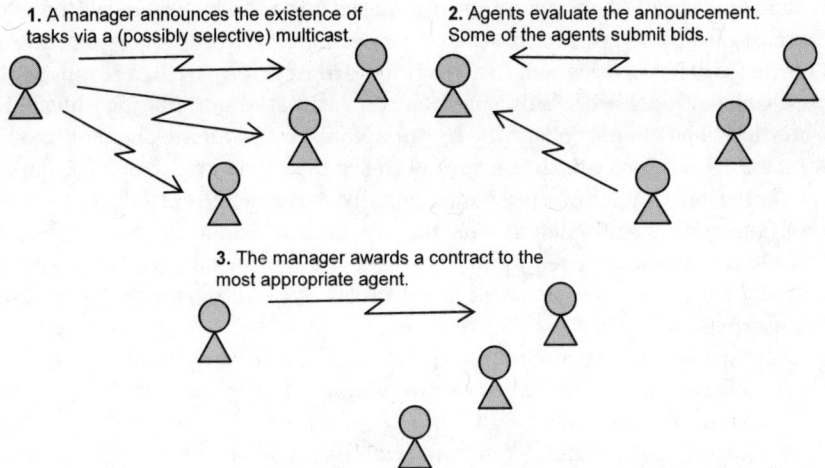

Fig. 2.5. Stages of the contract net protocol

Agents in the system can switch from manager agents to contractor agents (and vice versa) to meet the system's requirements. The stages of the CNP are depicted in Figure 2.5 [8]:

(1) The negotiation is initiated when an agent realizes that a new task has to be executed. It then acts as a manager agent and announces the properties of the new task suitable for remote execution to the other agents.

(2) Agents in the net are listening to the task announcements. They evaluate their own level of interest in the announced task with respect to their specialized resources by using task evaluation procedures. If an agent has resources for the completion of a task it offers the service in a message to the manager agent.

(3) The manager agent waits a specified time interval for bids to be submitted after broadcasting the task offer. After the interval expires, all submitted bids are evaluated and the best bid is determined. The best bidder is sent an award message. This message contains additional information that is needed for the winner agent to process the task. If the bidder agent receives the bid acceptance it becomes contract agent and fulfills the task.

The CNP itself is not an economically-oriented protocol. However it can easily be modified for economic bargaining processes by including information about W2P and reservation prices in the messages.

Malone et al. (1988) present a distributed resource allocation protocol *Enterprise* based on the CNP that uses a sequence of announcement and

---

[8] Figure in allusion to Huhns & Stephens (1999, p. 102).

award actions. Clients announce requests for bids by broadcasting a description of the required task and an estimate of the processing time together with an assigned task priority. Contractors with free processing capacities reply with bids that contain estimated completion time information. The clients collect the incoming bids and evaluate them at fixed intervals. The tasks are then awarded to the best bids, rewarding the earliest completion time. Enterprise additionally allows the mutual pre-selection of contractors and clients. Contractors as well as clients can list feasible contract partners. The system enhanced the performance of the Grid-like structured computer network significantly, as was shown by simulations. Enterprise as a very early development of such system management, does not provide explicit pricing information. The allocation control is done by the generation of artificial priorities assigned to the task, that can be seen as a kind of currency.

With *Mariposa*, Stonebraker et al. (1994) present the first purely economic model for the allocation of resources in a distributed database system using the CNP. The Mariposa model consists of clients that request database queries, server agents who hold the (fractioned) data and a service infrastructure, as well as broker agents who manage the database queries in the network. The clients are endowed with a fixed amount of a virtual currency which they can spend for each query. Clients pass their query request to the broker agents. These agents send a bid for the query service combined with the budget restriction given by the clients to those server agents that seem to be capable of fulfilling the requests (indicated by a yellow pages service integrated into the system) (step 1 of the CNP). Due to the fractioned data in the system, requests can be divided into sub-requests if a single server agent can not provide the complete data. Server agents themselves can use the budget offered to structure sub-requests for the missing data. After a while, the server agents respond to the query bids with an offer containing the price for the (partial) query and the estimated processing time (step 2 of the CNP). The broker agents combine and select the best offers and award the query tasks to the server agents (step 3 of the CNP). The Mariposa system enables brokers to use different profiles of W2P, e.g. declining W2P in time, and uses a sophisticated price management for the server agents. Server agents can e.g. broadcast advertisement offers or coupons to promote their services.

A recent application (candidate) for the CNP in distributed computing applications is *Nimrod-G* (Buyya et al. 2001, Abramson et al. 2002). Nimrod-G provides a usable economic-inspired resource allocation system for a Grid computing system designed for computationally complex simulations. Nimrod-G provides manager agents located on the peer machines that organize the economic-based scheduling and accounting of the job execution. At the moment Nimrod-G provides only a budgeted cost-based and time constraint adaptive scheduling mechanism that minimizes (Buyya et al. 2000):

- execution time with a deadline and a cost budget constraint,
- execution cost with a deadline and a cost budget constraint,
- neither execution time nor cost with a deadline and a budget constraint.

Nevertheless the introduction of CNP-based management for resource reservation in the Nimrod-G is one of the next milestones in the project. The *CATNET* approach by Eymann (2001) addresses the "catallactic" self organization process in markets described by F.A. Hayek. CATNET provides a simulation system for a distributed resource allocation process in an application layer network. The model aims to compare a decentralized allocation approach with a centralized allocation approach (Eymann et al. 2003). In the CATNET scenario there are four types of agents:

- *client:* a computer program on a host, that tries to access a web service,
- *service:* an instantiation of a general application function,
- *service copy (SC):* an instance of the service hosted on a resource,
- *master service copy (MSC):* a dedicated service coordinator which is known to the individual service copies.

In the *centralized variant* of the CATNET model, called the baseline model, a client requests a service on a resource. The resources route the request via SC to the MSC. The MSC initiates a CNP-like negotiation process acting as manager agent and takes care of the completion of the task as well as for a money transfer from the client to the SC (or resource) in exchange for the service. In the *decentralized variant* of the CATNET model the negotiation process is established directly between clients and SCs (or resources). An additional feature in the catallactic version of the CATNET is the SC's ability to move around in the network according to local supply and demand fluctuations. The pricing mechanism in CATNET is simple: for each completed service transaction the SCs increase their bid price by one money unit, for each unsuccessful bid the service price claimed is lowered by one money unit. Eymann measures the efficiency of the centralized and decentralized allocation method in different types of resource network in several experiments employing the maximum social welfare criterion (Sandholm 1996). For networks with a high node density (high proportion of resource nodes related to the total number of nodes in the network) the decentralized solution outperforms the baseline approach. The relationship reverses if low density networks are used. Based on the CATNET system Reinicke et al. (2003) presents a real world application for bandwidth provision for mobile devices in cellular radio networks.

Despite a vast amount of theoretical literature concerning the topic of multilateral negotiation protocols in agent systems, see e.g. (Sandholm 1996) and (Jennings et al. 2001), the application of sophisticated multilateral negotiation strategies for resource allocation in distributed computer systems outside the CNP is rare. This may be because the domain of market-based multilateral negotiations has strong coincidences with purely game theoretically oriented allocation mechanisms.

- **Tâtonnement Procedures**

  As shown in chapter 1 the proof of the existence of a *general equilibrium* for a particular micro economic system and the development of algorithms for the calculation of the resulting allocation and prices, remain a challenging task for economists as well as for computer scientists. A general equilibrium is reached if a set of commodity prices is found such that supply meets demand for each type of good while the agents optimize their utility perceived from the use of the resources at the current price level (see 1.5.4 for a formalization). One way to calculate this equilibrium under feasible conditions (guaranteeing convergence) is to use *tâtonnement procedures* with a virtual auctioneer for each good, inspired by the original approach of Léon Walras.

  Kurose & Shima (1989) present the first resource allocation model in distributed computer systems that is inspired by tâtonnement procedures in general equilibrium theory. Their optimization aims to allocate files in a network of distributed data storage called *file allocation problem*. The transfer of the files in the network has communication cost and a processing delay. The model tries to discover the optimal allocation of files (fragments) in the system by solving the global optimization problem in a distributed fashion. As is usual in GEE, each node has an initial allocation of resources. Kurose & Shima perform the optimization by trading amounts of resources iteratively according to their marginal rate of substitution until an equilibrium point is reached.

  Wellman (1993) employs a *Walrasian tâtonnement process* in a decentralized market (a market with an virtual auctioneer for each good) to solve a task allocation and transportation problem. The *WALRAS* algorithm (Cheng & Wellman 1996) is based on a distributed gradient search algorithm directly based on the formulation of the GEE in section 1.5.4. Sandholm (1995, p. 227) presents such an algorithm designed for the general version of the GEE that is depicted in Table 2.5 (see section 1.5.4 for the explanation of the corresponding variables used in the algorithm's formulation). The WALRAS algorithm differs from the Sandholm formulation in some minor points:

  (i) The bidding process is done asynchronously: each agent that wants to change its production (consumption) set can do this at any time.
  (ii) Agents can bid only for subsets of the produced and consumed goods instead of bidding for the entire quantity vector that has changed.

  It can be shown that this modified process still converges to an equilibrium under certain conditions (Cheng & Wellman 1996). Wellman applies the algorithm to several *multicommodity flow problems* (Wellman 1994, Wellman 1995). As an example of the suitability of this tâtonnement procedure to distributed computer systems, Mullen & Wellman (1995) present a market for network information systems. The system optimizes the access to an ISIP system for weather information called *Blue Skies*. Three types of resources are managed by WALRAS in the Blue Skies model: trans-

Table 2.5. Pseudo code of a 'Walrasian' tâtonnement algorithm.

```
Algorithm for the virtual auctioneer:
```
SET $p_j = 1$ for all goods $g \in [1, \ldots, n]$
SET λ_g to a positive number for all for all goods $g \in [1 \ldots n - 1]$
```
    REPEAT
```
        Broadcast **p** to consumers i and producers j
        Receive a production plan $\mathbf{y}_j$ for each producer j
        Broadcast the plans $\mathbf{y}_j$ to the consumers
        Receive a consumption plan $\mathbf{x}_i$ from each consumer i
        FOR $g = 1$ to $n - 1$
$$p_g = p_g + \lambda_g\left(\sum_i (x_{i,g} - x_{e,g}) - \sum_j y_{j,g}\right)$$
```
    UNTIL
```
$\left| \sum_i (x_{i,g} - x_{e,g}) - \sum_j y_{i,g} \right| < \epsilon$ for all $g \in [1, \ldots, n]$
Inform consumers and producers that equilibrium has been reached

```
Algorithm for consumer i:
    REPEAT
```
        Receive **p** from the virtual auctioneer
        Receive a production plan $\mathbf{y}_j$ for each j from the auctioneer
        Announce a consumption to the auctioneer plan $\mathbf{x}_i \in \mathfrak{R}_+^n$ that
           maximizes $u_i(\mathbf{x}_i)$ given the budget $\mathbf{x} \cdot \mathbf{x}_i \leq \mathbf{p} \cdot \mathbf{e}_i \sum_j \theta_{ij} \mathbf{p} \cdot \mathbf{y}_i$
```
    UNTIL informed that equilibrium has been reached
    Exchange and consume

Algorithm for producer j:
    REPEAT
```
        Receive **p** from the virtual auctioneer
        Announce to the auctioneer a production plan $\mathbf{y}_j \in \mathbf{Y}_j$ that
           maximizes $\mathbf{p} \cdot \mathbf{y}_j$
```
    UNTIL informed that equilibrium has been reached
    Exchange and produce
```

portation (network), Internet (Web-service), machine (CPU, DSK). These resources are managed by five producer agent types: carriers that provide data transport on the network resources (i), manufacturers of CPU time and disk space providing computational power for the Web-services(ii), transport arbitrageurs (iii) and delivery arbitrageurs (iv) that distribute and bundle the ISIP services required by the consumers, as well as mirror site providers (v). The consumers in the model are the users of a particular Blue Skies site. The approach follows the GEE formulation in section 1.5.4 and uses a *constant elasticity of substitution* utility function for the description of the consumers' preferences. As described in the GEE model (1.5.4), the consumers own shares of the producers to close the economy. In different test runs the economy showed an oscillating behavior, which is attributed to non-substitutionalities in the underlying resources, causing convergence problems in the algorithm.[9]

Ygge (1998) use a market-based allocation approach to implement a distributed computational system for temperature control in a building. Like the Wellman approach, the Ygge & Akkermans market model for resource allocation is based on a conventional tâtonnement procedure. The temperature control algorithm executes mainly three steps (Ygge & Akkermans 1999, p. 321):

(i) Each agent submits a net demand function to the virtual auctioneer, describing what change in allocation is desired by the agent at price p.

(ii) The virtual auctioneer computes an equilibrium by calculating a price p^* such that excess demand converges toward zero.

(iii) Each agent receives its required resources as calculated at the obtained market equilibrium price p^*.

The focus of Ygge & Akkermans' work is to demonstrate that a decentralized market solution using local information can be as efficient as a conventional control algorithm using global information. The outcome of the study proves the hypothesis that market-based resource allocation is at least as efficient as centralized resource control resulting in the conclusion: "*local information + market communication = global control*".

The most recent approach to tâtonnement-based resource allocation in a distributed computer system by Wolski, Plank, Bryan & Brevik (2001) strongly relies strongly on applying GEE algorithms designed for use in economic theory. They present a GEE-based Grid model with hard disk capacity (DSK) and processor time (CPU) resources. The algorithm used for the equilibrium tâtonnement process is deducted from a *Newton search method* proposed by Smale (1975). Unlike to previous GEE inspired models in distributed computer systems, Wolski, Plank, Brevik & Bryan do not use a barter economy with a closed-loop currency circuit. Consumers have iteratively refreshed budgets of monetary units to spend for the acquisition of computational power and disk space. In this model supply and demand functions are calculated on historical load data. To achieve a convergence of *Smale's algorithm* despite discontinuities of the supply and demand functions, these are approximated using polynomial approximation functions. For the sake of performance comparability, Wolski, Plank, Brevik & Bryan (2001) additionally propose a second price auction that is simultaneously performed for each resource type, processor time (CPU) and disk space (DSK), as an alternative to Smale's algorithm. This parallel auction type tries to address the problem of resource complementarities in ISIP applications. The resulting performance of this auction approach, however, is poor due to the fact that bidders have to pay for acquired resources, even if they do not receive the de-

---

[9] Mullen & Wellman (1996) present a related model for the provision of ISIP services in a digital library, there are, however, no experimental results for this work.

sired complementary resource. In various simulations Wolski, Plank, Brevik & Bryan compare the performance of both systems under different load conditions. In lower load regimes Smale's method performs well, especially in situations with sharp load changes. Prices are much more stable than in the auction system. In the higher load regimes, however, multiple price equilibria occur using the tâtonnement procedure, while the auction process delivers unique pricing results. According to the authors this may derive from the complementarties in the modeled two goods' economy.

- **Auctions**
  Auctions are a widespread market mechanism for resource allocation in distributed computer systems (see section 2.1.2). This is may derive from the fact that auction protocols can be *highly standardized*, but also provide *good adaptability* to various application scenarios. Two general types of auction protocols have to be distinguished in this context: auctions that do not take substitutionabilities and complementarities into account, like *conventional forward, reverse, and double auctions*, as well as auctions that are able to do so, like *combinatorial auctions*.

  – *Forward, Reverse, and Double Auctions*
  Ferguson (1989) employs an auction approach for the allocation of CPU time and communication bandwidth in a processor network. Unlike other distributed resource allocation approaches (e.g. Spawn), the processor network consists only of homogeneous production capacities. The model uses two types of agents, jobs and processors. Jobs reside on the processors and bid for their completion on the particular processor. Jobs are endowed with a budget while having to pay for task completion. To accelerate job completion or to save money, processors can migrate to processors with lower load and lower prices respectively. Migration to other processors is costly. Processor agents auction the execution time slots according to the price or time preference of the agents. Competition between the revenue maximizing processor agents determines the price equilibrium of the system. Actual execution prices on the particular processors are published on a bulletin board, ensuring information equivalence in the computational economy. After performing distributed resource allocation simulations using several auction variants, like the English auction, a modified Dutch auction, and a sealed bid auction, Ferguson finds increased modularity, limited complexity, and a good load balancing performance as prominent system properties.

  With *Spawn* Waldspurger et al. (1991) present a distributed allocation approach for computational resources based on sealed-bid second-price auctions. The system uses a top-level application manager process that represents the ISIP application. This top-level manager process spawns the main application into subtasks (application workers) each of them coordinated by a local manager process. In analogy to the Contract-Net protocol, manager agents can initiate the generation of new (child) manager and application worker agents if a further decomposition of tasks seems to be useful. Local managers bid for task execution in auctions that are performed by the resource managers, one for each CPU. The bids hold three types of information: length of requested time slot, quantity of funding available for the requested task and a brief task description. Auctions are characterized by minimum and maximum task length that is acceptable to the auctioneer and a strategy function for the bidders comprising slot length, current bids, and other market parameters. These parameters depend on market conditions, like e.g. discounts for longer tasks in situations with lower resource loads. The price finding process in the Spawn economy for computational resources is controlled by a central fund. All processes are funded from a central budget. Spawn agents hierarchically allocate shares of the incoming fund stream to their child managers in order to acquire time slots for task completion. Manager agents employ several strategies to allocate funds for resource acquisition to their children, e.g. depending on relative workload, task completion progress, or cost effectiveness of the child processes. The system turns out to provide a stable price equilibrium even in the case of exogenous demand fluctuations and a small number of agents. With Spawn Waldspurger et al. additionally demonstrate that funding policies can be used to control task priority in a distributed computational economy with heterogeneous nodes.[10]

A computational resource allocation approach presented by Stoica et al. (1995) proposes an auction-like mechanism for a P2P system. In a first-price scheduling mechanism, bidders can reserve CPU time for the execution of their tasks. The system calculates the estimated price for the total task execution (users pay per CPU time) and releases the task immediately, if the agent has the highest W2P and enough virtual currency to pay for the entire task (tasks are non-preemptive). As in many P2P-based market-oriented resource allocation systems, users can collect the monetary units for paying for CPU time procurement while providing CPU time themselves (barter economy). Due to the stochastic task arrival process, the Spawn scheduling auction has the properties of a YM system when considering the opportunity cost of non-executed jobs. See section 4.5 for a further discussion of YM-related allocation mechanisms for multiple machines.

Miller et al. (1996) present a resource allocation system using a forward first-price auction for the allocation of CPU time and network bandwidth in a distributed computer system. The system, which is very close to a previous computational auctioning approach by Drexler & Miller (1988), provides video delivery services in an ISIP scenario.

---

[10] Heterogeneous means mainly different CPU capacities in this context.

Both resources are simultaneously auctioned without explicitly regarding resource complementarities. A specialty of the Miller et al. model is the use of an hierarchical accounting system for the payment of the acquired resources that should prevent bidders from using strategic behavior, like the submission of untruthful bids with reduced W2P.

Regev & Nisan (1998) proposes an electronic exchange (*Popcorn*) to buy and sell CPU time using alternatively a Vickrey, a sealed-bid double and a clearinghouse double auction. The Java-based system enables users to open accounts that gather or spend monetary units by selling or buying computing time, as well as providing or consuming other valuable services like interesting information on Web-pages. If an agent participating in the Grid-like distributed system wants to acquire CPU power, it submits bids to the other CPU time offering agents. Depending on the auction type supply and demand is met (Regev & Nisan 1998, p. 151):

(i) in the *Vickrey auction* the bidder with the highest W2P is selected from among the current requesting agents and receives the CPU time for the second-highest bid.

(ii) in the *sealed-bid double auction* each bidder (seller/buyer) submits a minimum and a maximum W2P with a bid increment. The auctioneer matches supply and demand by stepwise increasing selling/offering bids in the predefined ranges until an equilibrium price is found.

(iii) the *clearinghouse double auction (CDA)* is similar to the sealed-bid double auction except for the fact that for each round (in intervals) more offers and bids are collected and a supply/demand curve is calculated. The matching price for the CPU time is the intersection of the calculated supply/demand curves.

Regev & Nisan measure the allocation efficiency (social welfare) of the three mechanisms by comparing simulation results (uniformly distributed bid/ask valuations, Poisson arrival process for the bids) with a result calculated comprising all bids occurring in the simulation interval. The performance of the CDA is highest (up to 98% of optimum), whereas Vickrey and sealed-bid double auction are less effective.

Levy et al. (2001) presents a reverse auction model *(MAJIC)* for the allocation of distributed ISIP services on the Internet. Unlike traditional auctions, this mechanism operates by using production cost functions submitted by the suppliers and utility functions submitted by the sellers. The service searches for the optimally matching seller-buyer combination if a potential buyer sends a service request. The allocation efficiency of MAJIC is tested by using a first and second-price allocation mechanism, while a load balancing effect of the system could be demonstrated.

A similar approach to the auction-like scheduling mechanism proposed by Stoica et al. (1995) is presented by Chun & Culler (2002). Unlike

Stoica et al. this approach does not employ budget constraints and pre-emptive jobs. Chun & Culler (2002) compare a first-price auction for the economic scheduling of CPU time with a traditional non-economic scheduling method. The results of their simulations are clearly in favor of the economic-driven scheduling system.

With *Tycoon* Lai et al. (2004, 2004, 2005) present a distributed first-price (or second-price) auction for resource allocation in a distributed computer system. The system uses balanced bank accounts for each participant to control the allocation policy. They discriminate between two cases: an open-loop and a closed-loop system. In the open-loop system the the bidding and resource consuming agents receive an allotment of funds when they join and at set intervals. In the closed-loop (P2P) funding policy, users themselves bring resources to the system when they join. After an initial funding, these agents have to earn their funds by enticing other users to pay for their resources (Lai, Huberman & Fine 2004, p. 5). Additionally, Tycoon tries to address the problem of resource allocation in the presence of complementarities (see next section). The resulting proposal, however, handles this point as unsatisfactorily as other simple auction approaches.[11]

A number of further auction-related resource allocation and scheduling approaches in distributed computer systems exist, like 'lottery' (Waldspurger & Weihl 1994) and 'stride' scheduling (Waldspurger & Weihl 1995, Waldspurger 1995). Some of them, also integrate elements of tâtonnement procedures and negotiation mechanisms (Ernemann et al. 2002). These models are not unambiguously identifiable in the context of this taxonomy and lie beyond the focus of this thesis.[12]

- ### *Combinatorial Auctions*

   Combinatorial auctions are a suitable tool for optimally allocating interdependent resources according to the W2P of the participants.[13] A number of recent resource allocation approaches use combinatorial auctions to find the optimal task allocation in different application domains. This is due to the fact that the process of producing information services on distributed resources comprises an allocation problem with strong complementarities. If an information product has to be processed on two different computers at two different points in time (t_1, t_2), the acquisition of CPU capacity at t_1 without complementary CPU power at t_2 is worthless. The value for the user is only obtained by

---

[11] It is intended to let users bid for complementary resources in the case of not achieving the acceptance of the desired time slot allocation, however, bidders do not have to pay for the current service.

[12] See Tucker & Berman (1996) and Chun & Culler (2000) for further examples.

[13] See section 5.1 for a detailed explanation of the nature of combinatorial auctions.

receiving the resources in the correct combination. Interestingly the application of combinatorial auctions as a resource allocation mechanism in distributed computer systems has so far not been developed very far (in practical terms), despite its special suitability for this domain. This may be due to the computational complexity of the allocation mechanism.

In a recent approach Chun et al. (2004) present a combinatorial auction-based mechanism for resource allocation in a *SensorNet* testbed. *SensorNets* are networks of wireless coupled sensors for the collection of observation data, like seismic and structural data for geological research or environmental data in ecology. The sensor devices have different capabilities that can be used in various combinations to perform successful observations. This results in non-linear valuations when the users are booking the sensors for their observation tasks. Chun et al. implemented a Web-based, periodically performed, sealed-bid combinatorial auction in their *microeconomic resource allocation system (MIRAGE)*, where SensorNet users can submit combinatorial bids for the electronic sensor resources. The users have accounts based on a virtual currency. They can spend virtual currency units to acquire SensorNet resources and earn virtual currency units by selling their own SensorNet capacity in the closed combinatorial economy. The combinatorial allocation mechanism of MIRAGE, however, is strongly simplified to achieve sufficient real time performance.[14]

As a logical continuation of the SensorNet work, AuYoung et al. (2004) propose the application of combinatorial auctions as a resource allocation mechanism in a Grid computing environment called *Bellagio*. The approach relies on Berkeley's combinatorial auction-based allocation scheme *SHARE* (Ng et al. 2003*b*). The different resource types available are identified using a Grid specific *scheme for auction-based resource allocation (SWORD)* (Oppenheimer et al. 2005). The bidders can select the desired resource combinations from the set of resources provided by SWORD and place their bids. Each bidder has a budget of a virtual currency for the payment of the tasks. The allocation of the resources is made by a kind-of second-price combinatorial auction, which is considered to be strategy proof in this context (Ng et al. 2003*a*). In several experiments the system has been tested with respect to scalability, efficiency and fairness. Due to the simple greedy algorithm used in the system, scalability is good. However, the efficiency and fairness of the resulting resource allocation are subject to further research in the Bellagio project. This results from the fact that smaller bidders are often edged out in the allocation process, if computational load is high.[15]

---

[14] A similar approach for radar sensors is presented by Ostwald & Lesser (2004).

[15] Bubendorfer & Hine (2005) propose a mechanism for Grid resource allocation that is very close to a combinatorial action, however, it is not specified exactly

2.4 Multi-Agent Systems and Automated Resource Allocation

The main objective of this thesis is the development of stable revenue and yield maximizing mechanisms for the optimal allocation of resources in ISIP production tasks. Besides this objective, allocation and coordination mechanisms presented in this work strive for stability, low response time to supply/demand fluctuations, minimized processing time and scalability. With respect to the goals of increased fault tolerance, scalability, and robustness a disjoint decomposition of the resource allocation system seems to be a good choice (Ferguson et al. 1996, p. 158).

2.4.1 Distributed Artificial Intelligence and Multi-Agent Systems

John McCarthy coined the term *'artificial intelligence' (AI)* in 1956, providing the following definition: "AI is the science and engineering of making intelligent machines, especially intelligent computer programs. It is related to the similar task of using computers to understand human intelligence, but AI does not have to confine itself to methods that are biologically observable" (McCarthy 2004). Following the exhaustive discussion of AI capabilities in the eighties, the definition was extended to the term *distributed artificial intelligence (DAI)* in the nineties: "DAI is concerned with the study and design of systems consisting of several interacting entities which are logically and often spatially distributed and in some sense can be called autonomous and intelligent. Two primary types of DAI systems can be distinguished: distributed problem solving and planning systems, where the emphasis is on task decomposition and solution synthesis, and multi-agent systems, where the emphasis is on behavior coordination" (Weiss & Sen 1996). Lesser (1995, p. 340) provides a more precise definition of *multi-agent systems (MAS)*: "In general, multi-agent systems are computational systems in which several semi-autonomous agents interact or work together to perform some set of tasks or goals. These systems systems may involve computational agents that are homogeneous or heterogeneous and may involve activity on the part of agents having common goals or goals that are distinct." The building blocks of an MAS are the *software agents* themselves. They are sometimes defined as *intelligent agents*, in accordance with the AI's attempt to engineer intelligent machines (Wooldrige & Jennings 1995).

The characteristics of an intelligent agent in the sense of MAS are (Schwind & Grolik 2003):

- *Reactivity:* This property describes the agent's ability to perceive its environment and to react promptly to changes without abandoning the goals it is aiming for.

---

enough to provide further insight into its usability for the resource allocation in a Grid.

- *Proactivity:* The performance of goal oriented actions initiated by the agent itself.

A couple of software architectures try to provide the property of intelligence from the AI point of view: *Deliberative, reactive, hybrid* and *believe desire intention (BDI)* agent architectures.[16] Additionally the intelligence property of agents often includes the ability to learn. Reinforcement learning is a common approach at this point. This topic will be elaborated in detail in chapter 4.

The usual definition of a *software agent* is established between *game theoretical factors*, *DAI properties*, and common *software engineering* considerations: "A software agent is a programmed object within a software system (MAS) that is able to interact autonomously with its environment to achieve the goals specified by the user and owner of the agent. A software agent has full control of its state and actions" (Schwind & Grolik 2003, p. 340 seq.).

Sycara (1998) presents a catalog of MAS properties that demonstrates their advantages and indicates factors favoring their deployment in industry processes:

- *Computational efficiency:* MAS provide parallel computing power and enable the successful application of 'divide and conquer' strategies in the design of algorithms to enhance system performance, as long as communication complexity remains in bounds (communication complexity vs. computational complexity). MAS can be configured to conduct *distributed problem solving (DPS)* strategies (Durfee 1999) by using simple negotiation mechanisms like e.g. the CNP or more sophisticated game theoretic oriented task allocation approaches developed in "Rules of Encounter".[17]
- *Reliability and Robustness:* Due to their architecture, distributed systems have a higher resistance to environmental and technical interferences. Redundant parts in an MAS can take over functions of failed system components (Schillo et al. 2001).
- *Extensibility and flexibility:* Software agents can be directly attached to the IT systems which they control and administrate. This makes it easy to collect current operation data from system devices for control purposes. The modularity of system components in an MAS enables the quick and flexible altering of functions according to environmental states. Extensions of an MAS in terms of scale can be made simply by replicating software agents. Additionally, flexibility of MAS is backed by the self-organization property and the mobility of individual software agents.[18]
- *Maintainability and reusability:* The modularity of MAS also implies easy maintainability of the system. Individual components can be changed in

---

[16] See Schwind & Grolik (2003, p. 341) for a further discussion of these approaches.

[17] See Rosenschein & Zlotkin (1994) and section 2.3.1.

[18] Agents can migrate between host computers on the Internet in most MAS implementations.

the running MAS system. Additionally the construction of new MAS applications is eased by reusing already developed components.

- *Responsiveness:* Due to their decentralized structure and flexibility, MAS can swiftly response to changing environmental conditions.

Regarding the arguments given in the previous section, MAS can be considered as a suitable tool for performing price controlled resource allocation in distributed computer systems. Software agent engineers focus on creating MAS as flexible self-organizing communities where complex behavior emerges from social interaction. MAS are ideally suited to reproduce economic behavior and markets as they behave in nature. For this reason a couple of the economically oriented resource allocation systems enumerated in section 2.3 are based on MAS implementations. In contrast to this Grid, computing aims at "resource sharing and coordinated problem solving in dynamic, multi-institutional virtual organizations" (Foster et al. 2004, p. 8). The goals of MAS engineering and Grid engineering, are very close but differ in one crucial point: whereas Grids are more related to bundle computational power that is mostly not very task specific, MAS have a very differentiated system of DPS and task allocation mechanisms. For this reason Foster et al. (2004, p. 11) concludes: *"Clearly, neither situation is ideal: for Grids to be effective in their goals, they must be imbued with flexible, decentralized decision making capabilities. Likewise, agents need a robust distributed computing platform that allows them to discover, acquire, federate, and manage the capabilities necessary to execute their decisions. In other words, there are good opportunities for exploiting synergies between Grid and agents. One approach to exploiting such synergies might be a simple layering of the technologies, i.e., to implement agent systems on top of Grid mechanisms. However, it seems more likely that the true benefits of an integrated Grid/agent approach will only be achieved via a more fine-grain intertwining of the two technologies, with Grid technologies becoming more agent-like and agent-based systems becoming more Grid-like."*

Combining all these arguments, the experiments in this thesis mainly employ MAS to simulate the allocation of resources for ISIP provision in a distributed computer system (Grid). The type of allocation system used in this thesis has the following structure which uses a communication protocol near to the CNP (see Figure 2.6):

- Resource consuming *task agents* acting as bidders in the allocation process submit bids for the use of resource bundles (network capacity (NET), computing power (CPU), volatile memory (MEM) and non-volatile memory (DSK)) for specific points in time associated with a W2P attached to the bundles.
- One or more *mediating agents* receive the bids and calculate an allocation profile for the available resources managed by the *resource agents* according to the chosen allocation mechanism. In the framework of this thesis, this is either a *learning yield management process* or a *combinatorial auction mechanism*. After a successful allocation the *bid acceptance* is

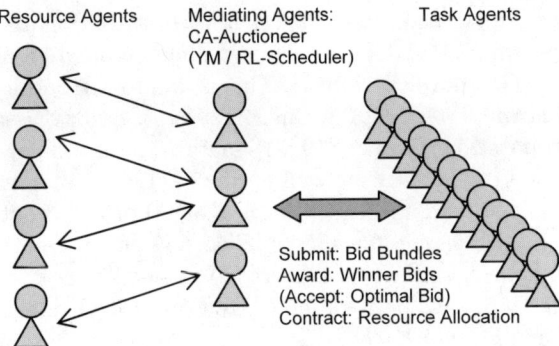

Fig. 2.6. Agent architecture of the automated resource allocation system

reported to the bidders. The bidder agents then confirm the bid allocation in a *contract*.

- The *resource agents* know the actual status and the future booking profile of the resources they manage. They provide this information to the *mediating agents* at short intervals to keep them up to date and confirm the acceptance and fulfillment of the tasks accepted by the mediator agents. The mediator agents are enabled to allocate the free resources reported by the resource agents to the bidders in charge of the resource agents without call back. Allocated resources are *reserved in advance* by the allocation agents until task execution.

2.4.2 Ontologies for the Coordination of Multi-Agent Systems

One important issue of MAS, which has not yet been discussed, is the use of *agent communication languages* for the coordination of agents' actions. A theoretical concept for the design of such communication protocols including reasoning capabilities within the communication language is *ontologies*. Sowa (2000) defines ontologies as follows:

"*The subject of ontology is the study of the categories of things that exist or may exist in some domains. The product of such a study, called an ontology, is a catalog of types of things that are assumed to exist in a domain of interest $\mathfrak{D}$ from the perspective of a person who uses language $\mathfrak{L}$ for the purpose of talking about $\mathfrak{D}$. The types in the ontology represent the predicates, word senses, or concept and relation types of the language $\mathfrak{L}$ when used to discuss topic in the domain $\mathfrak{D}$. An uninterpreted logic, such as a predicate calculus, conceptual graphs, or KIF, is ontologically neutral. It imposes no constraint on the subject matter or the way the subject may be characterized. By itself, logic says nothing about anything, but the combination of logic with an ontology provides a language that can express relationships about the entities in the domain of interest.*"

In most cases, logic used in connection with ontologies as a vehicle for *knowledge representation* and *knowledge interchange* in MAS is the classical *first-order logic(FOL)* developed in AI. This is due to the ontological neutrality of FOL (Guarino 1995). The definition of an ontology as a conceptualization (also proposed by Gruber (1995)) enables the use different ontologies as a representation of the same concepts, lik the use of different natural languages. One widely spread implementation of an AI oriented ontology in FOL is the *knowledge interchange format (KIF)* proposed by Genesereth (1998). The underlying concept of KIF is a predicate logic that is intended to serve as an interchange language between heterogeneous knowledge databases. Unlike *conceptual graphs*, which were designed for the direct mapping of natural languages, KIF was designed for ease of mapping to and from computer languages. A problem which accompanies the design of ontologies for various domains is the standardization and synchronization of ontology sets. The use of ontology servers for the provision of ontology dictionaries upon request is one approach to dealing with this topic that is widely discussed in the literature (Wendt et al. 2002). The KIF concept is realized within the commonly accepted *foundation for intelligent physical agents (FIPA)*[19] standard for the implementation of MAS in industrial and experimental environments. FIPA offers a generalized *agent communication language (ACL)* as an extension of KIF (FIPA 2000a, FIPA 2000c, FIPA 2000d). In connection with ACL message definitions and MAS engineering a derivative of UML, called *agent unified modeling language (AUML)*, was developed (Odell et al. 2000, Odell et al. 2001). AUML sequence diagrams are used to model communication acts within the FIPA standard definitions of negotiation protocols like the CNP (FIPA 2000b).

The FIPA specification of the CNP has been slightly adapted. Figure 2.7 shows the AUML sequence diagram of the FIPA-CNP, which is proposed for bilateral negotiations. The manager agent in the classical CNP interaction protocol (see Figure 2.5) is called *initiator* and the potential contractor is named *participant*. Like the original CNP, the initiator announces the existence of tasks via a multi-cast of a *call-for-proposal (cfp)*. A number of interested participants may respond with a *propose* message to the multi-cast task announcement, while the other agents send a *refuse* message. The initiator then continues by *accepting* any number of proposals. The remaining agents are informed about the rejection of their proposals *(reject-proposal)*. Unlike other interaction protocols, the FIPA-CNP allows for preference specification by all negotiating agents and also permits each party to discontinue the negotiations process.

Several FIPA compliant MAS implementations of the FIPA protocol exist. One of these implementations amongst various others like JOT, LEAF, etc., is the *Java agent development framework (JADE)*.[20] Due to its widespread

---

[19] www.fipa.org
[20] http://jade.tilab.com

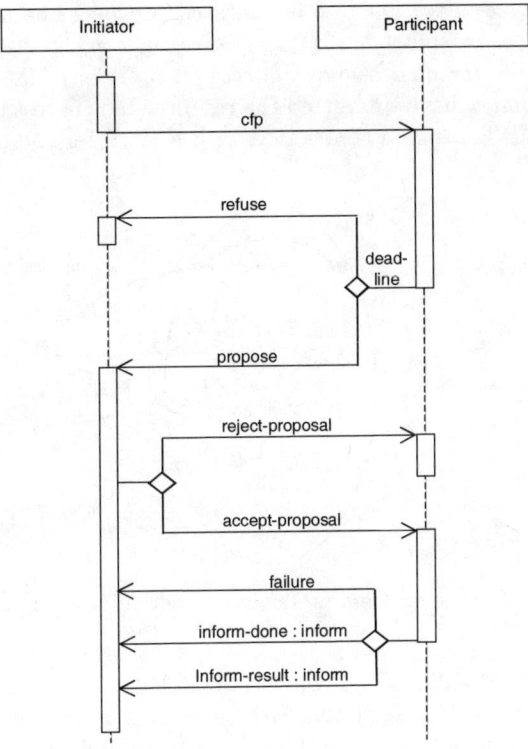

Fig. 2.7. AUML-FIPA definition of the contract net protocol

use in the MAS community, broad support and its continuous further development, JADE is used for the implementation of an ISIP provision framework in a distributed computer system within this thesis. Additionally an FIPA compliant software engineering tool for the development of ontologies, called Protégé[21], is employed for the implementation of the software applications presented within this thesis.

2.5 Dynamic Pricing versus Automated Resource Allocation

The use of resource prices as lead variables for optimal allocation raises the question of the existence of a unique and stable *equilibrium price*. This problem has been briefly discussed in the light of mathematical economics in section 1.5.4, however, for practical application purposes it is a good idea to decompose the interdependent optimization problem of *dynamic pricing* and

---

[21] http://protege.stanford.edu

automated resource allocation into an iterated process. The automated allocation of resources q_o controlled by a *price variable* p_o can be performed by using a separate *automated resource allocation function(ARA)*. This is done after *ISIP consumers* have submitted the resource bids necessary for the completion of their ISIP tasks in connection with a W2P for these bids.

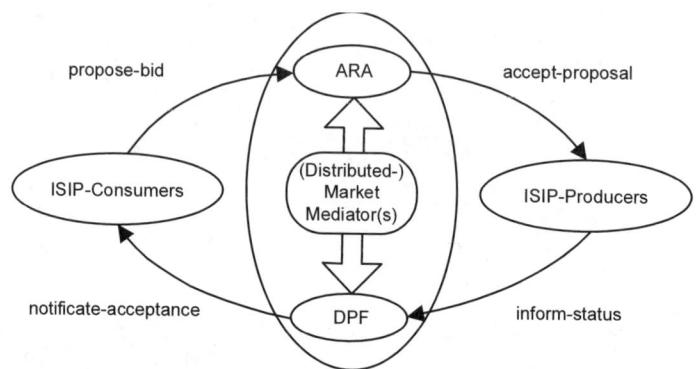

Fig. 2.8. ISIP consumer and producer interaction via market mediator

The function $ARA(p_o) = q_o$, which is coupled to the *market mediators*, can generally represent one of the various market-based allocation mechanisms enumerated in section 2.3.2, like auctions or multilateral negotiations. After a successful allocation process by the ARA mechanism the *ISIP producers* are informed about the *acceptance* of the *bidders' proposals* enabling them to reserve the required resources in advance. The ISIP producers for their part *inform* the market mediators about their actual resource load and the *status* of the allocated task completion. This data is used to calculate a notification of the ISIP consumers about the *acceptance* and status of their bids. Additionally, information about the current prices of the individual resources is provided by the *dynamic pricing function(DPF)*. The pricing function $DPF(q_o) = p_o$ is directly coupled with the ARA function for most resource allocation methods. The interaction of ISIP consumers and ISIP producers creates a bilateral optimization problem including $ARA(p_o)$ and $DPF(q_o)$ standing in an antagonistic relationship:

$$p_o \uparrow (\downarrow) \Rightarrow DPF(q_o) \uparrow (\downarrow) \qquad (2.1)$$

and

$$q_o \uparrow (\downarrow) \Rightarrow ARA(p_o) \downarrow (\uparrow) \qquad (2.2)$$

meaning simply that scarce resources imply rising prices (vice versa) and rising prices produce a falling resource load (vice versa). A well constructed allocation and pricing mechanism should lead to a stable self-controlling system (see Figure 2.8). This mechanism is sometimes described as a decentralized market mechanism, due to the fact that the pricing decisions are met in a decentralized way by the participants employing information interchange via the market mediator. Two methods from the set of *dynamic pricing* mechanisms proposed in section 2.1 (RP, auctions, NBP, DPD, YM) have been selected for further investigation in the context of this thesis: *yield management* and *combinatorial auctions*. In the following the decision for these mechanisms is explained in connection with the criteria presented in the previous sections.

2.5.1 Yield Management Systems

As introduced in section 2.1.5 *yield management* is concerned with revenue optimization for perishable assets through price discrimination. ISIP resources, like CPU time and network bandwidth (NET), can be considered perishable assets in our distributed computer system. Additionally, resource capacities in such a system are only gradually changeable, a fact which is another argument for the application of YM systems.[22] To find an optimal load utilization, market segmentation by a slow changing price discrimination scheme, seems to be a suitable approach. later in this process, the inherent self-selection property of YM prevents the emergence of consumer dissatisfaction. Due to the use of precalculated pricing schemes, the response time to a resource request of the YM allocation and pricing system should be to be short compared to other dynamic pricing approaches.

The YM system, presented in this thesis, provides the direct acceptance and assignment of single tasks and task bundles in the moment of the request's occurrence. The YM system is capable of dealing with different length of the *consumable* resources bundles, while the resources are assumed to be *divisible* into *discrete* units. Each task is assigned an *economic weight* (W2P) transforming the YM acceptance problem into an *economic allocation* problem. Furthermore the bid bundles of the approach presented include multiple resources, transforming the economic allocation problem into a *combinatorial allocation* problem. Using a learning YM system makes especially sense if requests arrive stochastically.[23] For this reason the YM system presented in this thesis is endowed with a *learning capability*. The functionality of the economic resource allocation mechanism is taken over by an adaptive YM process acting as mediator agent in a *non-cooperative allocation game*. The allocation game is performed by the $ARA(p_o)$ function, which optimizes the task fulfillment locally, considering short term information about the production status and a learned request profile. Pricing is only performed indirectly in the YM system by omitting an explicit $DPF(q_o)$. The ISIP consumers are simply notified about the acceptance or rejection of the submitted bid bundles. The construction of the YM resource allocation system allows for the design of a decentralized self-organizing system.

---

[22] This will be extensively elaborated in section 4.2.

[23] *Markov* property is assumed here for the underlying stochastic arrival process.

2.5.2 Combinatorial Auction Mechanisms

Combinatorial auctions (CAs) deal with the optimal allocation of bid bundles including complementary resources. A CA starts with the collection of bundle bids in a *bid acceptance* phase. After closing the bid submission process the calculation of the optimal allocation is performed. The inclusion of task bundles, which differ in pricing from the simple sum of prices bid for the single items exhibiting complementarities, produce a *combinatorial auction problem (CAP)* that is $\mathcal{NP}$-complete (see section 5.1.6). As indicated by the name, CAs are a method of dealing with *combinatorial allocation* problems. CAs are capable to represent all problem types depicted in Figure 2.2. The particular adequacy of CAs for resource allocation in distributed computer systems has already been discussed in section 2.3.2. In the scenario presented in this thesis combinatorial resource allocation is performed by employing a resource allocation function $ARA(p_o)$ representing a CAP optimizer working as a *global scheduler*. A pricing function $DPF(q_o)$ is directly coupled to the $ARA(p_o)$ providing price estimates for the particular system resource types for market formation. By performing the CA iteratively ISIP consumers have the opportunity to adjust their bid prices according to the auctioneer's task acceptance/rejection decisions and their W2P. In terms of the response time of the allocation process, CAs can be time consuming for big problems due to the $\mathcal{NP}$-completeness of the CAP. This problem, however, will be addressed by using specifically designed heuristics guaranteeing a real-time allocation process which is suitable for e.g. Grid applications. In contrast to the YM system introduced in the previous section, the CA optimization mechanism has no learning property. Due to the existence of a pivotal auctioneer, CA mechanisms are often regarded as a *centralized* market-based allocation mechanism and therefore provide a contrast to a distributed YM system. However, as already mentioned above, pricing decisions are met and revised by the individual market participants based on the auctioneer's acceptance or pricing information, leading to a categorization as decentralized allocation mechanism, if the CA is carried out iteratively.

Both systems seem to be especially appropriate for dynamic pricing and automated resource allocation in information systems due to their ability to capture the resource complementarities occurring in distributed computer systems. However, each approach has its advantages and drawbacks, especially due to their strongly differing construction. The comparison of a CA and a YM system for dynamic pricing and automated resource allocation in a qualitative way and the performance exploration of the CA system in a quantitative way, employing the aspects and criteria mentioned in this chapter will therefore be a main aspect of chapter 7.

3

Empirical Assessment of Dynamic Pricing Preference

With the emergence of e-commerce in late nineties dynamic pricing increasingly came into the focus of applied market design, especially for B2B and B2C applications (Bichler & Werthner 2000). However, the e-commerce bubble burst that followed in the years 2002 and 2003 put a huge setback to automated pricing acceptance. At this time the attitude of business and consumers, who were originally open-minded toward a new kind of market making, turned into a sceptical view of dynamic pricing methods. To introduce this kind of method into new application areas, like the ISIP provision, where dynamic pricing seems to be highly adequate, it is necessary to evaluate the present attitude of companies toward the implications, application potentials, perceived hurdles and chances for dynamic pricing. With these arguments in mind it would be easier to react to the doubts while constructing economic and technical market solutions for ISIP solutions in the Internet. For this reason, a German industry survey on this topic was carried out in 2004 in the context of a series of surveys on enterprises' attitude to Internet use.[1]

3.1 Basic Data of the Study

The main objective of the *empirical study* within the context of this thesis was to evaluate the readiness of German industry to adopt dynamic pricing models. 12,000 mid and large sized enterprises beginning with a minimum of 10 employees were contacted in the course of the study. The selection of the

---

[1] The survey was done under the institutional aegis of PREMIUM ("Preis und Erlösmodelle im Internet, Umsetzung und Marktchancen" which is a joint research project funded by the German state organization (BMBF). The questionnaire was elaborated in a cooperation between the Johann-Wolfgang-Goethe-Universität and the Albert-Ludwigs-Universität Freiburg as joint research of the PREMIUM subprojects "EVENT" (Evaluierung des betrieblichen Einsatzes von Netzwerktechniken), "Reverse Pricing", and "Dynamic Bundle Pricing" (Sackmann & Strüker 2005).

contacted enterprises was done by employing the "Hoppenstedt Database"[2] for mid and large size enterprises (edition 2/2003). The layer size of the industrial sectors was chosen according to NACE[3] categorization in connection with branch data of the German Federal Statistical Office[4] (Statistical Yearbook 2003, p. 115). 506 questionnaires were returned either by mail or by the completion of an online form. The distribution of the industrial sectors according to the answers to question Q 1.1[5] in the returned questionnaires can be seen in Figure 3.1. Figure 3.2 shows the result of a branch aggregation into three main industrial sectors, production, commerce and services according to the NACE categorization, to give a better overview. The distribution of these sectors correlates close with the aggregated branch distribution of the contacted enterprises.[6] In a second step for sample segmentation, the participating corporations had to indicate their main customer group according to their self assessment (see Q 1.4). Three groups were counted, business-to-business at 61.1% (n = 277)[7], business-to-consumer at 32.2% (n = 146) and business-to-administration amounting to 6.6% (n = 30) over the entire questionnaire sample.[8] The size of the participating enterprises was evaluated in terms of employees (see Q 1.2) and annual income, - balance sheet total for banks and insurances respectively - (see Q 1.3), employing eight classes each. The class structure and the distribution of both measures for the whole sample data (n = 506) can be seen in Figure 3.4 and Figure 3.5 respectively, indicating that about the half of the enterprises involved had more than 100 employees and an annual income of more than 100 million €.[9] Different accounting methods for banks and the insurance industry in contrast to the remaining companies led to an increased turnover of companies in the segment 'greater than 500 million €'. This might be a problem when evaluating empirical survey data.

---

[2] www.hoppenstedt.de

[3] Statistical systematic of economic branches in the European Community (NACE Rev.1) according to the order of the EU council (EWG) Nr. 3037/90 dating from the 9. of October 1990 (EG Nr. L 293 p.1).

[4] www.destatis.de

[5] The entire questionnaire is depicted in appendix A. In the following, questions are referenced by employing Q and the corresponding number in the German original.

[6] The data was tested for representativeness against the distribution of the aggregated number of employees that are subject to social insurance contribution in the industrial sectors according to the German Statistical Yearbook 2003 (p. 115). Chi-square testing could not reject representativeness of the sample at a significance level of 0.95.

[7] In the following the data in brackets denotes the number of valid answers without don't know and missing values.

[8] Multiple classification was not allowed in this analysis.

[9] Comparing the distributions of the employees number and the companies' annual income with the corresponding figures in the Statistical Yearbook 2003, as the branch segmentation case, no significant deviation from the basic population could be diagnosed (Sackmann & Strüker 2005, p. 14).

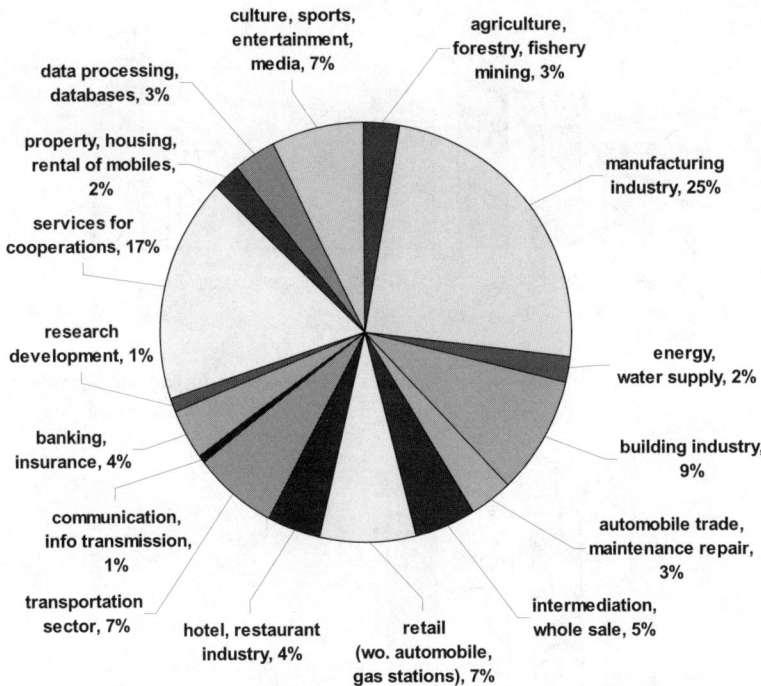

Fig. 3.1. Detailed sector distribution of the participating enterprises

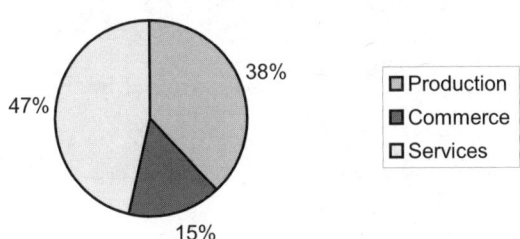

Fig. 3.2. Distribution of the participating enterprises according to aggregated industrial sectors

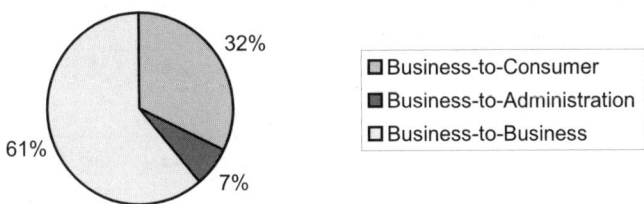

Fig. 3.3. Distribution of the participating enterprises according to their main customer segment

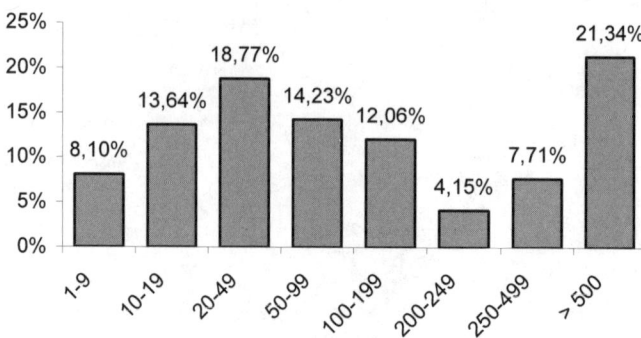

Fig. 3.4. Sample distribution of the number of employees in the participating enterprises

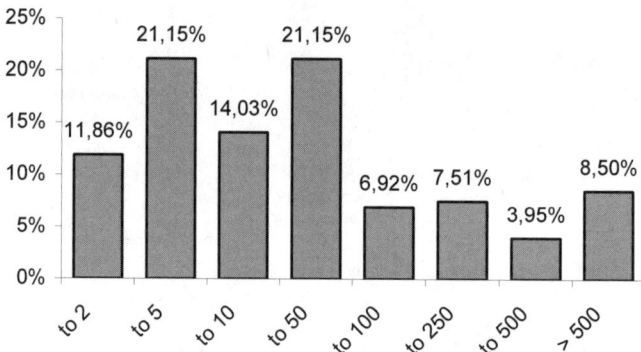

Fig. 3.5. Sample distribution of the turnover attained by the participating enterprises in million €

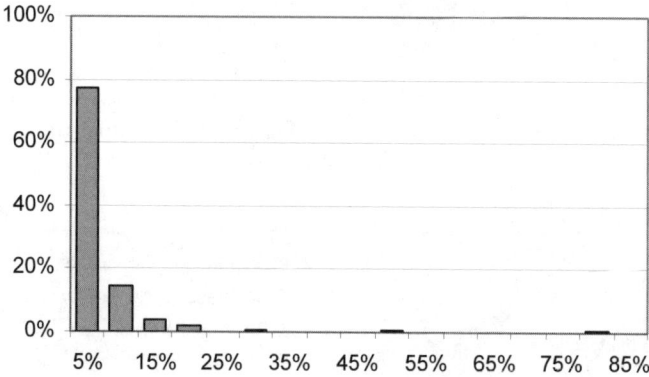

Fig. 3.6. Distribution of the companies' IT expenditure in relation to the total costs

However, the option of investigating the banking and insurance sector separately to resolve this problem, was not used in this study, due to the fact that the statistical analysis presented here does not make extensive use of turnover related data. The last item in the general part of the study considered here is related to the IT expenditure. The measurement of firms' IT expenditure' impact on productivity is a heavily discussed topic in the management of information services domain (Brynjolfsson 1993). He argues that the widely observed weak and sometimes negative correlation of the firms' IT expenditure and their measured productivity, that is often called *"productivity paradox"*, is mainly caused by the difficulty of measuring the direct impact of IT expenditure and firms' productivity. Nevertheless, the percentage of IT costs in relation to the total costs was asked in the questionnaire, providing a log-normal distribution with mean expenditures of 4.9% of the total costs ($\sigma = 9.59$).[10] The linear IT expenditure distribution (n = 351) is shown in Figure 3.6.

3.2 Reference Groups Used in the Evaluation

The analysis presented here uses three main types of criteria to group the data for the evaluation of the preparedness for and attitude of German enterprises towards the application of dynamic pricing methods. Table 3.1 depicts the segmentation into evaluation groups according to the criteria. In the following sections these groups will be introduced in connection with a discussion of the associated criteria.

Table 3.1. Reference groups used in the evaluation

Group	Digital Business	Digital Yield			Dynamic Pricing
Subgroup		e-yield information sales	e-yield contact sales	e-yield product/ services sales	
Criteria	Enterprises provide digital products/ services or distribute digital products	Enterprises that reach more than 1% Internet-generated turnover in one of the yield sectors: information sales, contact sales, and products / services			Enterprises ranking the importance of at least one of dynamic pricing methods 'rather high'
Subcriteria		minimal 1% turnover in information sales	minimal 1% turnover in contact sales	minimal 1% turnover in product/services sales	

---

[10] All calculations for the statistical evaluation of the empirical data in this chapter have been performed using SPSS 11.5.

3.2.1 Digital Business Group

To find out whether a company in this sample is operating in *digital business* either by the direct provision of digital products and services (ISIP provision) or by the production of *digital goods* as a part of their Internet activities the following two questions (Q 2.7, 2.8) were asked:

- "Does your enterprise create digital products or provide digital services (e. g. software, news-ticker, ring tones)?"
- "Is the direct distribution of digital products (e. g. software download) accomplishable via your Internet presence?"

81.5% of the participants sayed yes to this question, 17.1% replied no to the first question, and 1.5% indicated a planned provision of digital services or products within the next two years. For the second question the results were positive at 82.1%, negative at 12.4% and "a realization is planned within the next two years" at 5.5% (see Figure 3.7).[11]

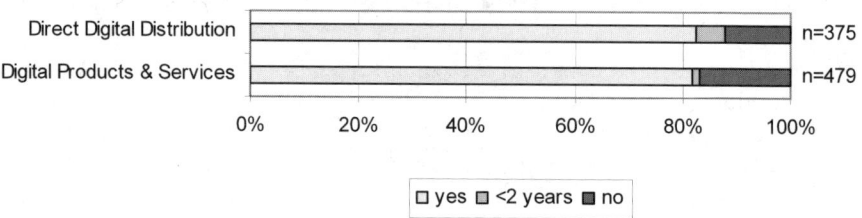

Fig. 3.7. Enterprises providing digital products and digital services

The high proportion of more than 80% with direct download capabilities might be explained by the fact that many companies use download features as a simple instrument to make their Internet presence more appealing to the customers. The first segmentation group in this study is therefore the cluster of enterprises actually providing digital services and products or directly distributing digital products.[12] It will later be called *"digital business group"* further on (see Table 3.1) and is very close to the ISIP industrial sector discussed in this thesis.

3.2.2 Digital Yield Group

The simple question of digital download capability and the provision of digital products and services represents only an indicator for the engagement of

---

[11] In the following figures with an aggregated percentual representation of answers the total number of answers will be given behind the corresponding bars.

[12] Potential adopters that answered "a realization is planned within the next two years" are not incuded in this group.

a company in the ISIP sector. A further question (Q 2.11) therefore addresses the percentage of total turnover attributed over the Internet distribution channel. The resulting distribution of the Internet sales revenue in relation to total sales revenue over the complete survey sample is shown in Figure 3.8. The mean value of the Internet return taking the valid answers of all contacted enterprises amounted to 3.49% with a standard deviation of $\sigma = 10.21$.

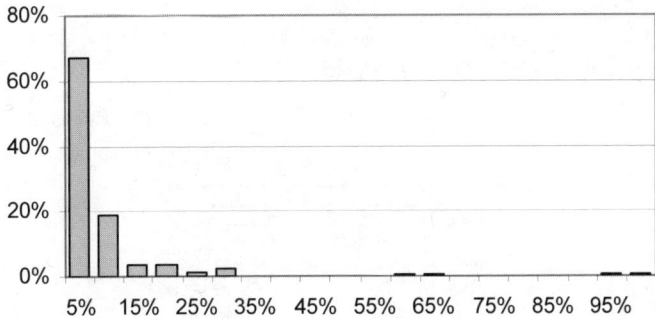

Fig. 3.8. Distribution of companies' fraction of turnover generated by Internet sales

In a second subsequent question (Q 2.13) three segments of the Internet yield sources were specified and prompted both for Internet and conventional business:

"Which part of return contributes to the total return for the following sources?"

- sales of products /services
- sales of contacts
- sales of information

The resulting yield distribution over the whole survey sample is depicted in Figure 3.9 for conventional (n = 404) and Internet sales (n = 191). Conventional sales thereby cover the turnover achieved with sales made without customer interaction in the Internet, whereas Internet sales subsume the return obtained by employing the Web. The corresponding turnover for conventional and Internet sales is then broken down into the product categories sales of products /services, sales of contacts, and sales of information by question (Q 2.13). It can be seen that the reward achieved from information sales in the Internet is 4.82% of the total Internet sales, which is higher than 1.41% in conventional business with a significance of t = 0.006[13]. This is not very surprising because information is especially intended to be sold as a digital good.

---

[13] In this study the degree of significance is partitioned into three classes: weakly significant where $t < 0.1$, significant where $t < 0.05$ and highly significant where $t < 0.01$.

Interestingly the turnover made from the sales of contacts over the Internet shows at 3.37% no significant difference to the return that is achieved by the sales of contacts (2.48%) over conventional distribution channels. The stake of product and services sales is 96.11% of the total return achieved when distribution over conventional channels is used. For the distribution of products and services over the Internet channel it is 91.81% of total return according to the companies' anwers to question Q 2.13.

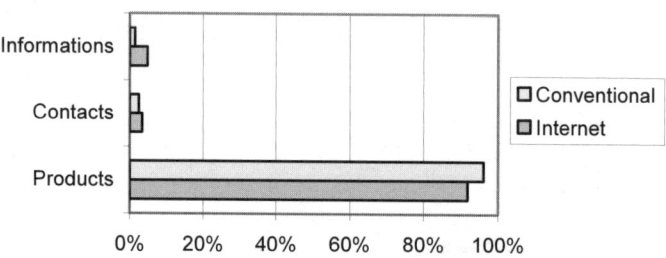

Fig. 3.9. Distribution of yield sources in conventional and Internet business sectors

The cluster of enterprises reaching more than 1% turnover[14] in the yield sectors, information, contacts, and products/services will be called *"digital yield group"* in the analysis that follows in this survey. According to the assessment of question Q 1.13, the digital yield group is subdivided into the sectors *"e-yield information sales"*, *"e-yield contact sales"* and *"e-yield products/services sales"* (see also Table 3.1). In analogy to this, companies reaching more than 1% of total return in the digital revenue assessment of the preceding question are subsumed under the *"digital revenue group"*.

3.2.3 Dynamic Pricing Group

To study the relevance of dynamic pricing methods for Internet applications from the companies' point of view the following pricing methods were considered: fixed pricing, bilateral price negotiations, English auctions, Dutch auctions, combinatorial auctions, bundle pricing and reverse pricing. For this purpose the participants in the study were asked to answer the following question (Q 5.25) in the range of a five point Likert scale[15]:

"How do you assess the application potential of the following pricing models in the Internet for the sales division in your enterprise?"

---

[14] This low treshold of 1% was set to avoid the dropout of too many elements from the evaluation sample in the grouping process.

[15] Like all assessment questions in this study, participants could answer in the following graduation: high, rather high, rather low, low, none.

- fixed pricing (e. g. catalog pricing)
- bilateral price negotiations
- English auctions (highest bid wins)
- Dutch auctions (price declines, first bid wins)
- combinatorial auctions (auction for bundles)
- bundle pricing (price reduction for bundles)
- reverse pricing (seller sets price, first accepting buyer wins)

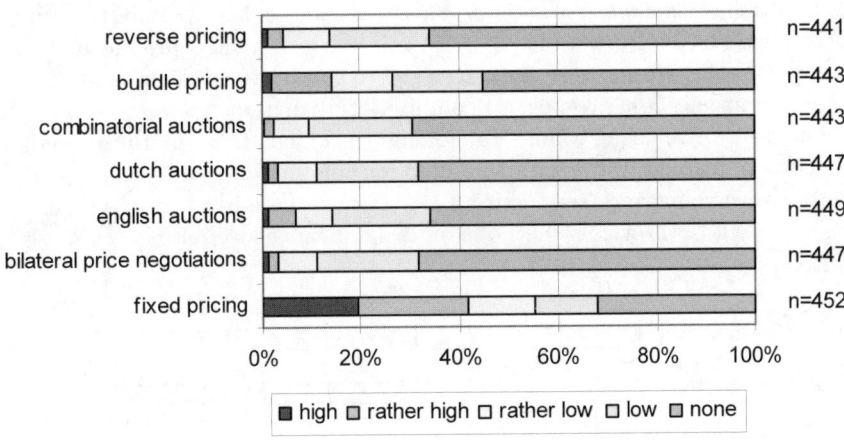

Fig. 3.10. Assessment of dynamic pricing methods for all sectors

The answer distribution for fixed and dynamic pricing methods for all returned questionnaires can be seen in Figure 3.10. As might be expected, fixed pricing has a maximum acceptance rate of $\mu = 3.16$, where acceptance rate is defined as the mean of the application potential rating.[16] By definition, fixed pricing does not belong to the class of dynamic pricing methods and serves for comparative purposes in this question. The second highest acceptance ($\mu = 3.68$) is found for bilateral price negotiations, which is the first representative of dynamic pricing methods.[17] This does not come as a surprise, because bilateral price negotiations are widely accepted in the business-to-business domain. The third highest acceptance is found for bundle pricing with $\mu = 4.13$. This is not very astonishing either, because the widely accepted practice of rebates and gross discounts can be subsumed under the bundle pricing

---

[16] (high = 1, rather high = 2, rather low = 3, low = 4, none = 5)

[17] The order follows the mean of the rating values, however, due to small differences the ranking can not always be proved. The order gives only a hint about the ranking for the group of auctions and reverse pricing: English auction and RP, as well as Dutch auction and CA could not be distinguished on a 0.05 significance level by employing a paired samples t-test.

mechanism. Bundle pricing is followed by the group of classical dynamic pricing methods, like the English auction ($\mu = 4.43$), Dutch auction ($\mu = 4.53$), and combinatorial auction ($\mu = 4.57$) in the acceptance rate ranking. Reverse pricing, which is related to auction mechanisms, lies in the middle field of the acceptance rates ($\mu = 4.46$) perceived for the auction domain.

To investigate the properties of Internet dynamic pricing applications, a reference cluster is formed out of the companies that rank the importance of at least one of dynamic pricing methods (bilateral negotiations, auctions, reverse-, bundle pricing) equal to or higher than "rather high". This reference group is called *"dynamic pricing group"* in the subsequent part of the study.

In an additional question (Q 5.26) related to dynamic pricing methods the survey participants were directly asked to specify their assessment of consumer acceptance with respect to combinatorial auctions and reverse pricing in general.[18] The mean values range slightly higher than in the preceding question, but are still low with $\mu = 4.33$ for combinatorial auctions and $\mu = 4.24$ for reverse pricing (see Figure 3.11). In fact there is no difference in the assessment of both pricing methods on a significance level of 0.05.

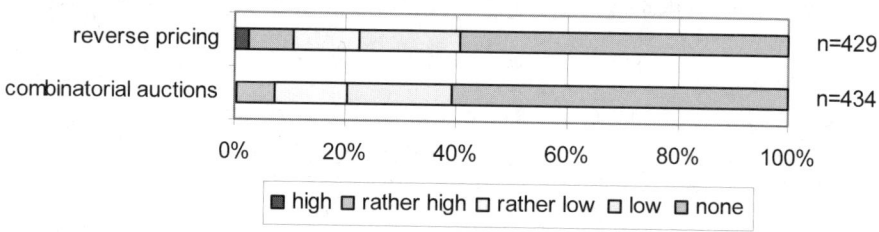

Fig. 3.11. Assessment of combinatorial auctions vs. reverse pricing customer acceptance

3.3 Detailed Findings on Dynamic Pricing Preference

The application potential of dynamic pricing methods in the Internet is strongly coupled to the benefit of online price variations perceived by vendors and sellers as well. As a first hypothesis, one could claim that Internet technology-related sales enable higher price variability compared to pricing in conventional distribution channels:

H1: If Internet is used, higher price variability will be the result, compared to pricing in conventional distribution channels.

---

[18] The answer scale for all assessment question following in this survey is identical with the five point Likert scale presented in the description of question Q 5.25.

3.3.1 Conventional vs. Internet Pricing Behavior

For this reason one of the first questions (Q 5.9) investigates conventional vs. Internet pricing behavior with respect to the conventional competitors:

"If your company provides price information in the Internet, do you vary Internet pricing "

- more often than your competitors in the offline world?
- more individually than your competitors in the offline world (e.g. price differentiation between customers)?

The answers unambiguously indicated a strong tendency towards higher price variation on the Internet (see Figure 3.12). 89.37% of the survey's participants claimed to use price variation more often than their competitors using a conventional distribution channel (1.33% plan to do so within the next two years, 9.30% answered no). Slightly fewer, 87.16%, of companies indicated more individual Internet pricing *(INP)* compared to their conventional selling competitors (1.69% plan to do so within the next two years, 11.15% answered no)[19].

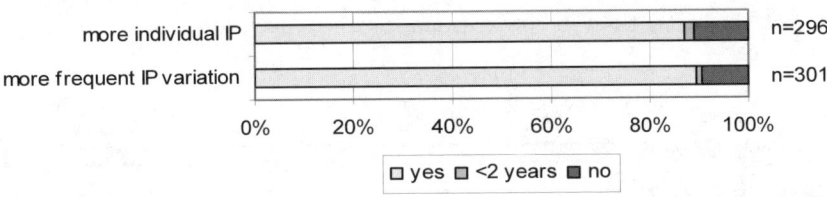

Fig. 3.12. Comparison of competitors' with respect to Internet pricing behavior

To get a more detailed view of the types and scope of Internet price variation and differentiation the following characteristics were evaluated (Q 5.10):

"If your enterprise has other distribution channels besides the Internet,..."

- do you vary prices on the Internet more often than on other distribution channels?
- do you vary prices on the Internet in smaller steps than on other distribution channels?
- do you grant specific Internet rebates, premiums, etc. that would not be given otherwise?
- do you grant rebates, premiums, etc. on the Internet more often than on other distribution channels?

---

[19] Whether the companies' self assessment matches econometric evidence in INP is a matter of further research and beyond the scope of this thesis (Brynjolfsson & Smith 2000).

- do you grant rebates, premiums, etc. on the Internet which are higher than on other distribution channels?
- do you form prices on the Internet more individually than on other distribution channels (e.g. price differentiation between customers)?

Figure 3.13 shows the resulting distribution of answers for the INP characteristics questions over the whole query sample. Again a strong tendency towards higher variation and differentiation of INP occurs. Companies indicated a more frequent INP variation than on conventional distribution channels (90.88% yes, 0.8% planned, 8.32% no) combined with a finer graduation of pricing steps (93.24% yes, 0.81% planned, 5.95% no) and a more individual treatment of pricing (92.02% yes, 0.8% planned, 7.18% no). Concerning the price level, participants in the study indicated a more frequent (92.53% yes, 1.87% planned, 5.6% no) and higher (94.96% yes, 1.33% planned, 3.17% no) granting of rebates and premiums. In addition, the existence of special exclusive rebates and premiums is reported with a slightly lower occurrence of 86.0% (2.64% planned, 11.35% no). The observations confirm empirical studies on Internet pricing behavior performed by Brynjolfsson & Smith (2000). Price variations are found to be more frequent and of finer graduation in Internet-related business, as an econometric study of book and CD retail prices shows. This could mainly be attributed to lower costs for price changes in the Internet *(menue costs)* than in conventional business according to the authors of this study.

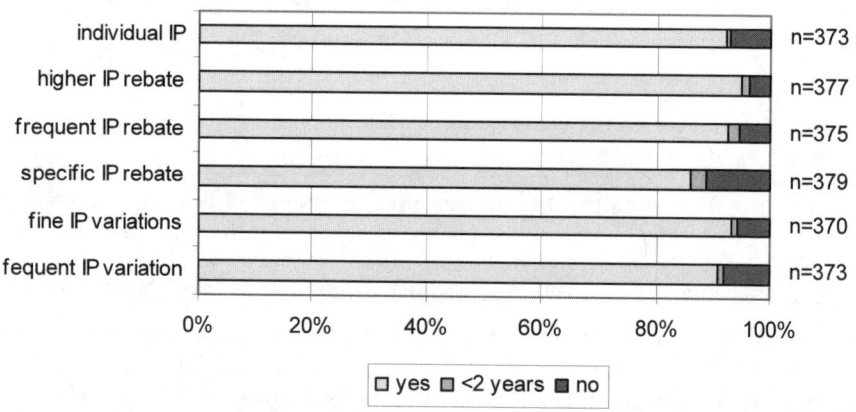

Fig. 3.13. Comparison of Internet and conventional pricing behavior in terms of variation, individualization and rebates

3.3.2 Automated Pricing Acceptance

A first step towards assessing the acceptance of automated pricing processes is to evaluate how far enterprises and customers are prepared to provide and use automated pricing and product information. For this reason, participants in the study were asked the following question (Q 5.14):

"How much importance do you assign to the following barriers to the automated communication of price and product information from your company's view?"

- low buyer or customer acceptance
- costly integration into existing IT infrastructure
- costly integration into existing work flow organization
- costly processing, maintaining and administration of the information base (e.g. ERP system)
- opportunity to easily reuse the provided information (e.g. price comparison)
- costs exceed profit
- peril of price wars with competitors
- other barriers

The resulting distribution of answers for the overall survey sample is shown in Figure 3.14.

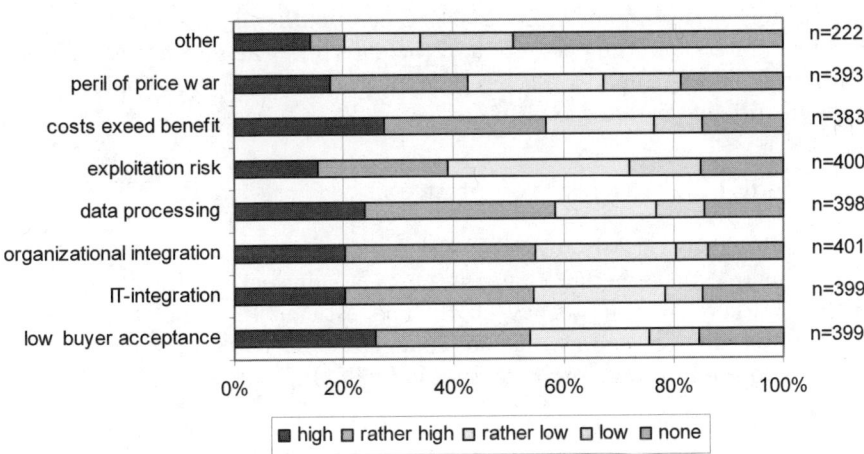

Fig. 3.14. Barriers to the application of automated communication of price and product information with respect to the customers' acceptance

A predominant argument ($\mu = 2.54$)[20] against the use of automated pricing and price information in Internet business is the enterprises' estimation of costs exceeding benefits. Another often-cited argument used to justify the rejection of automated pricing and product information systems is the expectation of costly processing, maintaining and administration of the related information base ($\mu = 2.56$). Interestingly this perception contradicts the low menu cost argument presented in Brynjolfsson & Smith (2000). Consequently this argument is reinforced by organizational ($\mu = 2.58$) and IT integration concerns ($\mu = 2.62$) on the part of the enterprises' decision makers. Market risks like lower buyer acceptance ($\mu = 2.6$) or the exploitation of pricing information by arbitrage makers ($\mu = 2.89$) play a secondary role in the entrepreneurs' considerations. The lowest importance is attributed to the likelihood of price wars ($\mu = 3.91$) initiated by the application of dynamic pricing methods and the automated publication of price information on the Internet. Besides menu costs, there are many arguments about INP behavior in econometric evaluations with respect to the frequency and level of price changes in B2C. The literature relates to classic marketing arguments like price points (e.g. 2.99€), price discounts and relative price. On the side of non-pricing elements, product popularity, shipping costs and product information are considered to be control variables in B2C INP. Other more remote impact factors for the INP are the specific Internet market structure (monopolistic tendencies, high market transparency), Internet characteristic information asymmetry (product quality, lower search costs) and consumer demand behavior (less inventory-oriented INP policy, greater influence of demand shocks) (Kauffman & Lee 2004). These marketing arguments seem to be of less interest in the perception of the enterprises as shown in the survey. Technical aspects come to the foreground, leading to the non-application of dynamic pricing methods.

3.3.3 Individual Pricing Acceptance

A subject which is closely related to the acceptance of dynamic pricing methods is the question whether companies expect sufficient yield from the introduction of such systems to pay off their investment in automated dynamic pricing systems. The survey thus put some questions concerning the potential barriers to the introduction of flexible/individual pricing systems (Q 5.11):

"How much importance do you assign to the following barriers to flexible/individual pricing in the Internet from your company's point of view?"

- no turnover increase attainable
- costly determination of prices

---

[20] As in the previous question no significant ranking of the assessments could be given based on μ. For this reason μ is given only for orientation purposes in the discussion of the following questions.

- cost of price variation (menu costs) too high
- costly processing, maintaining and administration of the information base (e.g. ERP system)
- costly integration into existing work flow/organization
- low buyer or customer acceptance
- opportunity to easily resell goods (arbitrage)
- expected negative customer reaction due to perceived "inequality"
- expected negative customer reaction due to "price uncertainty"
- legal aspects
- peril of price wars with competitors
- other barriers

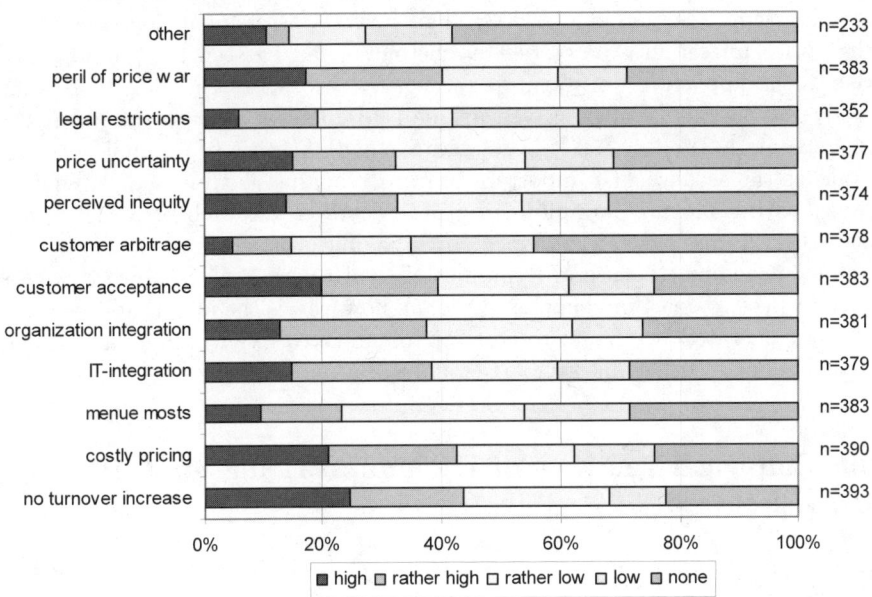

Fig. 3.15. Assessment of barriers to the application of individual / flexible pricing methods

As an outcome of this subject unit, a very prominent barrier to the application of individual and flexible pricing methods (first degree price differentiation using information about the individual buyer) is the companies' expectation ($\mu = 2.86$), that no turnover increase can be achieved from using this methods. Costly determination of prices plays another important role in the rejection of individualized dynamic pricing methods ($\mu = 2.98$). These two arguments are followed by the companies decision makers' presumption that consumers will not accept individual price differentiation ($\mu = 3.04$), especially if they recognize the price inhomogeneity (Bergen et al. 2003, Wirtz et al. 2003). Another market related argument for avoiding dynamic pricing form the companies' perspective is the danger of price wars occuring. The mean of sample for this impact factor was $\mu = 3.11$. Interestingly in disagreement with the answers to the above question regarding the acceptance of automated pricing and price information, technical reasons do not play the major role in the context of the adoption of dynamic and flexible pricing on the Internet. IT integration ($\mu = 3.16$) and organizational integration ($\mu = 3.14$; n $= 381$) both rank low, like the marketing aspects recognized as a barrier to introducing Internet dynamic pricing methods. The remaining influential factors asked about in this complex can be considered to have low relevance within the assessment group of technical factors that are responsible for hindering the use of flexible pricing technology. Price uncertainty on both consumer and vendor side is an often quoted argument against dynamic pricing (Bergen et al. 2003). The survey's assessment does not show that price uncertainty perceived by customers can be seen as a major barrier to the application of dynamic pricing by enterprises ($\mu = 3.29$). In this study the possibility of perceived inequality ranks low as a potential barrier to dynamic INP ($\mu = 3.32$). Legal scrutiny with respect to dynamic pricing ranks as the second least important argument against flexible and individual prices with a mean of sample of $\mu = 3.7$. See Weiss & Mehrotra (2001) for a further discussion of legal aspects of dynamic pricing in US markets. An argument against the use of dynamic INP with very low relevance $\mu = 3.9$ with respect to question complex (Q 5.11), however, is the idea that arbitrage can be generated by the customers' opportunity to resell individually lower priced goods to other market participants.

3.4 Empirical Analysis of Dynamic Pricing for ISIP Provision

After discussing the empirical assessment of dynamic pricing by German enterprises in general, this section will look more closely at the empirical data to analyze the application potential of flexible pricing in information production and information services industries. As shown above, the structure of this industry is closely related to the digital business group segmentation defined for our survey sample.

An interesting hypothesis associated with the assessment of flexible pricing, could be the assumption that a significant increase in acceptance for dynamic pricing methods can be expected for companies engaged in the digital business group (see Figure 3.16).

H2: Firms engaged in digital business will have a higher acceptance rate for dynamic pricing approaches.

For this reason the mean acceptance rate for dynamic pricing methods of the digital business group was tested against the mean attitude of com-

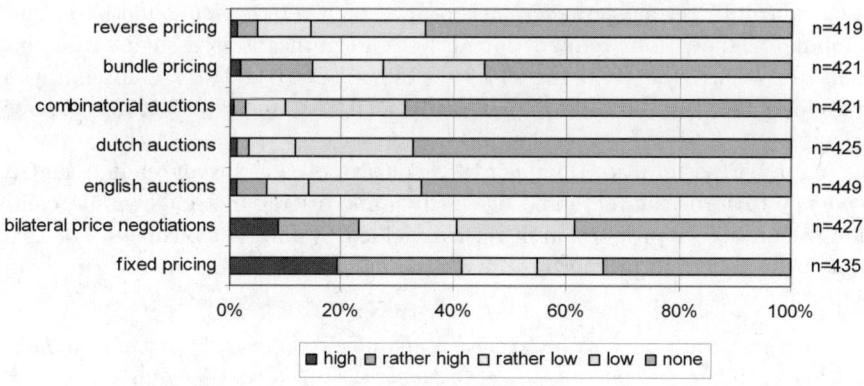

Fig. 3.16. Assessment of dynamic pricing methods for digital business group

panies not engaged in digital business, employing a t-test for the equality of means and Levene's test for equalities of variances. The result was that the acceptance rate for most dynamic pricing methods is significantly higher for the digital business group: Bundle pricing had the highest acceptance rate increase ($\bar{r} = 4.59; \Delta\bar{r} = 0.49; sig_F = 0.006; sig_t = 0.012$)[21] followed by reverse pricing ($\bar{r} = 4.82; \Delta\bar{r} = 0.37; sig_F = 0.000; sig_t = 0.000$) and Dutch auctions ($\bar{r} = 4.86; \Delta\bar{r} = 0.35; sig_F = 0.000; sig_t = 0.000$). The acceptance registered for combinatorial auctions was only slightly increased ($\bar{r} = 4.86; \Delta\bar{r} = 0.31; sig_F = 0.000; sig_t = 0.001$) in the survey sample. No other dynamic pricing method in the digital business group showed a significantly increased rating of applicability in the opinion of the interviewed companies.

As one might expect, the same observations hold for the direct comparison of combinatorial auctions and reverse pricing (see Figure 3.17).

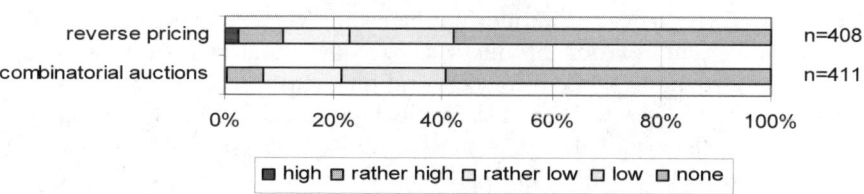

Fig. 3.17. Assessment of combinatorial auctions vs. reverse pricing customer acceptance in the digital business group

---

[21] The first value $\bar{r}$ is the average ranking of replies excluding the specific property defined for the test segment, the second value $\Delta\bar{r}$ is the difference to the mean rankings including the group's property, the third number sig_F denotes the F-test significance level of the Levene test for equal variances and the fourth value sig_t is the t-test significance level for equality of means (Hartung et al. 1999, pp. 531).

In absolute terms, however, acceptance of reverse pricing methods and combinatorial auctions remains low despite a significant assessment value increase ($\bar{r} = 4.78; \Delta\bar{r} = 0.48; sig_F = 0.000; sig_t = 0.003$) for combinatorial auctions and ($\bar{r} = 4.62; \Delta\bar{r} = 0.37; sig_F = 0.037; sig_t = 0.065$) for reverse pricing.

A further assumption made in the context of this survey claims that a tendency towards the increased use of dynamic pricing methods should coincide with a higher proportion of return yielded by Internet business. For this reason the dynamic pricing group was tested for higher mean returns from digital business.

H3: Firms that have an increased preference for dynamic pricing methods will have a higher proportion of return yielded by Internet business.

The result of the test astonishingly showed a significantly lower return share yielding from the Internet business (2.34% of total revenue share) for the firms voting for dynamic pricing methods than for the enterprises which were not in the dynamic pricing group (4.3% of total revenue share) ($sig_F = 0.03; sig_t = 0.055$). This may result from problems occurring when using dynamic pricing. Negative impact for dynamic pricing assessment could result from customer perception (and consequently the companies' view) of price uncertainty and inequality discussed previously (see section 3.3.3). The different assessment results between the digital business group and the dynamic pricing group can be explained by the broader definition of the digital business group. Only a small proportion of the companies in the digital business group are generating yield by employing the internet as their distribution channel. These findings however should be examined in further studies.

Another interesting view of dynamic pricing is the industrial segment structure associated with the firms engaged in information service and information product provision (see Figure 3.18). One of the first clusters considered in this context is the class of enterprises engaged in selling information over the Internet according to the e-yield information services group defined above. Enterprises engaged in information selling via the Internet mostly distribute similar information products using conventional channels as well.[22] Enterprises providing services for corporations hold a prominent position in this group. The segment share in the e-yield group (n = 22) is 49% compared with 17.4% in the overall distribution and 23.8% in the services business operating with conventional distribution channels (c-yield group, n = 21). As one

---

[22] In analogy to the "digital yield group", which is subdivided into the sectors "e-yield information sales", "e-yield contact sales" and "e-yield products/services sales" (see Table 3.1), yield type classes for conventional distribution channels were formed. Like the "digital yield group", the "conventional yield group" is subdivided into "c-yield information sales", "c-yield contact sales" and "c-yield products/services sales". According to the e-yield definition firms belong to theses classes, if more than 1% of total return in conventional business derives from information-, contact- or product services sales correspondingly.

would expect, the media related sector shows increased activity in information selling, whereas Internet distribution quantity is equal to conventional sales volume (e-yield 9.1%; c-yield 9.5%) compared to the overall segment distribution (7.3%).

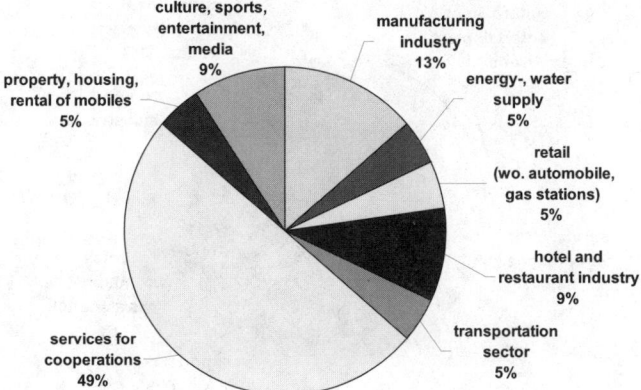

Fig. 3.18. Sector distribution of enterprises belonging to the information services e-yield group

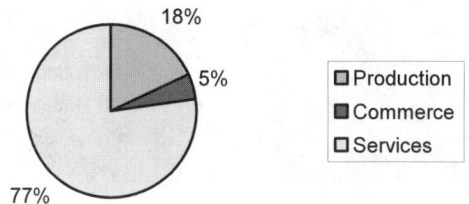

Fig. 3.19. Aggregated sector distribution of enterprises belonging to the information services e-yield group

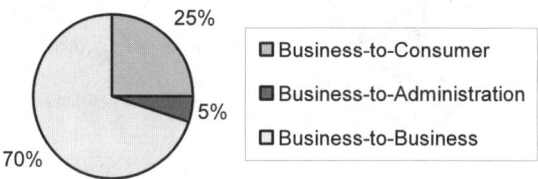

Fig. 3.20. Distribution of enterprises' main customer segments in the information e-yield group

Another area which is traditionally related to information-dependent yield management is the hotel industry. Therefore, it is not surprising that this sector shows higher revenue values for the e-yield (9.1%) and c-yield (9.5%) segmentation, compared to 4.3% in the overall distribution.

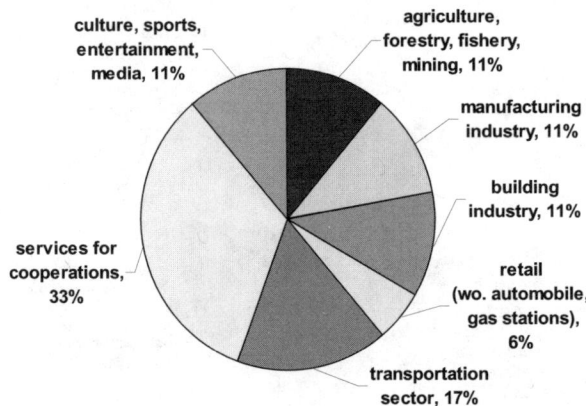

Fig. 3.21. Sector distribution of enterprises belonging to the contact selling e-yield group

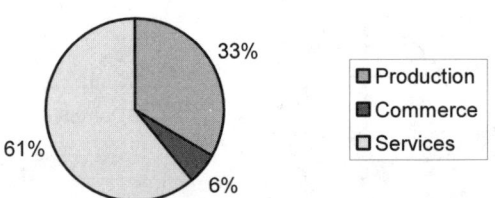

Fig. 3.22. Aggregated sector distribution of enterprises belonging to the contact selling e-yield group

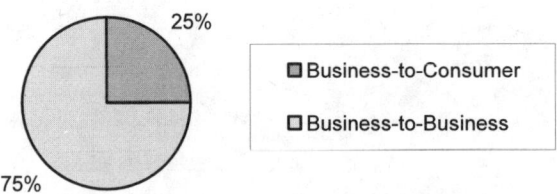

Fig. 3.23. Distribution of enterprises' main customer segments in the contact selling e-yield group

All other industrial segments occurring in the information sales area of this study have less importance than in the overall sector segmentation. The aggregated industry distribution (Figure 3.19) consequently underpins the service character of information provision (77% services compared to 47% in the overall sample Figure 3.1) and the B2B character of this activity (70% e-yield vs. 61% overall).

Following the same procedure that has been applied in the case of the information services group, the selling of advertising contacts is also analyzed for the e-yield contact segment. This business can be regarded as a member of the ISIP business cluster due to the information character of advertising. As in the information services group, the industry sector of the enterprises yielding more than 1% of their Internet return by selling advertising contacts (n = 18) has a more than proportional share (17,4% overall distribution) of firms in the companies' services business sector (e-yield 33.3%; c-yield 22.2%). Additionally, Figure 3.21 depicts the increased importance of contact selling in the media sector (e-yield 11.1%; c-yield 13.9%) compared to the overall distribution (7.5%). Interestingly, the transportation sector plays an outstanding role in the Internet contact selling business (e-yield 16.7%; c-yield 8.3%) compared to 6.5% in the whole questionnaire sample.

A look at the aggregated industry sector distribution underpins the services character of the contact selling business with a 61% share of service industry (see Figure 3.22). This confirms the findings for the information services group. The proportion of B2B customer segment reaches the high level of 75% (Figure 3.23) compared to 64% in the overall distribution (Figure 3.1). E-yield group for services and products is not considered in this section, due to the lack of differentiation properties of conventional and digital products distributed over this channel. Products and services covered by this part of the inquiry are not guaranteed to be typical of the ISIP segment. The empirical assessment therefore leaves this topic aside, considering information services and contact selling groups as sufficient representatives of ISIP business.

3.5 Empirical Implications of Dynamic Pricing for ISIP Provision

Over 80% of companies covered by the questionnaire analyzed here, provide the Internet-based download of digital products and information as core business or a free additional customer service. Sales of information and advertising contacts amounts to 8% of the overall Internet return, generating business besides the yield achieved by selling real world and digital products as well as services over the Internet. This gives a good impression of the importance of ISIP-related systems for e-commerce. Dynamic pricing is closely related to ISIP business because the yield management aspect of information and contact sales (see section 2.1.5 and 4.2.1) requires flexible valuation of the information in general and the dynamic allocation of their provision resources in particular. Internet pricing has already been shown to be more flexible and volatile than conventional business, a finding which is enforced by this analysis (Brynjolfsson & Smith 2000).

The survey, however, reveals some important arguments against the use of dynamic pricing in ISIP applications from the companies' point of view. The main argument against automated provision price information to customers is the expectation that cost of dynamic pricing applications will exceed the reward. The same argument holds for the application of flexible pricing strategies in the Internet business, mainly due to the companies' belief that IT integration and maintenance of dynamic pricing applications, - including the price finding process -, will generate high expenses but not higher reward. These are the main technical objections to dynamic pricing on the part of the consulted companies. Marketing concerns are the other crucial set of reservations about the application of dynamic INP. Interestingly, this group does not play the dominant role with respect to the arguments against the application of flexible INP. The widely accepted argument that the introduction of individual and flexible prices will offend customers because of a feeling of inequity and price uncertainty rank far lower in the enterprises' awareness than it would be expected in this context.[23]

In general, the assessment of the utility of dynamic pricing is low, showing the highest reception for more traditionally oriented price negotiation mechanisms. As a result of eBay's success, auctions have received more attention from the companies' in terms of reward increasing dynamic pricing methods, leading to the second best ranking for this flexible INP group. The participants in the survey are not, however, particularly aware of differentiated auction mechanisms and it seems to be in some doubt whether they really recognize the innovative potential of applying e.g. the combinatorial auction mechanism. As a further result the study showed that the main business domain of flexible ISIP pricing lies in the B2B oriented service sector, including especially the media and the hotel industry. As a consequence of the empirical survey results, the necessity of "easy to handle and integrate" dynamic pricing IT system development has a high priority in convincing enterprises to adopt flexible pricing methods. Configurable mechanisms for automated price finding have to be designed to help companies while meeting yield maximizing pricing decisions. Moreover the economic advantages of sophisticated pricing mechanisms like combinatorial auctions have to be brought in the awareness of corporate managers to eliminate the "no-increasing-return" argument. The theoretical work in this thesis should give some advice about which direction this IT and mechanism design should take.

---

[23] Previous arguments confirmed in this study, are in full agreement with two postulates made by Reinartz (2001) that should guarantee successful dynamic pricing strategies by claiming that "costs of segmenting and pricing must not exceed revenue due to customization" and "notions of perceived fairness must not be violated".

4

Reinforcement Learning for Dynamic Pricing and Automated Resource Allocation

The beginning of *reinforcement learning (RL)* as an AI discipline dates back to the early days of cybernetics and work in statistics, psychology, neuroscience, and computer science. In an RL problem an agent has to learn *reward maximizing (punishment minimizing)* behavior by *trial-and-error* interaction with a dynamic environment. A crucial advantage of RL compared to other AI approaches like DPS, is the fact that with RL there is no need to specify how the task is to be achieved (Kaelbling et al. 1996).

In the first section of this chapter an introduction to the methodology of RL is given and after a short survey over classical YM the application of RL to YM, dynamic pricing, and allocation tasks is discussed. Two models of the use of RL in the dynamic pricing and automated resource allocation context are then presented. The first model addresses the yield maximizing scheduling of ISIP tasks on a single resource, whereas the second model enables the yield maximizing allocation of interdependent ISIP tasks (exhibiting complementarities). Both approaches are compared with alternative AI-related scheduling and allocation approaches and their advantages and drawbacks are discussed at the end of this chapter.

4.1 Basics of Reinforcement Learning

The selection of reward maximizing actions by RL agents is closely linked to optimal decision making in an environment that can be described as a stochastic process.

4.1.1 Markov-Processes in Decision Trees

Decision problems, in principle, can be described as a tree structure, where the edges represent the individual decisions or actions of the agents respectively $(a_1, a_2, ...a_t \in A: action\ space)$ and the vertices denominate the system states $(s', s'', s''' \in S: state\ space)$. In the framework of *decision process optimization*

(DPO) an additional parameter is required to describe the reward for an action $(r_1, r_2, ...r_t \in R_t$: *reward function*). In this context the expected value of the next decision's reward could be written as:

$$\mathcal{R}^a_{ss'} = E\left\{r_{t+1} \,|a_t = a \,, s_t = s, s_{t+1} = s'\right\} \tag{4.1}$$

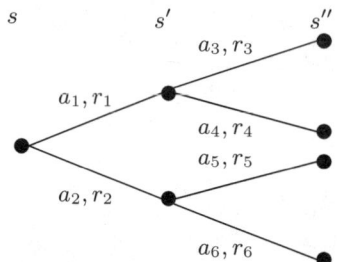

Fig. 4.1. A decision tree with states actions and rewards

An optimization pass through the *decision tree* should lead to a maximum sum of all rewards r_t. In every node of the tree a decision about the selection of the next edge is taken.

Additionally one can assign an a-priori probability to the selection of the actions or the edges:

$$\mathcal{P}^a_{ss'} = \Pr\left\{s_{t+1} = s' \,|s_t = s, a_t = a\right\} \tag{4.2}$$

A further parameter π is employed for this purpose. This parameter is defined as policy, because the strategy used to select the edges, called the decision policy, has a direct impact on the probability of running through a particular path in the tree.

For the techniques discussed in what follows, it is essential to be aware, whether the route passed through has significance for the subsequent states or not. To facilitate the analytical handling of the problem while using π, the *Markov property*[1] is assumed. It implies that the probability of the transition from state s to state s' etc. in the particular stages of the tree is independent of the previous system states and decisions. The Markov property allows a successive, stepwise solution of the problem, employing an iterative procedure based on *dynamic programming* methodology, as realized in RL. The basic idea of RL will now be presented in the following paragraphs.

---

[1] Due to the outstanding importance of the Markov property in connection with decision problems, a whole class of problems in literature is named '*Markov decision problems*' (Puterman 1994).

4.1.2 Idea of Reinforcement Learning

When considering the base elements of decision optimization –states, actions and reinforcements– from a system-theoretic perspective, the following model could be implied together with the interpretation as a decision tree.[2]

- The reinforcement learning agent (RL-agent) is connected to its environment via sensors.
- In every step of interaction the agent receives a feedback about the state of the environment s_{t+1} and the reward r_{t+1} for its latest action a_t.
- The agent chooses an action a_{t+1} representing the output function, which changes the state s_{t+1} of environment and thus leads to state s_{t+2}.
- The agent receives new feedback from reinforcement signal r_{t+2}.
- Objective of the agent is to optimize the sum of the reinforcement signals in the long run.

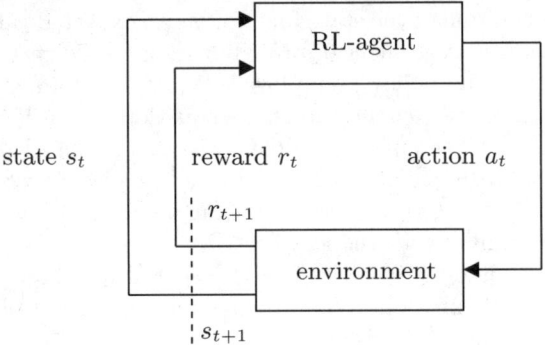

Fig. 4.2. A reinforcement learning agent interacting with its environment

Now the question arises of how such an agent, modeled as a subsystem of the environment, could find its way through the decision tree. The fundamentals essential for this process are the subject of the discussion in the following section.

4.1.3 Stochastic Dynamic Programming

To determine the optimal path through the decision tree, it is usually necessary to evaluate all edges in the tree. Using a myopic greedy decision strategy, – while selecting the next edge with the highest reward r_t –, is not a guarantee of obtaining the optimal path in the decision tree. A branch from node

---

[2] The following sections follow the argumentation in Sutton & Barto (1998) as well as the introduction to reinforcement learning given in Schwind (2003).

s' to node s'' which seems favorable first, could prove disadvantageous in the further progression of the chosen subtree. This problem, resulting from the sequential structure and non-reversibility of the decisions, leads to the complexity of the optimization problem, which grows exponentially in terms of state space dimensions.[3] Bellman (1957) solves the problem of wrong branches by calculating recursively the optimal decisions in every state layer s. For this purpose, we introduce two new terms:

$$V^\pi(s) = E_\pi \left\{ R_t | s_t = s \right\} = E_\pi \left\{ \sum_{k=0}^{\infty} \gamma^k r_{t+k+1} | s_t = s \right\} \qquad (4.3)$$

- The state value $V^\pi(s)$ of a policy π is defined as the value which is expected to result from the discounted state value $V^\pi(s')$ of the next state s' summed with the expected reward r in $t+1$. As can be seen in the example, depicted in Figure 4.3, three possible actions a could be taken from state s, which can produce two different rewards r_{t+1} as environmental reaction, while migrating into one of two states s' in the next decision layer. The definition of the state function as an expected value can be seen as a result of the assignment of probabilities π to the particular decisions a depending on the policy π employed. While determining the new state values, following the *Bellman equation* (equation 4.3), it is important that the state values $V^\pi(s')$ of all possible decisions are considered with weight π in the recursive calculation of $V^\pi(s)$. We can execute recursion through the decision tree by forward induction starting at the root or by backward induction beginning with the leaves.
- The use of a discount factor γ is not essential for classic dynamic programming, but it turns out to be valuable for convergence of reinforcement learning in the following discussion.

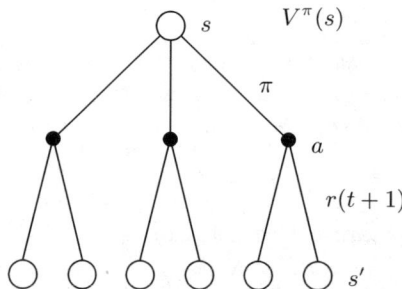

Fig. 4.3. The state diagram for $V^\pi(s)$

---

[3] Bellman's curse of dimensionality: "At high dimensions all objects are far from each other."

$$Q^{\pi}(s,a) = E_{\pi}\left\{\sum_{k=0}^{\infty}\gamma^{k}r_{t+k+1}\middle| s_t = s, a_t = a\right\} \qquad (4.4)$$

- The action value $Q^{\pi}(s,a)$ of a policy π, in addition, depends on the state value of an action chosen in s. As can be seen in Figure 4.5 for the transition from s to s', and is apparent from equation (4.4), action a determines the selection of one edge in the decision tree, whereas the edges which are not selected do not account for the calculation of $Q^{\pi}(s,a)$. Having arrived at decision layer s', the Bellman equation is employed to calculate the state value of the decision subtree which remains in s'.

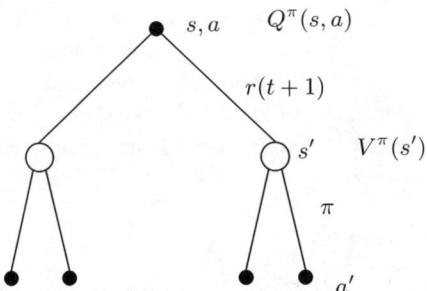

Fig. 4.4. The state diagram for $Q^{\pi}(s,a)$

The objective of reinforcement learning is to find a path through the decision tree which *maximizes the sum of rewards*. For this purpose we need a definition of the optimal state value and the optimal action value respectively.

$$V^{*}(s) = \max_{\pi} V^{\pi}(s) \qquad (4.5)$$

$$V^{*}(s) = \max_{a}\{r(s,a) + \gamma V^{*}(s')\} \qquad (4.6)$$

$$Q^{*}(s,a) = E\{r_{t+1} + \gamma V^{*}(s_{t+1})| s_t = s, a_t = a\} \qquad (4.7)$$

- The optimal state value can be calculated recursively, based on the Bellman equation (4.3), differing from the normal state value in one respect. Not all decision alternatives are taken into account here, but only the alternative which provides the highest value adding the discounted optimal state value of the lower decision layer and the reward $r(a,s)$ of the optimal action a when the optimal action is the action yielding the highest reward and the optimal state value $V^{*}(s')$ of the lower decision layer has been calculated the same way one step before (equation 4.6 and 4.7).

- In analogy with equation (4.5) the optimal action value $Q^*(s, a)$ of a policy π is the discounted optimal state value of the selected action a plus reinforcement signal $r(a, s)$ (equation 4.8):

$$V_{k+1}(s) = \max_a E \left\{ r_{t+1} + \gamma \sum_{s'} V_k(s_{t+1}) \middle| s_t = s, a_t = a \right\}$$
$$= \max_a \sum_{s'} \mathcal{P}_{ss'}^a \left[\mathcal{R}_{ss'}^a + \gamma V^\pi(s') \right] \qquad (4.8)$$

An alternative way of writing equation (4.7) is given in equation (4.9). $V^*(s)$ is computed by calculating $V_{k+1}(s)$ repeatedly. This procedure is known as value iteration. We will return to the algorithmic formulation later. So far, we have assumed that the decision policy is given initially and will be kept to the end of the evaluation. The calculation is performed by the iterative approximation of the state values following the modified Bellman equation (4.3):

$$V^\pi(s) = E_\pi \left\{ r_{t+1} + \gamma r_{t+2} + \gamma^2 r_{t+3} + \dots \middle| s_t = s \right\}$$
$$= \sum_a \pi(s, a) \sum_{s'} \mathcal{P}_{ss'}^a \left[\mathcal{R}_{ss'}^a + \gamma V^\pi(s') \right] \qquad (4.9)$$

The iteration takes place until the difference between the previous and the new calculated state value falls below a threshold σ:

$$\max_{s \in S} |V_{k+1}(s) - V_k(s)| \le \sigma \qquad (4.10)$$

As an alternative to value iteration one can think of a policy adjustment during the evaluation to avoid the necessity of exploring the whole decision tree to rate a policy. Such a technique is called policy iteration in contrast to the value iteration, where the state values and not the decision policies are changed. At first it must be ascertained in the context of the policy iteration, whether a change from policy π to policy π' in state s leads to a refinement of $V^\pi(s)$ and whether this improvement in policy π' is valid for all further decision situations as well. In this case the policy improvement theorem tells us.[4]

If the following holds for two deterministic policies:

$$Q^\pi(s, \pi'(s)) \ge V^\pi(s) \qquad (4.11)$$

Then the subsequent argument holds too:

$$V^{\pi'}(s) \ge V^\pi(s) \qquad (4.12)$$

---

[4] A short proof can be found in Sutton & Barto (1998, p. 95).

That means that a change in policy π could be evaluated in terms of an improvement for all potential actions a in state s, as far as the value function $V^{\pi}(s)$ of policy π is known at node s. Selecting the optimal policy in every visited state s and regarding only the next step, a greedy strategy for permanent policy improvement results:

$$\pi'(s) = \arg\max_a Q^{\pi}(s,a) = \arg\max_a \sum_{s'} \mathcal{P}^a_{ss'} \left[\mathcal{R}^a_{ss'} + \gamma V^{\pi}(s')\right] \qquad (4.13)$$

$$V^{\pi'}(s) = \sum_{s'} \mathcal{P}^a_{ss'} \left[\mathcal{R}^a_{ss'} + \gamma V^{\pi'}(s')\right] \qquad (4.14)$$

After optimizing a policy π yielding the improved policy π' by using $V^{\pi}(s)$, a new value function $V^{\pi'}(s)$ can be calculated to improve π' to π'' etc. The iteration of evaluation and improvement towards a greedy strategy depicted in Figure 4.5 is called policy iteration. Table 4.1 shows the related algorithm identifying the dichotomy between evaluation (equation 4.14) and improvement (equation 4.13).

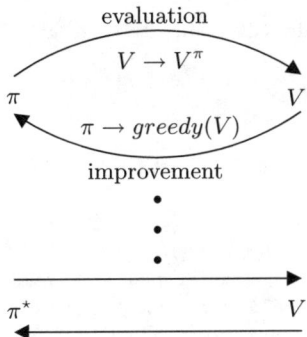

Fig. 4.5. Alternation between evaluation and improvement at policy iteration

A disadvantage of policy iteration is that states have to be passed in a multiple manner at every phase of evaluation to determine $V^{\pi}(s)$ with sufficient accuracy, while the exact value is obtained only as limit value. It can be shown that a policy evaluation converges, even if the state space is passed exactly once. Thus step 2 and 3 of the policy iteration method coincide, resulting in the value iteration method:

$$V^{\pi'}(s) = \max_a \sum_{s'} \mathcal{P}^a_{ss'} \left[\mathcal{R}^a_{ss'} + \gamma V^{\pi'}(s')\right] \qquad (4.15)$$

The value iteration algorithm is outlined in Figure 4.2 and can also be derived directly from equation 4.6.

Table 4.1. Policy iteration algorithm

1. Initialization
       Select $V(s) \in \mathcal{R}$ and $\pi(s) \in A(s)$ arbitrary for all $s \in S$
2. Policy evaluation
       Repeat
               $\Delta \leftarrow 0$
               For each $s \in S$:
               $v \leftarrow V(s)$
               $V(s) \leftarrow \sum_{s'} \mathcal{P}_{ss'}^{\pi(s)} \left[\mathcal{R}_{ss'}^{\pi(s)} + \gamma V(s') \right]$
               $\Delta \leftarrow \max(\Delta; |v - V(s)|)$
               until $\Delta < \Theta$ (small positive number)
3. Policy improvement
               policy-stable $\leftarrow$ *true*
               For each $s \in S$:
               $b \leftarrow \pi(s)$
               $\pi(s) = \arg \ \max_a \sum_{s'} \mathcal{P}_{ss'}^a [\mathcal{R}_{ss'}^a + \gamma V(s')]$
               If $b \neq \pi(s)$, then *policy-stable* $\leftarrow$ *false*
               If *policy-stable*, then stop; else
               go to step 2

Table 4.2. Value iteration algorithm

Initialize v arbitrary, e. g., $V(s) = 0$, for all $s \in S^+$
Repeat
       $\Delta \leftarrow 0$
       For each $s \in S$:
               $v \leftarrow V(s)$
               $V(s) \leftarrow \max_a \sum_{s'} \mathcal{P}_{ss'}^a [\mathcal{R}_{ss'}^a + \gamma V(s')]$
               $\Delta \leftarrow \max(\Delta; |v - V(s)|)$
               until $\Delta < \Theta$ (a small positive number)
Output: a deterministic policy π, such that:
               $\pi(s) = \arg \ \max_a \sum_{s'} \mathcal{P}_{ss'}^a [\mathcal{R}_{ss'}^a + \gamma V(s')]$

4.1.4 Monte-Carlo Methods

Another downside of the estimation procedure described so far when using Bellman dynamic programming to determine $V(s)$, lies in the fact that all states in the decision tree have to be passed to evaluate the value function of state s (*full backup method*). Because it is sufficient to estimate $V(s)$ coarsely, a *Monte-Carlo method (MC)* can be employed in most cases. For this purpose it is not necessary to traverse the entire decision tree, but only several traces, so called episodes, represented in Figure 4.6 as reward vectors (r_1, r_3, r_7) and (r_1, r_4, r_8). While passing through the episodes the mean value of the rewards which occurred on the edges are used as an estimator for the state values. Two fundamental proceedings have been established in this context:

- *First-visit MC method:*

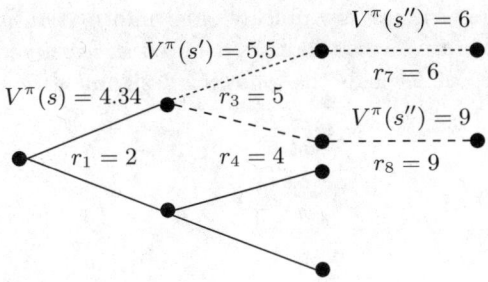

Fig. 4.6. Sample for the pass of two episodes using the first-visit MC method

Starting with the leaves of the decision tree, the episodes already selected are passed through, while recording the mean value of the previous rewards in each visited node. In example Figure 4.6 this is initially path (r_7, r_3, r_1) containing the values $(6, 5.9, 5.34)$. If a state s, which has already been visited, is passed while traversing an episode, the mean value $V(s)$ recorded at the first visit, remains preserved. In our example, this is the case for the second episode (r_8, r_4, r_1) including the state values $V^\pi(s') = 5.5$ and $V^\pi(s) = 4.34$. Table 4.3 demonstrates the algorithm for the first-visit method in detail.

Table 4.3. First-visit algorithm

```
Initialize
        π ←policy to evaluate
 V ←arbitrary state function
        returns(s) ←empty list, for all s ∈ S
Repeat infinitely:
        (a)Generate an episode using policy π
        (b)For each state s, placed on the episode:
        R ←reward on the first visit
                of s
                Insert R into returns(s)
                V(s) ←average (returns(s))
```

- *Every-visit MC method:*
  This method differs from the first-visit MC method only by the update-strategy for the state values. In contrast to the first-visit MC method, the every-visit method employs an update rule, e.g. the constant-α MC method, to improve the state value of an already visited state s:

$$V_{neu}^{\pi}(s_t) = V(s_t) + \alpha\left[R_t - V(s_t)\right] \tag{4.16}$$

The factor α determines how quickly older information is replaced by the value of recent states. It is therefore denoted as *learning rate*. For the example above $\alpha = 0.2$ yields the following assignment of state values:

$$V_{old}^{\pi} = 4.34 , \quad V_{old}^{\pi} = 5.5 , \quad V_{old}^{\pi} = 6$$

$$V_{new}^{\pi} = 4.47 , \quad V_{new}^{\pi} = 5.7 , \quad V_{new}^{\pi} = 9$$

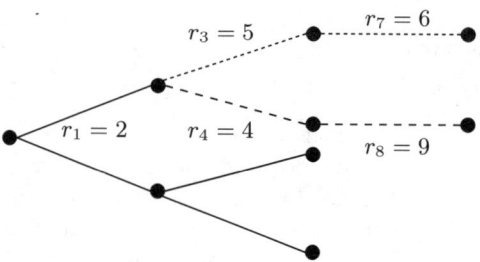

Fig. 4.7. Sample for the pass of two episodes using the every-visit MC method

Up to now nothing has been said about the selection of the episodes using the MC method. The selection strategy, however, is crucial for a relatively exact estimation of the state values, because the MC method randomly selects episodes from the decision tree for evaluation. The technique is therefore classified as a sample backup procedure. For the selection of the episodes, the approach that has already been used for the policy iteration should be applied:

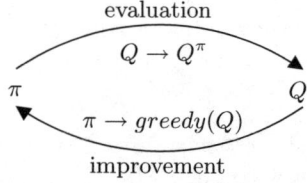

Fig. 4.8. Alternation between evaluation and improvement at Monte Carlo method

Starting with an arbitrary policy, an episode in the decision tree is generated during the evaluation phase. Along the trace of this episode action values are generated using the first- or every-visit method. Thereupon the improvement phase follows, yielding an optimization of the selected policy by adjusting the decision making process in the direction of the particular maximum action value. This process is iterated in analogy to the policy iteration (Figure 4.8).

An important aspect of the MC method is the way in which the policy is designed, which is responsible for an effectively generating episodes. While using deterministic decision policies, the path generated will always be the same. This would prevent a learning effect, because the necessary exploration does not takes place. To achieve this, the MC algorithm has to be endowed with a stochastic element. This can be done in two different ways described in the following section.

On-Policy Monte-Carlo Evaluation

This variant of the MC method is based on employing a stochastic element while selecting a path. It is based on the fact that the decision with the highest action value $Q(s, a)$ in state s will not be taken with probability $\varepsilon / |A(s)|$, but instead one of the suboptimal decision alternatives will be selected by the algorithm. Therefore, in the case of the so-called ε-greedy policy, paths with an initially poorer estimated action value can be chosen by the algorithm. The fact that the selection policy does not exclude any episode grants a pass of all the paths in the limit case and constitutes the alternative notation as ε-soft to illustrate the difference to a 'hard' selection of the best value. Table 4.4 shows the associated ε-soft algorithm for the evaluation phase. An important feature of this algorithm is the fact that the policy for evaluating the action value is identical with the improvement policy of the MC method. This property is why the method is called on-policy to the method and distinguishes it from the off-policy method, which uses a different policy to update the Q-values.

Off-Policy Monte-Carlo Evaluation

The advantage of the off-policy method lies in the possibility of using the results of a decision tree evaluation performed under a specific estimation policy π to explore and improve another policy π' without being forced to recalculate all the Q-values. The question arises whether the separation of estimation- and decision-policy allows the inference from the action values of one policy to the Q-values of another policy. Assigning the probabilities $p_i(s)$ (or $p_i'(s)$) to the nodes s of the decision tree in the visiting sequence determined by the estimation policy π (or the decision policy π'), an unbiased estimator for $V(s)$ resulting from either ε-soft policies[5], can be constructed from the ratio of the visiting probabilities:

---

[5] The policies may not have the visiting probability 0 for any state s, therefore they have to be ε-soft.

Table 4.4. ε-soft on-policy Monte Carlo algorithm

```
Initialize for each  s ∈ S, a ∈ A(s) = 0:
     Q(s, a) ← arbitrary
     returns(s, a) ← empty list
     π ←arbitrary ε-soft policy
Repeat infinitely:
     (a) Generate an episode using policy π
     (b) For each pair (s, a), occurring in this episode:
          R ←reward on the first visit of s, a
          Inserting R into returns(s, a)
          Q(s, a) ← average (returns(s, a))
     (c) For each state s placed on the episode:
          a* ← arg max_a Q(s, a)
          for all a ∈ A(s) :
```

$$\pi(s, a) \leftarrow \begin{cases} 1 - \varepsilon + \varepsilon/|A(s)| & \text{if } a \neq a^* \\ \varepsilon/|A(s)| & \text{if } a \neq a^* \end{cases}$$

$$V(s) = \frac{\sum_{i=1}^{n_s} \frac{p_i(s)}{p_i'(s)} R_i(s)}{\sum_{i=1}^{n_s} \frac{p_i(s)}{p_i'(s)}} \tag{4.17}$$

Normally, the visiting probabilities of both policies are not known. This does not matter because only their ratio is needed. Writing the visiting probabilities of node s at time $T_i(s)$ as product of the transition probabilities $\mathcal{P}_{s_k s_{k+1}}^{a_k}$ and the decision policy $\pi(s_k, a_k)$ of the states passed so far:

$$p_i(s_t) = \prod_{k=t}^{T_i(s)-1} \pi(s_k, a_k) \mathcal{P}_{s_k s_{k+1}}^{a_k} \tag{4.18}$$

By constructing the ratio of the transition probabilities it can be seen that it is independent of the decision tree transition probabilities and therefore independent of the system dynamics observed:

$$\frac{p_i(s_t)}{p_i'(s_t)} = \prod_{k=t}^{T_i(s)-1} \frac{\pi(s_k, a_k)}{\pi'(s_k, a_k)} \tag{4.19}$$

Thus, an off-policy MC algorithm (Table 4.5) can be formulated that calculates a decision policy π' using the estimation policy ratio ω with $\pi(s_k, a_k) = 1$.

4.1.5 Temporal-Difference Learning

While *stochastic dynamic programming* allows one to select an optimal action a for the next state s on a decision level k, as long as all actions in the tree have been evaluated until time t (full-backup), the MC method is based on the evaluation of at least one complete episode including state s to calculate the Q-values and to take optimal decisions. *Temporal difference learning (TD)*

Table 4.5. Off-policy Monte Carlo algorithm

```
Initialize for all  s ∈ S, a ∈ A(s):
```
$\quad Q(s,a) \leftarrow$ **arbitrary**

$\quad Z(s,a) \leftarrow 0$, **nominator and**

$\quad D(s,a) \leftarrow 0$, **denominator of** $Q(s,a)$

$\quad \pi \leftarrow$ **an arbitrary, deterministic policy**

```
Repeat infinitely:
```
(a) **Select a policy** π', **to generate an episode:**

$\qquad s_0; a_0; r_1; s_1; a_1; ....s_{T-1}; a_{T-1}; r_T; s_T$

(b) $\tau \leftarrow$ **last point in time, at which** $a_\tau \neq \pi(s_\tau)$

(c) **For each pair** s, a, **occurring in this episode**

$\qquad$ **at a time** τ **or later, put:**

$\qquad\qquad t \leftarrow$ **time of the first occurrence of** s, a,

$\qquad\qquad$ **so that it is true** $t \geq \tau$

$$\omega \leftarrow \prod_{k=t+1}^{T-1} \frac{1}{\pi'(s_k, a_k)}$$

$\qquad Z(s,a) \leftarrow Z(s,a) + \omega R_t$

$\qquad D(s,a) \leftarrow D(s,a) + \omega$

$$Q(s,a) \leftarrow \frac{Z(s,a)}{D(s,a)}$$

(d) **For each** $s \in S : \pi(s) \leftarrow \arg\ \max_a Q(s,a)$

tries to avoid the disadvantages of both methods and to combine their virtues. Beginning with the constant-α MC method, the horizon of an episode can be shortened to a one step decision relying on a further convergence of the procedure that is in fact guaranteed for sufficiently small α:

$$V^\pi(s_t) \leftarrow V(s_t) + \alpha[r_{t+1} + \gamma V(s_{t+1}) - V(s_t)] \qquad (4.20)$$

Table 4.6. Temporal difference(0) algorithm

```
Initialize V(s) and evaluation policy π arbitrary
Repeat for each episode
        Initialize s
        Repeat (for each step of the episode):
```
$\qquad\qquad$ **Perform action** a, **observe reward** r

$\qquad\qquad$ **and the next state** s'

$\qquad\qquad V(s) \leftarrow V(s) + \alpha\,[r + \gamma V(s') - V(s)]$

$\qquad\qquad s \leftarrow s'$

```
        until s is terminal state
```

Based on equation 4.20, the *temporal-difference-algorithm* (Table 4.6) can be formulated.[6] TD(λ) procedures will not be within the scope of our reflections, but some interesting insights can be gained by showing the direct relation between DP and MC methods from the mathematical point of view to elucidate the position of TD learning. Writing the state value $V^\pi(s)$ as the expected value of the average rewards in the MC procedure (equation 4.21), some transformation yields the update equation (equation 4.22) for the DP procedure.

$$V^\pi(s) = E_\pi \{R_t \,|\, s_t = s\} = E_\pi \left\{ \sum_{k=0}^{\infty} \gamma^k r_{t+k+1} \,|\, s_t = s \right\} \qquad (4.21)$$

$$V^\pi(s) = E_\pi \left\{ r_{t+1} + \gamma \sum_{k=0}^{\infty} \gamma^k r_{t+k+2} \,|\, s_t = s \right\} = E_\pi \{r_{t+1} + \gamma V^\pi(s_{t+1}) \,|\, s_t = s\}$$
$$(4.22)$$

TD methods use estimated values in two aspects: On the one hand, the expected value of $V^\pi(s)$ is calculated employing sample-backups (episodes), on the other hand an estimator for the actual state value V_t replaces the exact value V^π.

Until now, only the evaluation of a policy using the TD(0) method has been considered and nothing has been said about policy improvement. The same alternation between policy evaluation and improvement in the action space that has already been illustrated in Figure 4.8 in connection with the MC method can be applied within the TD procedure. Thus, let us write equation (4.20) using Q-values:

$$Q(s_t, a_t) \leftarrow Q(s_t, a_t) + \alpha \left[r_{t+1} + \gamma Q(s_{t+1}, a_{t+1}) - Q(s_t, a_t) \right] \qquad (4.23)$$

Writing a policy iteration algorithm for the TD(0) method should now be easy. Estimation policy and decision policy are the same in this context (ε-greedy), assigning the method to the class of on-policy procedures.

4.1.6 Q-Learning

The off-policy variant of the policy iteration algorithm can be considered the most important procedure for reinforcement learning. It was formulated by Watkins (1989) and is known as Q-learning. In this procedure the approximation to the optimal action value function takes place independently of the evaluation policy by only using the path with the greatest action value to calculate a one-periodic difference.

---

[6] Strictly spoken, the method turns out to be a 0-th order TD method because only a one period look-ahead is considered in contrast to the TD(λ) method.

Table 4.7. On-policy temporal difference algorithm

```
Initialize Q(s,a) arbitrary
Repeat for each episode
        Initialize s
        Select a from s by using a policy derived from Q
        (e. g. ε-greedy)
        Repeat (for each step of the episode):
                Perform action a and observe r, s'
                Select a' from s' using a policy derived from Q
                (e. g. ε-greedy)
                Q(s,a) ← Q(s,a) + α [r + γQ(s',a') − Q(s,a)]
                s ← s'; a ← a';
        until s is terminal state
```

$$Q(s_t, a_t) \leftarrow Q(s_t, a_t) + \alpha \left[r + \gamma \max_a Q(s_{t+1}, a) - Q(s_t, a_t) \right] \qquad (4.24)$$

A proof for the convergence of the procedure can be performed easily, which is a reason for the popularity of the method. For the algorithm described here (Table 4.8) the ε-greedy policy has been selected as estimation policy:

Table 4.8. Q-learning algorithm

```
Initialize Q(s,a) arbitrary
Repeat for each episode
        Initialize s
        Repeat (for each step of the episode):
                Select a from s using a policy derived from Q
                (e. g. ε-greedy)
                Perform action a and observe r, s'
                Q(s,a) ← Q(s,a) + α [r + γQ(s',a') − Q(s,a)]
                s ← s'; a ← a';
        until s is terminal state
```

After having discussed the basics of RL techniques that are supposed to be useful for learning yield optimizing strategies for the PCRA scenario of this thesis, we should look at the basics of YM.

4.2 Basics of Yield Management

In section 2.1 YM was classified as a branch of *dynamic price posting* strategies which are especially suitable for *perishable* goods and services. This section firstly outlines traditional YM methods and subsequently illustrates the reasons for using YM sensibly in ISIP environments. It will then be demonstrated that dynamic pricing for ISIP coupled with the *automated resource allocation* of the required IT capacities has many parallels to a classic YM task. Afterwards it will be shown how to transfer such optimization methods to the domain of ISIP provision.

4.2.1 Classic Yield Management and Dynamic Pricing in Information Services and Information Production

Traditional microeconomic theory (Debreu 1959, Malinvaud 1974, Hildenbrand & Kirman 1976) assumes that goods do not lose their utility until consumption takes place. However, this is not the case for a class of goods which can be considered as perishable, as it is the case for goods of fast declining value like information products. Pricing of durable goods has dominated the normal marketing literature (Nieschlag et al. 1994, Simon 1992) up to the present, but not much literature can be found dealing with pricing perishable goods to achieve a more efficient resource allocation. YM tries to close this gap by employing a combined *dynamic pricing* and *contingent allocation* approach in theory as well as in practice and has provided valuable optimization results for the airline industry for years (Chen et al. 1999).

The failure of conventional *pricing theory* in many industrial sectors stems mostly from the existence of *limiting factors* that could only be expanded in the short term at the price of prohibitively high costs. Production processes, however, which are characterized by *high costs for capacity resourcing* and *dynamic standby* mostly have *low variable marginal costs* for each additional activity unit (Bertsch 1990). This leads to *high contribution margins* enabling appropriate price/quantity management to provide a reasonable profit rising potential. This is due to the fact that each additional demand unit which has a positive contribution margin, leads to covering irreversibly predisposed capacity resourcing costs. The higher these costs of an additional capacity unit are, the higher the resulting profit increasing potential of YM methods is.

In this context, Belobaba (1989) subsumes all techniques for *integrated pricing* and *capacity control*, which maximize fixed cost coverage by attributing the appropriate capacity type to a corresponding customer type. Kimes (1989) identifies the following characteristics which make production processes eligible for YM methods:

- Widely fixed production capacities
- Possibility of market segmentation
- Non-storability and perishableness of product units
- Pre-production sales
- High volatility of demand

- Low marginal costs of additional product units within capacity constraints and high costs for capacity expansion

Common examples of perishable products can be found in the food, fashion, hotel, and airline industries.[7] Even if initial approaches to overbooking management have existed since the beginning of the seventies, a systematic development of YM methods first started with the deregulation of American aviation industry in 1979. Due to positive responses from practitioners in the airline domain the concept was transferred to the hotel, travel, logistics, and transportation sectors in the late eighties (Smith et al. 1992, Weatherford & Bodily 1992). Since the end of the nineties, applications have been increasingly developed for resource allocation in the telecommunication domain, based on YM methods (Humair 2001). Recent YM research directly addresses ISIP applications. Nair & Bapna (2001), for example, present a model where the availability of multiple access points of an *Internet service provider* is controlled by a YM system according to *quality-of-service (QoS)* classes. It is important to point out that, in connection with the provision of services, perishableness does not refer to the potential factor itself (rental car, hotel room, airplane seat), but to the potential of using the resource to generate returns in the current period. In the next period the potential factor is again available unchanged (Weatherford & Bodily 1992, Bertsimas & Popescu 2000).[8]

The ISIP definition employed in this thesis comprises computer-based information systems that are utilized to perform value creating communication actions as well as the mapping of information resources onto relevance information by employing methods of information work and information production, like the customized retrieval, replication or generation of customized stock chart data or the broadcast of a sporting event to paying viewers via TCP/IP protocol etc. (see section 1.4.2). Compared with the supply of pure physical services, ISIP provision exhibits the following differences that can be recognized with respect to YM applications (Schwind & Wendt 2002):

- Processes of ISIP can be considered as interruptible, the state of processing can be buffered and the process resumed if resources are idle again. In the traditional YM domains this is not possible.[9]
- Whereas variable costs of production are low compared to fixed costs in the physical YM domain, they are negligible for ISIP applications.[10]

---

[7] An extensive overview of various YM applications and methods can be found in van Ryzin & McGill (1999).

[8] Sometimes YM is also called 'Perishable Asset Revenue Management', 'Profit Management' or 'Revenue Management' for this reason.

[9] A hotel guest will not accept the passing of a room booked for a specific day to a prioritized guest, even if he would be compensated the next day with a half rate.

[10] Additional personel in a booked-up hotel generates significant variable cost, additional server load would mostly enhance power consumption.

- Duration of ISIP tasks can not be predetermined as exactly as is the case in classic YM environments. The problem has to be solved by employing stochastic distributions for estimating duration.[11]

4.2.2 Solving Yield Management Problems Using Stochastic Dynamic Programming

The *YM problem* considered in the context of this thesis deals with the time orientation of the task acceptance process. The YM process is divided into two phases: The acceptance phase, which is primarily considered in our YM context, and the execution phase. The yield manager's goal is to accept as many tasks as possible at the highest possible price. The YM procedure therefore requires the return-maximizing booking of tasks, such that the maximum resource load is reached when the task execution phase begins. For this reason, YM can be considered as a decision problem that bears two types of risks due to the uncertainty generated by the stochastic process of the arrival of the tasks:

- *Risk of return loss*: Each non-accepted task is a potential loss in the YM process unless a new task with a higher reward arrives until the task acceptance process closes at expiration time. See Figure 4.9 (A) for a visualization of the return loss problem in the YM process.
- *Risk of return extrusion*: The acceptance of tasks early before expiration time bears the risk of return extrusion. Tasks with a higher yield-to-capacity ratio might appear before the YM acceptance phase is closed. Tasks with lower return that have been accepted before will then inhibit the acceptance of the higher yielding tasks, as depicted in Figure 4.9 (B).

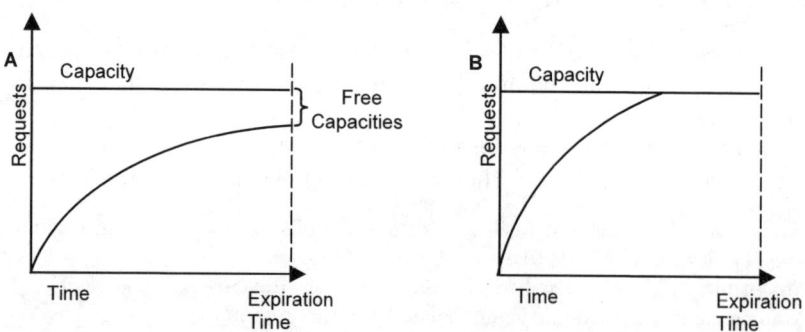

Fig. 4.9. Yield management decision risk: return loss and return extrusion

---

[11] Time necessary for retrieving and sorting data to build a chart for a Web browser, depends on the query. The duration of a flight can normally determined by looking at the flight schedule.

Techniques for solving this YM problem, which can be found in the literature, can be differentiated into *heuristic approaches* and *exact optimization procedures*. The best known representatives of the first group are *nested booking classes (NBC)* (Hornick 1991) and the concept of the *expected marginal seat revenue (EMSR)* (Belobaba 1987, Belobaba 1989, Hornick 1991). Both concepts are based on the division of available seats to booking classes, which have to be determined in size before the booking process starts and have to be adjusted during the allocation process if necessary. In the course of the booking process, requests with a higher contribution margin can make use of capacity from classes with lower contribution margins if the contingent of the higher booking class is already used up. Whereas class borders are determined by simulation in the NBC process, the EMSR procedure detects the point where the EMSR of an additional capacity unit falls below the EMSR of the next lower class. If the EMSR calculation includes the recent historic probability data of booking requests within these classes, the calculated points indicate the optimal size of the classes.

Static solutions like *linear programming* do not lead to an optimal solution (Kimes 1989) of the YM problem because it represents a cyclic dynamic decision problem under uncertainty. For this reason *stochastic dynamic programming (SDP)* is often used to find exact solutions (Alstrup et al. 1986, Bellman 1957). All decision problems, which are relevant in the YM context, can be described by a *tree structure*, where edges represent the individual decisions (*actions a*) and the vertices denominate the system states (*states s*).[12] The Markov property can also be assumed for the stochastic process of the arrival of the book requests.

In the framework of *decision process optimization* an additional parameter is required to describe the reward (return in economic terms) for an action (*reward r*). In every node of the decision tree a decision about the selection of the next edge is taken. Additionally, one can assign an a-priori probability to the selection of the actions or the edges. A further parameter π is employed for this purpose. This parameter is defined as *decision policy*, because it describes the strategy used to select the edges, while passing through the decision tree. Subsequently the *Markov* property implying the independence of the state transition probability from s to s' etc. has to be assumed to apply SDP (Puterman 1994).

In the following, a simple example of solving the time-dependent YM problem using SDP is presented. We assume a contingent of 6 seats available in the first class of a commercial aircraft, flying from A to B at a specific future time. Seats can be sold singly or as a bundle, implying that the remaining capacity can take the following values, spanning the state space $S = \{0, 1, 2, 3, 4, 5, 6\}$.

The YM system receives capacity requests for this flight, which can be either accepted or rejected and differ in number of seats and price. In the following for simplicity we assume only three types of seat requests (F1, F2

---

[12] See e.g. Figure 4.1 for illustration.

Table 4.9. Exemplary distribution of booking requests

Type	Probability	Requested Seats	Reward
F1	0.5	1 Seat	1 MU
F2	0.3	2 Seats	4 MU
F3	0.2	3 Seats	9 MU

and F3) which are specified in Table 4.9. In addition we presuppose that the number of incoming requests is known at decision time. Therefore we assign a back-counting index k for each request, which tells us that $k-1$ decisions will follow the kth request.

To achieve optimal acceptance at stage k it is not sufficient for a decision maker to know the type of request k and the residual capacity i: he also has to be aware of the amount of the remaining $k-1$ requests for the case of rejection. Subsequently it is necessary that the value $V_{k-1}(i-f_k)$ of request $k-1$, for capacity i reduced by the required seat capacity f_k, is available in case of acceptance. It could then be tested, whether

$$r_k + V_{k-1}(i - f_k) \geq V_{k-1}(i) \qquad (4.25)$$

is satisfied, where r_k denotes the reward for the request at stage k. If equation 4.25 is true, the request will be accepted, otherwise it will be rejected. Starting with $V_0(i) = 0$ for all possible residual capacities i, the optimal acceptance policy for stage $k = 1$ and as a result the value of $V_1(i)$ can be calculated using the following instruction:

"Accept any request which is less or equal to the residual capacity if there are no further incoming requests. The expected value $V_1(i)$ can be calculated for each remaining capacity i employing the probability distribution given in Table 4.9."

Based on this result we can determine the optimal decision policy for stage $k = 2$. At this stage the question of saving capacity for stage $k = 1$ arises for the first time. The expected return on stage $k = 2$ consists of the expected return $r(i, \pi_a)$ on this stage and the value of the remaining residual capacity $V_1(i')$ on stage $k = 1$, where i' denotes the reduced residual capacity:

$$V_2(i) = \max_{\pi_a}[r(i, \pi_a) + \sum_i \mathcal{P}_{ii'}^{\pi_a}(i, \pi_a, i')V_1(i')] \qquad (4.26)$$

By comparing the capacity value for all possible states i and each acceptance policy π_a, the optimal acceptance policy for each state i is evaluated. In this context $\mathcal{P}_{ii'}^{\pi_a}$ denominates the probability of transition from capacity i to capacity i' under acceptance policy π_a. Iterating this process leads to an optimal YM policy for each stage k. Figure 4.10 shows the asymptotic progression of the maximum value of a given residual capacity i subject to the number N of remaining capacity requests according to the data given in Table 4.9.

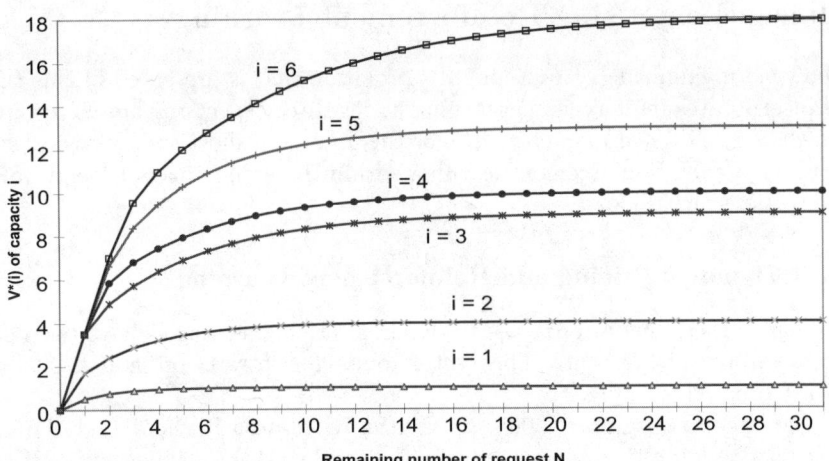

Fig. 4.10. Value of a given residual capacity vs. number of remaining requests

Despite the optimality of the SDP approach, the computational complexity prevents its widespread implementation. This applies in particular to the so-called *'network yield management' (NYM)* problem, where the customers request not only the use of one resource, but also a resource bundle (e.g. a multi-leg flight or a hotel stay for several days (Bitran & Mondschein 1995)). This is due to the combinatorial complexity deriving from the *Markovian* state space which is spanned by the probabilistic use of multiple resources. Other approaches which ignore this complexity, like the linear addition of several resource prices by means of a so-called *'bid pricing'* procedure, however, lead to suboptimal results (Talluri & van Ryzin 1998). The literature concerning complexity reduction in *Markovian* state spaces only takes account of general clustering mechanisms and does not regard the particular *multidimensional NYM problem* (Nollau & Hahnewald-Busch 1978, Doerninger 1984). The classical booking request acceptance problem of YM literature, such as that described above, is also referred to as the *stochastic knapsack problem (SKP)* with finite horizon in the recent *Operations Research* literature (van Slyke & Young 2000). Papastavrou et al. (1996) extend the SKP for the multidimensional case where more than one resource needs to be managed. This case is called the *multidimensional stochastic knapsack problem (MSKP)* in this branch of research and is the equivalent of the NYM problem in classical YM literature. In a recent article Kleywegt (2001) emphasizes the link between dynamic pricing and YM in the context of the SKP. He proposes a mechanism that allows one to solve the optimal acceptance problem in connection with discovering the yield maximizing posted price classes that are offered to the consumers to choose from. Kleywegt does this by analyzing demand structure, determining the usable resources and then evaluating the corresponding consumption behavior.

4.3 Dynamic Pricing, Scheduling, and Yield Management Using Reinforcement Learning

The yield maximizing system for ISIP provision that is presented in this thesis incorporates reinforcement learning under three aspects: *dynamic pricing, scheduling,* and *yield management*. For this reason, a short survey of applications employing reinforcement learning within these domains will be given in the sections that follow.

4.3.1 Dynamic Pricing and Reinforcement Learning

A couple of learning approaches to dynamic pricing strategies have been described in recent literature. This section only concentrates on models dealing with RL-based learning.[13]

Carvalho & Puterman (2005) provide an RL approach where a retail manager tries to learn the optimal setting of a good's price to achieve an optimal reward over a time period T by a day-by-day decision. If the manager agent knows the shape of the demand function, the problem can be solved using a stochastic maximization approach. However, this is not the case in the situation studied by Carvalho & Puterman. The model uses a simple regression model to approximate a logarithmic demand function. By introducing RL, the manager agent can dynamically learn the demand of the previous $t - 1$ days making an exploration versus exploitation trade-off by using a higher or lower ε for the pricing decisions. In a comparison of three different settings a myopic policy π with random exploration (high ε) performs better than a TD(1) RL policy with a low learning factor λ, where the performance is measured in terms of summarized reward after period T. The best performing strategy identified by Carvalho & Puterman is a high λ TD(1)-policy.

Hu & Wellman (1998) propose a *Nash Q-learning* method for general-sum stochastic games. Their approach does not address a pricing game directly, but describes a game theoretical solution to a game where multiple agents try to gain maximum reward. This approach is the core of the subsequent Q-learning pricing models.[14] In a stochastic game, each agent's reward depends on the reward for the joint action of all agents and the current state. The game is modeled as a decision process where state transitions obey the Markov property. The general sum property of the games constructed by Hu & Wellman allows the agents' rewards to be arbitrarily related, while the Nash equilibrium is the basic solution concept for general-sum games. The Nash Q-learning algorithm proposed by Hu & Wellman tries to learn the Nash Q-values through repeated play. Based on the Q-values learned the agents can

---

[13] Narahari et al. (2005) present a detailed survey of learning in dynamic pricing environments including other learning methods than RL.

[14] See Hu (1999), Hu & Zhang (2002), Hu & Wellman (2003) and Schwind & Meyer (2001) for a further detailed discussion of this method and Littman (1994) for an introduction to multi-agent RL.

derive the Nash equilibrium and choose their action accordingly. The Nash Q-algorithm begins with an arbitrary guess for the Q-values of the strategy sets of the concurrent agents. Following the classic RL procedure, Nash Q-learning updates the Q-values according to the actions and rewards that can be observed for the other agents in the game. At the end of the learning process the Nash Q-learning agent maintains a model of the other agents' Q-values and uses that information to update its own Q-values. The Q-learning agent chooses the next actions by using this information to derive a Nash equilibrium. Hu & Wellman theoretically prove the convergence of the algorithm under restrictive conditions. Additional experiments show that the algorithm converges even under relaxed constraints in a game with multiple equilibria despite the violation of theoretical assumptions.

Kephart & Tesauro (2000) study the aspects of multi-agent Q-learning in a market model with two identical, competing seller agents (pricebots) that try to price a commodity such that the maximum yield is generated. Buyers generate purchase orders at random times and acquire commodities from the sellers if their valuation lies above the sellers' offers. For this purpose the buyers compare the prices of some randomly picked sellers and select the lowest offer. The number of the sellers that are compared by the buyers differs from one single buyer to all buyers within seller groups. Acting in such a way, buyers form a distribution ω with different behavior γ that can be segmented by the sellers. Two main situations are observed for the seller agents, a myopic behavior called *myoptimal pricing* and a *Q-learning pricing* strategy. In the myopic optimization case without learning the system behaves as follows: Each seller agent sets the price p_m a profit maximizing monopolist ('best-response Cournot') would charge given the knowledge of the current competitors' price setting behavior. A price war develops between the seller agents who are fighting for a dominant market share of buyers and causing a linear price fall to a critical threshold p_t. If the threshold price is reached, some seller agents decide to give up the market share, because the share ownership does not compensate for the very low profit margin reached at this point. The seller agents jump back to the initial monopolists price p_m and the price war cycle begins again. In the Q-learning case, agents avoid a total price decline in their bidding strategy and return to the initial p_m earlier by regarding the total discounted profit for the next steps instead of immediate return. Besides symmetric agent behavior, asymmetric agent policies are modeled leading to oscillating equilibria in the Kephart & Tesauro (2000) model. This pseudo-convergent asymmetric solution has no analogy in ordinary Q-learning. The authors analytically compute the shape of such solutions, and numerically map out the conditions under which such solutions occur. Kephart & Tesauro (2000) suggest that the behavior thus observed should also be found in more general studies of multi-agent Q-learning.

In a modified version of the previous model, Kephart et al. (2000) attempt to understand the strategic pricebot dynamics in a multi-seller environment, where two additional price setting strategies are introduced. In total, they examine four price-setting strategies: game theoretical pricing, derivative following, myoptimal pricing, and Q-learning. *Game theoretical pricing* works by calculating a mixed-strategy Nash equilibrium in which each seller chooses prices randomly from a distribution $f(p)$ that can be computed from the number of sellers and the buyers' parameters ω and γ. The game theoretical pricing algorithm calculates $f(p)$ and periodically chooses a different random price from the distribution. In the game theoretic setting, the pricing decisions of the sellers are met without knowing the competitors' pricing. *Derivative following pricing (DFP)* simply experiments with increments (or decrements) in price, continuing to move the price in the same direction until the observed return falls and the strategy is reversed.

Kephart et al. (2000) observe various combinations of the different pricebot strategies in the same simulation environment. In homogeneous settings, when all the pricebots use the same pricing algorithm, a derivative following approach is shown to outperform game theoretic pricing and myoptimal pricing. In a market with heterogeneous pricebots, myoptimal and game theoretical pricing show better results than DFP, while the Q-learning strategy outperforms all others.[15]

Raju et al. (2005) present a dynamic pricing model for an electronic retail market, where multiple sellers stand in a price competition for buyers. The sellers use pricing agents (pricebots) that learn price schemes with the objective of maximizing the discounted cumulated profit or long run average profit per time unit. For this purpose pricebots adjust the sellers' prices at random intervals based on factors, like demand and supply information with respect to their own stock replenishment policy, that have crucial impact on the sellers' yield. Raju et al. (2005) investigate two learning situations in their model:

- The first scenario assumes that none of the retailers has information about the number of potential customers, inventory levels, and prices of the other retailers. Q-learning is employed to learn the optimal replenishment and pricing policy based on the utility function of the buyers.
- The second scenario models a situation where information about customers, inventory levels, and prices is available to learning agents.

Raju et al. (2005) compare the income of the Q-learning pricebot with the results of an agent using the DFP strategy in the non-information case. As one might expect, the Q-learning strategy performs better than the DFP strategy. Based on the setting in the non-information case, Raju et al., construct a pricing game for the partial-information case which works in analogy to the Hu & Wellman Q-learning approach. However, they use another type of RL to the search for Nash equilibria in the game. The algorithm shows a stable

---

[15] Further similar RL approaches and variants of the sketched multi-agent pricing games are presented in Sridharan & Tesauro (2000), Tesauro & Kephart (2001), and Tesauro (2001).

convergence to an equilibrium. Under a special parametrization, price setting agents reach a market segmentation resulting in a duopoly where one agent serves the low price customers and the other agent skims the high price market segment.

4.3.2 Scheduling and Reinforcement Learning

Resource allocation for ISIP processes as used for yield maximization in the PCRA environment of this work has a time dimension. This characterizes the problem as an economic scheduling problem (see section 2.2.3). For this reason, the following sections give a short survey of RL in connection with scheduling and problems. Two main approaches can be identified in the RL-oriented literature on scheduling and load balancing which is capable of dealing with the computational complexity of combinatorial optimization:

- *RL controls the optimal policy of local search methods or heuristics*, but does not try to solve the combinatorial optimization problem itself.
- *RL is applied as a distributed optimization technique*, operating only on the locally known states of the individual RL scheduling agent, hoping that emergent system behavior will generate feasible results.

The following RL approaches address *policy optimization in connection with local search*:

Zhang & Dieterich (1995) use a TD(λ) algorithm to learn the heuristic evaluation function for the process time minimization of a task allocation problem with precedence constraints. The RL model is situated in the application domain of planning the NASA shuttle payload processing. Resources have to be assigned to the payload processing tasks in a given order under several resource constraints. Starting with an imperfect schedule planned by a critical-path algorithm, the system incrementally repairs resource constraint violations. This is done by using a *reassign operator* that changes the resource assignment for one of the task's resource requirements and a *move operator* that shifts a task to a different execution time. After shifting a task, the move operator reschedules all temporal dependencies of this task by using the critical-path method to avoid resource constraint violations. The original version of the algorithm, proposed by Zweben et al. (1994), accomplishes this by employing a *simulated annealing*[16] approach called *iterative repair (IR)*. The RL variant of the IR algorithm uses reassign and move operators as well. Zhang (1996) introduces a *resource dilatation factor (RDF)* to provide a scale independent measure for the length and resource load of the schedule, that can be used to reinforce the shortening of total processing time without violating resource constraints by 'overallocation'. The objective of the learning is to select scheduling actions such that resource overload and processing time are reduced. The RL system uses an *artificial neural network (ANN)* for state

---

[16] Simulated annealing will be further discussed in sections 5.2.2 and 6.1.2.

space representation. Eight input factors are supplied to the ANN, e.g. the RDF and the percentage of task processing time in a schedule that causes a resource constraint violation in the actual schedule. The evaluation of the scheduling system for the payload processing planning problem proposed by Zhang & Dietterich (2000) shows that RL-controlled scheduling far outperforms the IR method of Zweben et al. (1992).

Boyan & Moore (1998) present an RL system called *(STAGE)*, that predicts the outcome of a *local search algorithm*. The system uses the value function $V^\pi(s)$ to learn the optimal starting point of a *hill climbing algorithm*. The states passed by a hill climbing episode (trajectory) are updated with the reward R of a local optimization function, if the optimization result is better than the result achieved in previous episodes.[17] State space compression is obtained by using a *polynomial regression network*, Boyan & Moore (2000) apply STAGE to *global optimization domains* like e.g. *bin packing problems*, *Bayes network structure finding* or *channel routing* and find that STAGE shows a better average performance than traditional simulated annealing and hill climbing algorithms. As Chang (1998) points out, this approach is also suitable for scheduling applications, due to its constructive goal of targeting combinatorial optimization problems.

Schneider et al. (1998) propose an approach that is close to the RL-YM system constructed in this thesis. The authors present a system that learns the optimal schedule for an eight week production scenario. The production program may only be changed at two week intervals and permitted variants of the scheduling plan are limited to only 17 configurations. State space compression is achieved by alternately employing first-nearest neighbor methods, locally-weighted linear regression, and global quadratic regression methods. Compared with SA and greedy optimization methods the performance of the Schneider et al. RL-YM system is superior.

Representatives of scheduling algorithms operating in *distributed and emergent RL systems* are presented here:

Schaerf et al. (1995) present a multi-agent framework in which RL agents learn processing properties for the execution of jobs in a distributed computational system with multiple homogeneous resources. The agents try to find an optimal load allocation policy (maximal use of resources in the entire system), that depends on the learned behavior of resources in response to allocation requests. The allocation problem addressed by the authors is very close to the 'El Farol' bar problem (Arthur 1994) described in section 2.3.1. Schaerf et al. test the impact of the system's learning behavior on allocation quality in different load situations and for various resource capacity availabilities. In most experiments conducted by Schaerf et al., agents are unable to communicate with each other. The authors determine the adaptiveness of the agents to varying situations in the system by altering the exploration rate ε. As one would expect, a low exploration is better in static situations and higher ε strate-

---

[17] R is the value of the objective function in the optimum of the local optimization.

gies perform better in situations where load or resource availability fluctuate. In some experiments, Schaerf et al. allow the agents to exchange experience about the response behavior of the resources amongst each other; however, this information interchange does not lead to a better system performance.

Brauer & Weiss (1998) present a multi-machine scheduling approach, where machines in a network-like production structure are represented by agents. The agents try to shorten the total processing time for all jobs on the machines. Employing Q-learning, the machines collectively learn and iteratively refine appropriate processing schedules. A major characteristic of the Brauer & Weiss approach is that individual machine agents conduct their learning activities in parallel and asynchronously. Several experiments show that after a longer training phase, Q-learning leads to a considerable reduction of total processing time compared with a 'random' scheduling approach. Brauer & Weiss additionally simulate the failure of some machines to test the adaptiveness of the learning strategy. The performance of Q-learning, however, is poor in this case in comparison with a simple adaptive scheduling heuristic.

Another example of complexity reduction by using emergence in RL-scheduling approaches is presented by Riedmiller & Riedmiller (1999). They employ Q-learning agents to learn optimal scheduling policies in a scenario with single and multiple resources as well. The optimization objective is to find a scheduling sequence for continuously arriving jobs such that the tardiness of the jobs is minimized for the entire system. To benchmark their approach, the RL scheduling results are compared with traditional scheduling heuristics, like earliest due-date, minimum slack, or lowest processing time. For both the single resource case as well as for the multiple resource case the RL approach outperforms heuristics in a setting where a set of ten different scheduling scenarios is used in the training phase. Additionally, the RL system shows satisfying generalization behavior for new generated problem sets.

In a scenario that is closely related to the YM-RL environment that will be described in this chapter, Snoek (2000) proposes an RL-based optimization that has the goal of finding a yield maximizing order of tasks. In the underlying problem, called job order problem, new jobs arrive continuously while a selection has to be made among the jobs offered due to limited resource capacity. This problem can therefore be considered as an instance of the economic job-shop scheduling problem (see section 2.2.3). The acceptance policy is learned by training an ANN using RL and the scheduling policy is based on a genetic search performed by the same ANN. This leads to the title *neuro genetic scheduling (NGS)* for the system. A comparison of NGS with standard scheduling heuristics like average and minimum-slack, shows that NGS provides a superior allocation (total yield for accepted tasks) in several machine/job constellations.

In a recent work, Vengerov (2005) examines an RL-based scheduling system for task execution on multiple resources in a distributed computer system. The approach is based on an economic scheduling paradigm that assumes a value function declining in time for task execution (see section 1.4.4). The system enables the inclusion of multiple resource constraints into the scheduling decisions. Q-learning is used for scheduling control by learning optimal allocation heuristics from a set of rules. As in the previous model, the system's optimization objective is to accept the randomly incoming tasks in such a way that the summarized utility of all executed jobs is maximized. Unlike the YM-RL job allocation models that will be proposed in this thesis, the Vengerov approach assumes preemptive jobs with variously shaped time dependent valuation functions (linear, concave, convex). The evaluation of the RL allocation system in a simulated Grid environment indicates superior allocation results (in terms of resource usage per time) for all types of time dependent valuation functions compared to the performance of fixed scheduling rules.

4.3.3 Yield Management and Reinforcement Learning

Gosavi (2004) presents an RL approach that is based on the classical airline YM model described in section 4.2.2. The objective of the system is to learn the optimal acceptance policy for flight booking requests that arrive in a Poisson distributed process. Their model uses three fare classes and additionally allows overbooking to compensate for no-shows, while the system takes penalty costs into account if capacity overbooking occurs. After a training phase Gosavi et al. (2002) compare a classical and a modified Q-learning algorithm to the EMSR heuristic. In an evaluation that measures total return after the booking process, the RL algorithms, especially a modified version of Q-learning, performed better than the EMSR method.

Martin (2004) proposes an RL-driven model for the placement of advertisements in television programs. In analogy to the airline case, the requests for advertisements occur in a stochastic arrival process and have a considerable cancellation risk. The capacity of TV promotion time is limited and, as in the airline case, the advertisement time is a perishable good. Following the classical YM task the advertisements are sold as packages of broadcasting time. Feasible broadcasting time is divided into several contingents with a transmission limit per hour. This package structure introduces combinatorial complexity into the acceptance decision for the advertising requests.[18] In a first step, the promotion packages that have different volumes and prices are allocated to the different time slots following a set of heuristics developed by (Benoist et al. 2001). The strategies are *best-fit*, describing the direct allocation of the incoming packages to the transmission hours with the lowest advertising load if time is remaining, *forward-sampling* that is derived from an MC method and analyzes multiple scenarios to find the best solution (high computational effort) as well as a YM strategy based on a continuous time version of SDP approach described in section 4.2.2. Martin (2004) implements an RL approach for learning an improved best-fit strategy by using the MC

---

[18] The resulting optimization problem is related to the SMKP problem of NYM.

method to calculate a reference solution. The benchmarking results of the best-fit strategy improved by RL show an increase of total reward by 7.7% for the accepted promotion tasks compared to the non-learning solution. Additionally, the response time for acceptance decision decreases by a factor of 1000, qualifying the system for real-time applications.

4.4 A Yield Maximizing Allocation System for a Single ISIP Resource

The following section develops and evaluates an RL approach to the yield maximizing scheduling of ISIP tasks in the context of the PCRA scenario as presented in section 1.4.3. In a first step, we concentrate on the application of learning in the single resource case to evaluate the feasibility of the proposed RL-YM concept. In a second step, a multi resource case is introduced and tested in the subsequent section, considering the experience that has been made in the single resource case.

4.4.1 Infrastructure of the Yield Optimizing System

According to the PCRA scenario described in section 1.4.4, task agents act on behalf of an ISIP provider and try to acquire the resources that are necessary to handle computational jobs. Task agents do this by submitting bids to the market mediator in the PCRA system as depicted in Figure 1.2. The market mediator, which is denoted as *yield optimizing system (YOS)* in our RL-YM case, accepts or rejects the incoming jobs according to the yield maximization objective described in the airline example in section 4.2.2. At the end of the task allocation period the total reward for the accepted jobs should have been maximized. The basic idea is to learn the properties of the incoming tasks and use this knowledge to identify high yielding jobs. This procedure helps to compensate for the uncertainty caused by the stochasticity of the arriving resource requests. Another objective of the yield maximization strategy is to ensure the effective use of resources. The resource level that is available to the mediating YOS agent for task allocation is derived from periodical reports that are provided by the resource agents (cf. Figure 1.2).

The RL-YM approach has the following properties:

- The resource requests are coupled with a reward for successful job execution according to the W2P of the task agents. The bids for task execution are connected to deadlines. A penalty is charged to the YOS for task execution after due-date.
- The system combines a deterministic scheduling heuristic with an RL-YM approach. The RL-YM algorithm has to decide whether a task should be accepted or rejected with respect to the total reward for the whole scheduling period T.

- If a task has been accepted by the RL-YM component, the deterministic scheduler puts the job into an execution queue. The heuristic of the scheduler tries to arrange the accepted jobs in the queue in such a way that they can be executed before due-date. If this cannot be achieved by the scheduler, the penalty has to be paied by the YOS agent. The deterministic scheduler informs the RL-YM about the success or failure of execution-in-time.

According to this function specification, the YOS agent (mediating agent in the PCRA environment) has the following infrastructure. It consists of three main components depicted in Figure 4.11:[19]

The first component is the *job providing interface (JPI)*, which receives the incoming ISIP job requests submitted by the task agents and holds these requests until a final acceptance/rejection decision is made by the RL system. Then the second module of the YOS agent incorporates the *reinforcement learning algorithm (RLA)*, that has to learn the yield maximizing job acceptance strategy. As a third module the *deterministic scheduling component(DSC)* arranges the accepted jobs in the production queue to avoid penalty costs caused by not-in-time-execution.

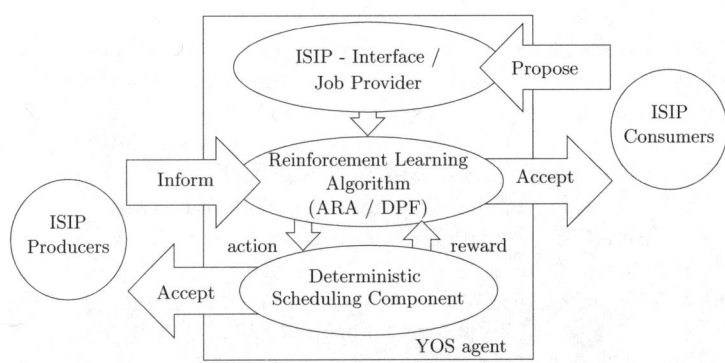

Fig. 4.11. Diagram of the yield optimizing system for ISIP provision

In the light of the ISIP consumer-producer interaction described in section 2.5 and depicted in Figure 2.8, the YOS agent acts between the consumers (task agents) and the producers (resource agents). In the YM-RL model depicted here, the interests of tasks and resource agents are clearly separated.

---

[19] The YOS presented here is to a great extent identical with that described in (Stockheim, Schwind & König 2003*b*). The system was constructed in collaboration with the Distributed Artificial Intelligence Laboratory (DAI-Lab) of the Technical University Berlin for a supply chain application scenario (Stockheim, Schwind, Korth & Simsek 2003, Simsek et al. 2004). The yield maximizing scheduling behavior implemented in this application, however, is completely compatible with the PRCA scenario of this thesis.

Task agents (ISIP-consumers) strive to acquire as many ISIP resources as required within the preferred production time. Resource agents are interested in accepting as many bids as possible from the offered requests, such that the total return from the accepted bids is maximized with respect to the available resource capacity. This optimization is performed by the YOS agent which acts as a mediator on behalf of the resource agents. The next sections will explain the core functionality of the YOS agent.

4.4.2 Properties of the Resource Requests

The task agents submit requests $q(o_j)$ for the resource capacity o_j that is required for execution of job j to the YOS agent in the job acceptance phase. These job requests $q(o_j)$ have the following attributes:

- b_j which is the bid-date of the resource request for job j (release time of the request),
- d_j which is the job's due-date (last date for job execution without necessity of a penalty payment),
- l_j denoting the job length (processing time),
- v_j indicating the value class of a job (class of returns for the fulfilled request) and
- p_j describing the contract penalty class of the job (class of penalty payments in case of job execution after due-date).

The bid-dates and due-dates are discrete in time. We use the notation of bid matrices (BMs introduced in section 1.4.4) to express the resource requests $q(o_j)$ for job execution in the time window between bid-date and due-date. To model this time window, requests are formulated as alternative BMs over the job length l_j. These BMs can only be selected exclusively (see Figure 4.12).

Bid Matrix: $q_1(o_j, v_j, p_j)$								
	Time Slot t							
Res.	1	2	3	4	5	6	7	8
o_j		1	1	1				

Bid Matrix: $q_2(o_j, v_j, p_j)$								
	Time Slot t							
Res.	1	2	3	4	5	6	7	8
o_j				1	1	1		

Bid Matrix: $q_3(o_j, v_j, p_j)$								
	Time Slot t							
Res.	1	2	3	4	5	6	7	8
o_j					1	1	1	

Fig. 4.12. Illustration of the single resource RL-YM requests for a job of length $l_j = 3$ (bid-date: $b_j = 2$, due-date: $d_j = 7$) as a combination of three BMs that are exclusively eligible by the YOS agent

4.4.3 Reinforcement Learning Algorithm

Before the functionality of the reinforcement learning algorithm is designed, state space, action space and state transition conditions have to be specified. The *state space* S of the RLA consists of three independent parameters,

- $g_f \in G$ representing the schedule queue filling,
- $v_j \in V$ denoting the value class of current bid for job execution j and
- $p_j \in P$ the contract penalty class for after due-date execution of current job j.

Consequently, the resulting state space amounts to

$$S = G \times V \times P \tag{4.27}$$

states and is of dimension m

$$m = |G| \cdot |V| \cdot |P| \tag{4.28}$$

Employing a moderate selection of parameter variables, the state space dimension is low enough to avoid the necessity of an ANN for state space compression in our YOS agent.[20] The *action space* is restricted to the *accept* and *reject* decision for the current job request.

$$A = \{accept, reject\} \tag{4.29}$$

A *state transition* $\delta(s_t, a_t); s_t \in S, a_t \in A$ is subdivided into several phases, of which only the second can be affected by the RLA (see Figure 4.13). The transition phases consist of

- δ_1: a scheduling phase for the ISIP production plan by the DSC and a classification of the requested job coupled with the propagation of the respective parameters to the RLA.
- δ_2: a decision phase concerning the acceptance or rejection of the requested job j_{t+1},
- δ_3: and a notification of the reward r_{t+1} coupled with the update of the RL system.

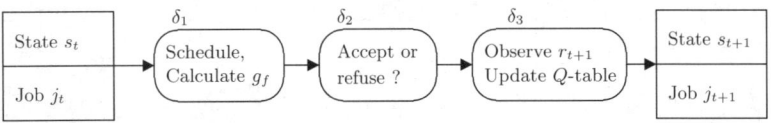

Fig. 4.13. State transition in detail

As a result of the previous state space design and the state transitions actions space $Q = \{S \times A\}$ is defined and a description of the RL-YM algorithm

---

[20] The versions of the YOS agent presented in the following section will make use of a state space dimension reduction system like an ANN, which is necessary to employ RL-YM scheduling for realistic problem dimensions.

can be given. The algorithm is based on the standard Q-learning depicted in Table 4.10. The characteristic feature of the RL algorithm designed here, is based on the fact that the reward is generated by the DSC's scheduling heuristic (line 7 to 8 of Table 4.10). The RLA learns a yield maximizing acceptance strategy based on the capability of the scheduling heuristic.[21]

Table 4.10. Reinforcement learning yield management algorithm

```
Initialize Q(sₜ, aₜ) arbitrary
Repeat (for a fixed number of episodes):
    Repeat for all jobs in job list of JPI
        Receive a request for job j from JPI at time bⱼ
        Determine actual state sₜ from pⱼ, vⱼ and gf
        Select an action aₜ for state sₜ (ε-greedy)
        Perform action aₜ and inform DSC about acceptance /
            rejection of job j
        Observe reward rₜ determined by the DSC
```
$$Q(s_t, a_t) \leftarrow Q(s_t, a_t) + \alpha \cdot [r_t + \gamma \cdot max_{a_{t+1}} Q(s_{t+1}, a_{t+1}) - Q(s_t, a_t)]$$

4.4.4 Deterministic Scheduler Component

After having outlined the RLA of the YOS agents, the DSC is now depicted in detail. The parameter g_f represents the degree of to which the production queue is filled on the assumption that the offered job is accepted while using the amount of provisional allocated tasks, job length l_j, bid-date b_j, and due-date d_j for calculation. The scheduler component sorts all jobs in descending order of their due-dates. To validate the resulting schedule/queue job ratios g_x have been composed by using equation (4.30).

$$g_x = \frac{\sum_{i=1}^{x} l_i - b_j}{d_x - b_j} \tag{4.30}$$

These job ratios g_x are employed to classify the schedule. Table 4.11 depicts the classifications that were used in the simulations.[22]

Based on these calculations the RLA decides on acceptance or rejection of the current job request. In case of acceptance, the DSC maps the job into the job execution plan.

4.4.5 Benchmark Algorithm for the RL-YM Scheduler

To benchmark the performance of the RL-YM scheduler a simple heuristic is employed, which accepts resource requests depending on the level to which the

---

[21] The RL scheduling method employed here falls into the first group of RL mechanisms described in section 4.3.2 for this reason.

[22] Tests showed that other classifications do not provide better results.

Table 4.11. Classification of jobs according to the time-to-due-date and due-date ratio g_f

	job class g_f				
	0	1	2	3	4
$max_x g_x$	$0.0 \ldots 0.2$	$0.2 \ldots 0.4$	$0.4 \ldots 0.7$	$0.7 \ldots 1.0$	> 1.0

production queue is filled (see Table 4.12). The benchmark algorithm should establish a job acceptance/rejection equilibrium providing a considerable return on ISIP resource provisioning.

Table 4.12. Benchmark algorithm for the RL-YM scheduler

```
Initialize scheduler with j = 0
Repeat for each requested job j
    If g_f ≤ 4 (job fits in schedule)
        accept
    Else
        reject
```

4.4.6 Performance of the RL-YM Scheduler

Using a JAVA implementation of the RL-YM algorithm, a test set consisting of 1000 scheduling requests produced by the JPI based on tasks with equally distributed length and a Poisson distributed arrival process was employed to learn a yield optimal acceptance policy. Subsequently the performance of the trained RL-YM system was evaluated using a separate set of 1000 scheduling requests. During the learning phase, the JPI passes a job request j taken from the training set to the RLA, which has to decide on job acceptance or rejection. After the RLA has accepted a task, the DSC tries to fit the job into the schedule avoiding the occurrence of penalty costs due to not-in-time delivery. Later on, when the scheduled production has been executed, the RLA gets feedback for the acceptance/rejection decision depending on a positive or negative production return. Thus the system learns yield maximizing acceptance decisions by updating the Q-values of the states corresponding to positive or negative rewards. Learning has been achieved in various intensities, including an average number of updating visits per state s, accounting from 10 to 200 epochs and varying the exploration intensity ε (see Figure 4.14). For each point of the diagram 40 series of requests were performed, measuring the minimum, maximum and average return after the simulation series. It turns out, that the RLA performs best for ε below 0.5 depending on the number of training epochs in terms of minimum and average income.

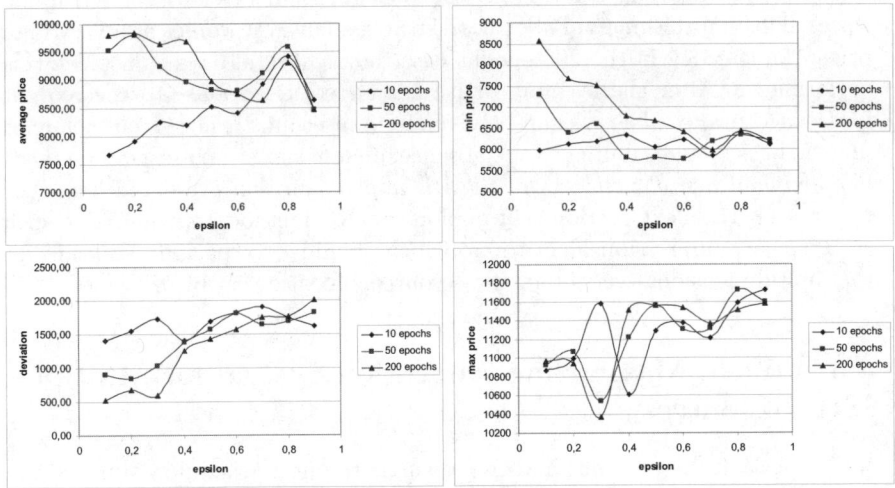

Fig. 4.14. RL-YM algorithms performance using a trained system after 10, 50 and 200 epochs of learning with epsilon increasing from 0.1 to 0.9 in steps of 0.1

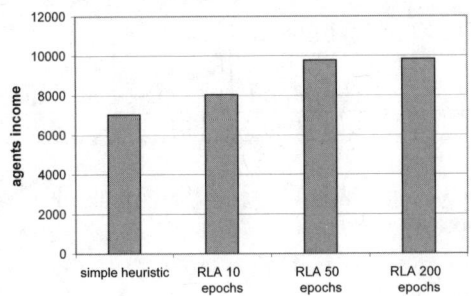

Fig. 4.15. Average income of RL simulation runs including 10, 50, 200 learning epochs compared to a simple heuristic

Interestingly maximum income is increased for high ε showing that learning with lower exploration rates cannot reveal all local yield optima in state space. Unsurprisingly this rise of maximum income occurring with higher ε has to be paid for with lower minimal income, generating an increasing volatility of return. Considering all output parameters of Figure 4.15, the 200 epochs learning case performs best in a range of ε between 0.1 and 0.5, showing high average income at low volatility.

By comparing the RL-YM schedulers results to the income gained with the benchmark algorithm the findings described above are confirmed. As can be seen in Figure 4.15, the RL algorithm outperforms the average income of the simple learning strategy in the benchmark heuristic for all training states. This improvement of allocation strategy, however, comes at a very high price: the learning in the 200 epochs case took a several days to achieve this performance. After the learning phase has ended, a response to the request for the acceptance of a task can be given in milliseconds. The system behavior may be basically attributed to the absence of *state space compression methods* like *artificial neural networks* in the RL implementation presented here. For this reason the next section will explore an RL method that employs such state space compression algorithms, which should be especially suitable for the multidimensional version of the resource allocation problem.

4.5 A Yield Maximizing Allocation System for Multiple ISIP Resources

In this section the previous YM-based single resource scheduling approach is extended to a multiple resource allocation case with complementarities. The work described in the following sections is published in Schwind & Wendt (2002). Unlike the single resource case, the multiple resource case does not allow for task shifts within time windows as illustrated in section 4.12.

Bid Matrix				
	Res. o_j			
Time Slot t_1	1	2	3	4
$q_1(o_j, t_1)$	1	1	6	4

Bid Matrix				
	Res. o_j			
Time Slot t_1	1	2	3	4
$q_2(o_j, t_1)$	6	6	9	3

Bid Matrix				
	Res. o_j			
Time Slot t_1	1	2	3	4
$q_3(o_j, t_1)$	6	2	5	

Fig. 4.16. Illustration of three RL-YM requests for multiple resources at execution time t as matrix representation

Figure 4.16 illustrates three alternative combinatorial resource requests $q(o_j, t)$ as matrix representations derived from the BM for the multiple ISIP resource allocation problem (see section 1.4.4). Differing from the previous bids in the single resource YM-RL example, the capacity requested in $q(o_j, t)$ per resource may be in a range from 0 to 9 resource units. For instance, in the above example, the first BM indicates that the resource user has a combined demand of one unit for resource o_1 and o_2, six units of resource o_3, and four units of o_4 at time t_1. The introduction of combinatorial bids into the standard YM setting turns the optimization problem that has to be solved by the market mediator into an even more demanding NYM problem.

4.5.1 Artificial Neural Networks for Value-Function Representation

A viable way to deal with the complexity of the NYM problem and the related evaluation of the optimal value function V^* of the classical YM problem is to use *artificial neural networks* (*ANN*) either to represent V^* or to map the decision function (Holland 1975). Both possibilities will be depicted in brief.

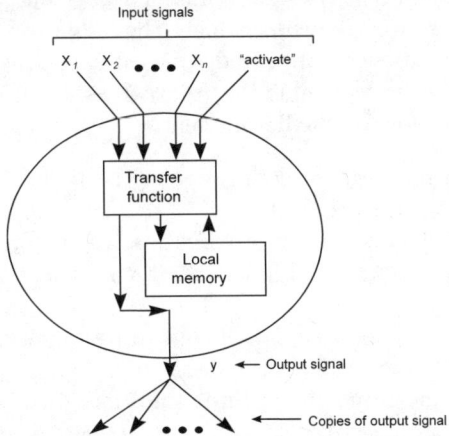

Fig. 4.17. Function of a neuron in an artificial neural network

In its basic construction, an ANN can be regarded as a *non-linear regression network* composed of elementary operators called *neurons*. The neurons are organized in a layer structure as depicted in Figure 4.18. As can be seen in Figure 4.17 each neuron simply weights w the input signal received from the neurons of the preceding ANN layer and propagates the output signal to the next ANN layer or output channel according to a transfer function stored in the *local memory* of the ANN neuron. The edges of the network thus act as unidirectional signal paths with identical output signals for all neurons in the next ANN layer. By switching a greater number of neurons together within a network, arbitrarily complex functions can be represented by the ANN. The system learns by feeding the network with the function to represent and adjusts the weights of the neurons according to the expected output function in a back-propagation algorithm.[23]

In the YM context presented here a *feed-forward ANN* (Principe et al. 1999, pp. 554 seq.) using three layers of neurons is employed for function representation. The following assignment of ANN layers to input and output signals seems to be advisable in the context of our NYM problem:

- Each resource available at valuation time t, represented as components of resource vector $q(o_j, t)$, is mapped by a neuron in the input layer. Besides the k *resource neurons*, a *time neuron* is required in the input layer to feed

---

[23] A detailed description of the *back-propagation process* can be found in (Bishop 1995, pp. 202 seq.). It is not discussed here because the methodology of ANNs is well known and focus of this work is not on this topic.

the amount of remaining future requests into the ANN. On the output side a single output neuron indicates the value class of the request. The structure of the resulting ANN is depicted in Figure 4.18.

- The ANN determines the output value of two resource vectors $q(o_j, t)$ in the case of a resource request, namely the value of residual capacity in the case of request-acceptance and residual capacity in case of request-rejection. The difference yields the *reservation price* for the request as a decision criterion for the mediator agent.

Employing this topology in connection with the back-propagation algorithm, the satisfying approximation of V^* depicted in Figure 4.10 can be achieved. Figure 4.19 shows the approximation results for the ANN representation of the classical YM value function after a diverse number of learning iteration steps.

Unfortunately it turns out to be difficult to use supervised learning in the YM domain, because the result of the decision process only appears after the entry of the last request and the optimal total marginal contribution for the particular request profile is not known until it has been calculated by employing SDP. One way to get out of this dilemma is to train different ANNs with the same sample of a value function, calculated by SDP and selecting the best performing ANNs, hoping that the generalization property of the network will provide satisfying solutions measured by the *average marginal contribution (AMC)*. In this way, the search for an optimal ANN representation of the value function can be seen as a parameter optimization problem. The *search space* is the vector space of all weights of ANN neurons, the goal function for a given parameter constellation is the AMC which is reached by the weights, if the particular ANN is applied.

4.5.2 Using Evolutionary Techniques to Improve the Value-Function

When considering the meta problem of global parameter optimization, the selection of reasonable optimization strategies depends strongly on the structural properties of the search space, which are not known in our example.

In the case of a *unimodal space* a gradient search could be applied. However, gradient search will be hopelessly inferior to techniques based on *evolutionary strategies* (Schwefel 1995) or *genetic algorithms (GA)* (Goldberg 1989a). These methods start searching at a multitude of points (here: individual ANNs) in search space and proceed with the evolutionary process by mutation, reproduction and selection. As the name suggests, GAs are an analogy to biological evolution, where natural selection leads to the adaptation of the genes of a species to their environment. The adaptation process can be abstracted and transferred to the optimization of combinatorial problems. The origins of the development of GAs were not aimed at seeking optimization per-se but to gain insight into natural processes (Holland 1975). GAs are a

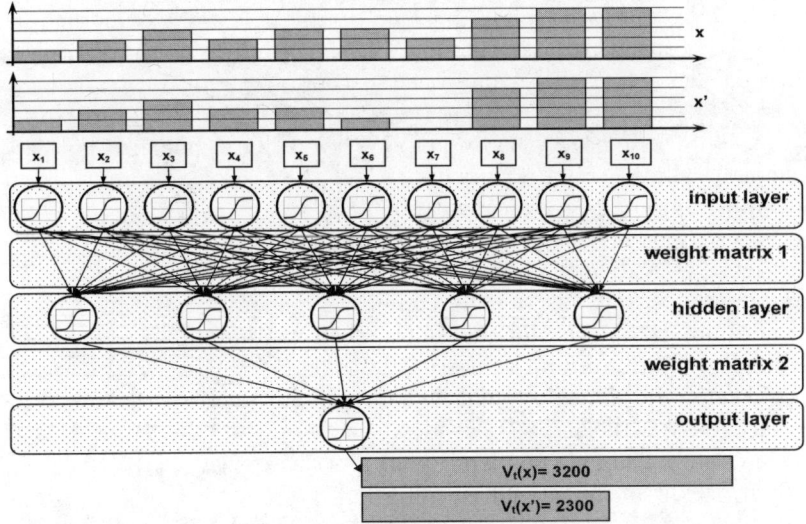

Fig. 4.18. Exemplary valuation of a reduction of resource capacity x_5, x_6 and x_7 by a requested order; the expected AMC amounts to 3200-2300=900 MU

class of adaptive search algorithms and as such a valid approach to numerical optimization. Adaptive search systems are defined by state space, operators that act on the state space, inputs from the environment and a mechanism that leads to the adaptation of the system to its environment (Holland 1975, p. 28). In GAs, the state space is coded as sequences of characters and numbers, while the adaptation mechanism is constructed in analogy to natural selection. A GA generally starts by constructing a population of candidate solutions. Each individual is then evaluated by assessing its value or fitness, according to the optimization criterion. The algorithm then enters a loop, which aims at the successive improvement of the population's fitness. In analogy to biological evolution each iteration is called a generation. Each generation has to

- generate new individuals by (1) recombining, (2) copying/reproducing or (3) mutating existing individuals and
- abandon old individuals by (4) incrementally replacing individuals or (5) selecting a subset from a generation.

A broad range of algorithms types can be constructed using different variants of these operators as well as hybrids of these operators. There is no general best performing combination of operators. This leads to a large number of techniques that provide approximate solutions to highly specified and often rather restrictive problems.[24]

---

[24] For a further treatment of the GA topic see sections 5.2.2 and 6.1.2.

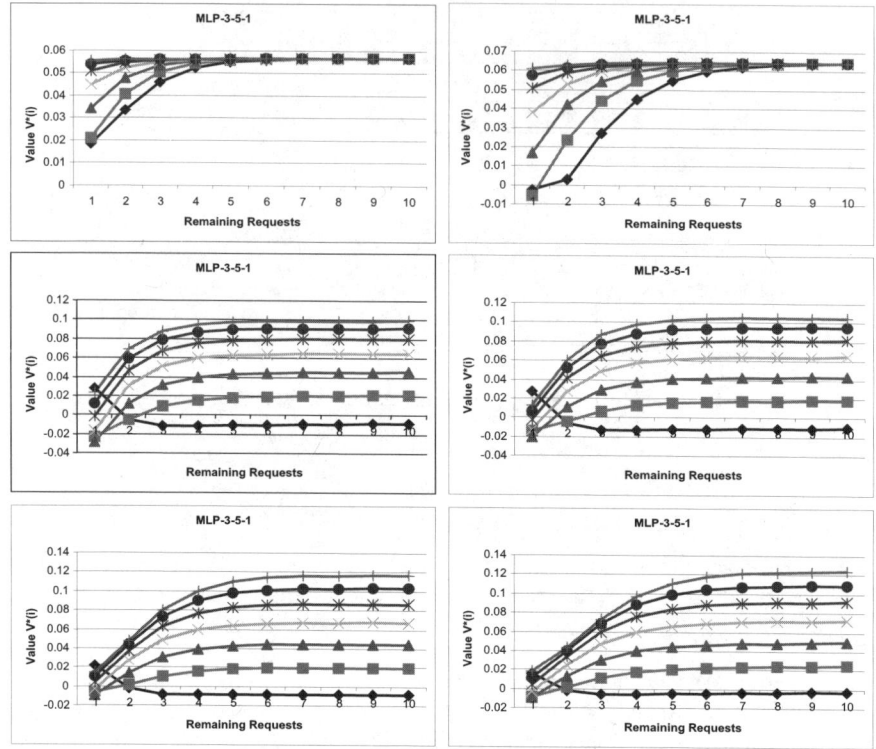

Fig. 4.19. Typical results of a back-propagation ANN 3-5-1 after 20,000, 40,000, 60,000 and 160,000 iterations

The application of GA-based local search strategies raises the question of optimal sample size. Fewer simulations carried out for a given ANN could lead to the repudiation of a superior vector of weights w in favor of an inferior vector w' due to estimation errors. Increasing the sample size helps in reducing this risk but implies a lower number of parameterizations to be evaluated per time span.

Taking this trade-off into account, several ANN topologies were tested allocating two resources only. This allowed the analytical calculation of the expected result using SDP and enables the rating of the ANN function approximation performance. The procedure led to a desired overall selection probability for high quality individuals. The ANN architecture used to do this was identical to the structure described above.

Breeding the ANNs employing the GA methodology was carried out according to the following procedure:

A population of 10 ANNs was initialized using random weight matrices. These were used to generate an offspring of 100 individuals. For the sake of efficiency any complex selection strategy was abandoned. The population was

simply sorted according to decreasing fitness followed by random selection
for the application of the *crossover operator (CO)* process. A mating partner
for an individual was then selected only from the subset of individuals with
lower fitness rankings. This led to the desired overall selection probability for
high-quality individuals. As CO a modified *1-point-crossover* was employed:
For each layer, the lines of the weight matrices were cut at a randomly chosen
column and recombined with the complementary lines in the matrices of the
mating partners that have been cut at the according columns as it is illustrated in Figure 4.20. The feasibility of the resulting value function remains
untouched by this process for all potential combinations of ANN weights w.

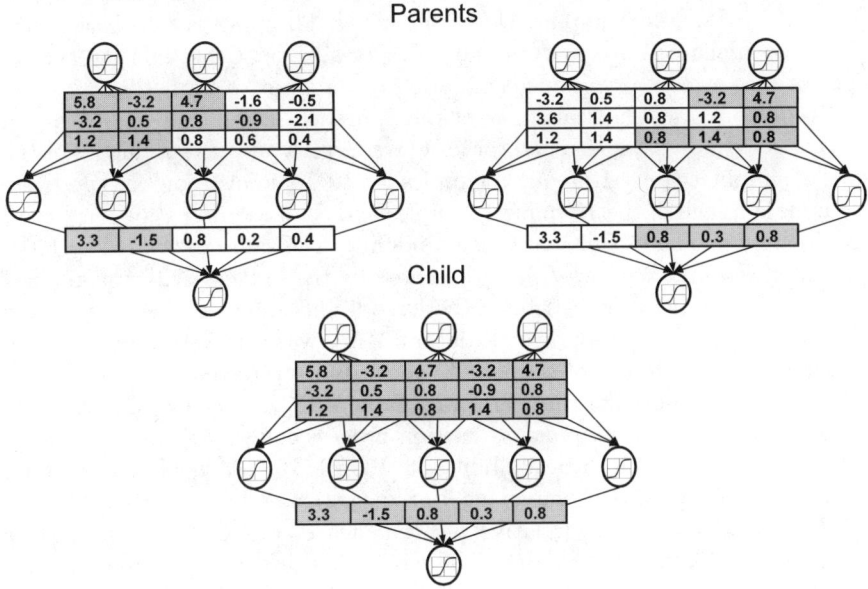

Fig. 4.20. Sample of a two ANN 1-point crossover

Preceding the CO process, 10% of the individuals were subjected to a
mutation procedure which selects a random column or line in the weight
matrix and superimposes the column or line with a stochastic offset vector
lying in an [-0.5; +0.5] interval. To determine the fitness of each particular
ANN in a population, 100 request processes were carried out, including 10
requests at a time. Finally the population was sorted according to their fitness
and reduced to the original size of 10 individuals.

Figure 4.21 depicts the impact of population size on the average fitness
of the value function that evolves after 100 generations. The sampling rate
denotes how many processes (10 requests each) are drawn from the expected
demand distribution, to reduce the estimation error. Each point represents the
result of a GA after 100 generations. The plain is the smoothed mean value

estimation of the AMC (or simply average return in the non-YM context) after 100 generations for each given population size and sampling rate. A positive effect of population size and sampling rate on the quality of the result can be recognized but may be misleading with respect to the necessary calculation effort.[25]

Employing the simple CO and mutation operators used in this context, the computational expense lies mainly in the evaluation procedure. Therefore the following diagrams no longer show the performance for a fixed quantum of 100 generations, but a budget of 100,000 and 200,000 evaluations as can be seen in Figure 4.22. It turns out that the best quality is achieved at small sampling rates and populations in the 100,000 evaluations study, whereas in the case of 200,000 evaluations larger populations are at an advantage leaving the benefit of smaller sampling rates untouched. That means calculation time should be allocated into greater populations or a longer GA evaluation period, but not into a more precise determination of the objective function.

To demonstrate the scalability of the approach in the context of NYM a stochastic request sample generator for O resource types was specified. For the sake of simplicity only 10 request samples for 10 different resource types were allowed. For each of the 10 request samples and each resource type a capacity requirement of 0 and 9 units is chosen randomly. The value associated with the requested resource bundles BM represented by the bidders W2P was defined to be likewise randomly distributed between 0 and 9. The overall availability of capacity (size of the market mediator's AM) was set to 10 units for each resource type. Such a pool of requests generated for 10 request types and 4 resources might look like that shown in Table 4.13. Using this problem size, the state space of the *Markovian* decision process can still be mapped easily using SDP (e.g. an initial availability of [10, 10, 10, 10] yields a maximum expected return of 13.067 employing 10 requests).

Figure 4.23 shows the relative performance of the GA-ANN procedure compared to the optimal solution. The results indicate declining performance for higher resource dimensions. This is less bad than might be supposed, in view of the fact that the computational effort increases exponentially whereas the evaluation budget remains constant at 500,000. The drop in efficiency can be explained by the difficulty of finding a promising value function in the search space of the weight vectors. Because the GA is able to learn quickly, that acceptance of all requests leads to a better performance than total rejection, an individual "recognizing" this replicates heavily and inhibits diversification as well as evolutionary progress.

---

[25] The use of population size 5 and sampling rate 10 per generation, implying 50 child-ANNs evaluated 5 times each, leads to $100 \cdot 50 \cdot 5 = 25,000$ demand streams, whereas a population size of 20 leads to $100 \cdot 200 \cdot 100 = 2,000,000$ evaluations.

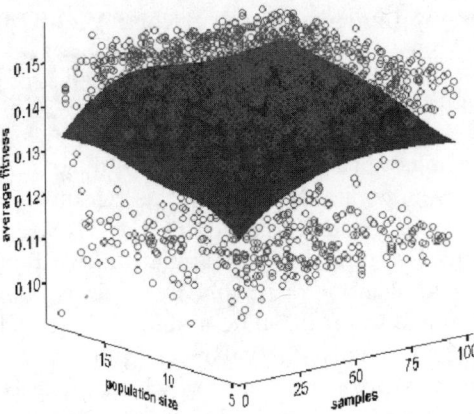

Fig. 4.21. Impact of population size and sampling rate on the quality of the value function (100 generations)

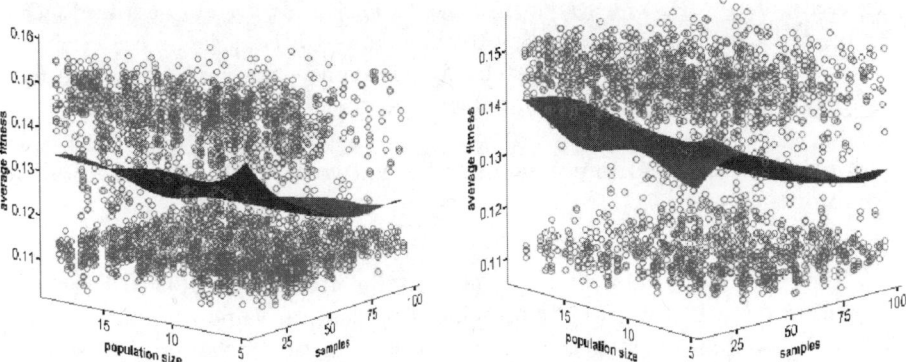

Fig. 4.22. Impact of population size and sampling rate on the quality of the value function (100,000

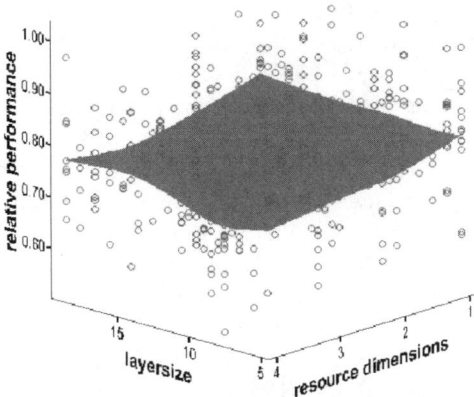

Fig. 4.23. Relative performance of the GA vs. layer size and number of resources

4.5.3 Reinforcement Learning for Network Yield Management Processes

The fact that domain specific knowledge about the structural properties of the value function (e.g. monotonicity) is not applied in the searching process and the quality of $V(s)$ can not be rated directly, leads to increasing computational expense at higher resource dimensions. For this reason the application of an adaptive learning system which replaces either the ANN representation of the value function or the GA procedure manipulating this representation seems to be indicated. The application of *reinforcement learning* seems therefore to be a promising approach to establishing a suitable feedback function in this context (Bertsekas & Tsitsiklis 1996). For this purpose the RL-YM system is connected to the YM process via a loop which gives feedback about the reward r_t for the acceptance of a priced resource allocation request in state s_t, after each optimization step (*action a_t*). The RL-YM system chooses a new optimal action a_{t+1} using the acceptance decision function and changes the state of the RL-YM system from s_t to s_{t+1}. The system gets feedback by receiving reward r_{t+2} according to the quality of the acceptance decision. The objective of the RL-YM system is to maximize the *sum of rewards* by learning the appropriate decision policy π_i as described in section 4.1.2.

Adapting Reinforcement Learning to Network Yield Management Problems

In order to adopt the NYM evaluation environment to the use of RL methods the SDP value iteration approach was replaced by the TD(0) learning algorithm described in Table 4.6. While the classical value iteration, starting with stage 0, calculates the complete value function for all possible states and then repeats this with all higher stages, TD(0) totally renounces this initialization, but replaces it rather with an arbitrary function (e.g. V(stage, state) $= 0$ for all stages and states) and then improves this incorrect function during the decision process. Based on this function most of the decisions are wrong, since for the example from Figure 4.10 and Table 4.9 for stage eight, $V(7.5) + 1 > V(7.6)$, as long as V equals zero for all stages and states. Nevertheless, an update of the value function occurs immediately after each decision, making the difference "less wrong" in the next run. If there are, for example, only requests of type F1, $V(8.6)$ could be set to max $[V(7.5) + 1; V(7.6)]$. For a decision that has to be made in the following stage (number nine) requiring the value function of $V(8.6)$, the TD(0) algorithm has then already learned from the previous RL iteration that the corresponding value of $V(8.6)$ must be greater than zero. Since there will normally be other requests with differing profit contribution besides F1, $V(8.6)$ will also be subject to a constant adaptation process. The selection of an appropriate learning rate α therefore turns out to be crucial in the evaluated experiments: if α is set too high, the system overreacts to single resource requests, if it is set too low, the adaptation process is not able to react adequately to significant demand fluctuations.

In the RL-YM system, implementation of the JAVA modules for the MDP and the ANN had to be adapted to represent the RL version of the decision process. An ANN is still necessary for state space compression since the RL algorithms have problems representing high-dimensional state spaces. Nevertheless, an important advantage results from the fact that the ANN will now be trainable by back propagation: the fact that the training samples will be as wrong as the value function represented by the ANN in the beginning only delays the RL training but does not disable its convergence, because appropriate reward information is fed back by the RL mechanism after every single ISIP task acceptance or rejection decision.

Results of Reinforcement Learning compared to Genetic Algorithms

To evaluate the performance of the proposed TD(0) approach, test instances with one to four resources o_j were generated and the results compared to the optimal solution calculated using the SDP method.[26] The test instance comprised a set of 10 randomly generated requests $q(o_j, t)$ for each simulation according to the BM structure depicted in Figure 4.16. The reward r associated with the successful execution of the tasks requested in the resource bid bundles was defined as a randomly generated contribution margin which is indicated in column one of Table 4.13.

Using this benchmark, the results shown in Figure 4.24 were obtained: after 20,000 evaluations a quality less than 5% from the optimum is reached on average and no negative correlation with the number of resources o_j was observed. For a value function represented by an ANN, however, 50,000 evaluations were still insufficient for an acceptable value function approximation. Even with 500,000 evaluations, training seems to have finished only for networks with small hidden layers. However, by using a higher learning rate α this convergence can be accelerated. If we compare the results in Figure 4.24 to the relative performance of the GA-ANN approach after 500,000 evaluations in Figure 4.23, the strong superiority of the TD(0) approach gets evident (see Figure 4.25).

The superiority of the TD(0) method can also be recognized from a direct comparison of Figure 4.25 and 4.23. Both diagrams present the performance of the different approaches for a fixed budget of 50,000 and 500,000 evaluation runs relative to the optimal solution that is achieved by using SDP.

The promising results of our TD(0) learning contradict the initial apprehension that direct learning can not be achieved as a result of decisions by an ANN in the YM context regarded here. The direct comparison with a GA-inspired optimization approach shows that our feed-back-oriented RL learning method outperforms the unsupervised GA learning method.

---

[26] This is possible since all states on all stages can still be represented directly by SDP in this example.

Table 4.13. Resource requests for a four-dimensional YM problem

	Req. Resource			
Reward	o_1	o_2	o_3	o_4
7.0	9	9	7	4
9.0	1	1	6	4
1.0	4	5	3	4
8.0	9	1	4	6
3.0	6	6	9	3
8.0	6	9	0	5
1.0	4	3	7	7
6.0	2	6	6	0
7.0	6	2	5	0
2.0	7	9	4	3

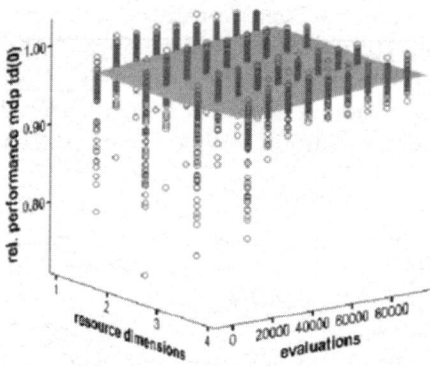

Fig. 4.24. Relative performance of reinforcement learning as a function of the number of resource dimensions and evaluations

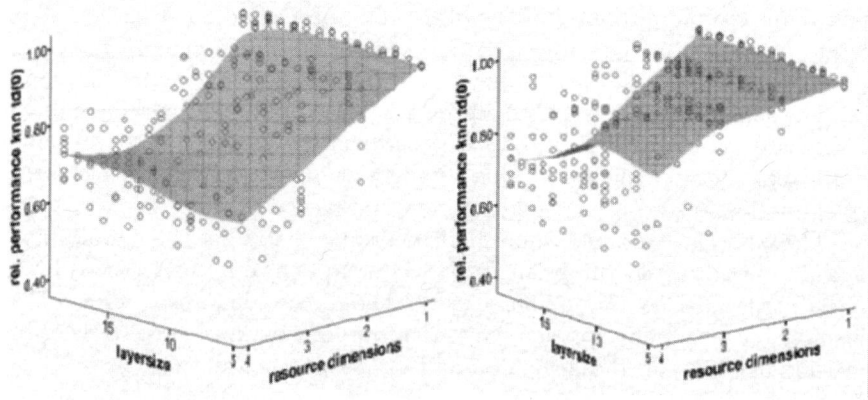

Fig. 4.25. Reinforcement learning based on a three layer ANN value function (left: 50,000 evaluations, right: 500,000 evaluations)

The findings of this chapter suggest the conclusion that it is reasonable to assume that dynamic pricing of ISIP processes for anonymous market demand shows high structural similarities to classical YM problems known from other service industries. Although thre are minor differences, customizing YM for ISIP provisioning processes seems to offer a more promising road than adapting classical price theory for dynamic pricing applications.

In the single resource case addressed in the first part of the RL-YM evaluation presented here, Q-learning surpasses a simple resource allocation heuristic in terms of income earned by the system's market mediator. However, the RL method used does not employ an ANN for state space compression, making the learning process very slow and hindering appropriate YN system reaction to market fluctuations. Similarly, the approach cannot take the complementarities of the renewable system resources into account. These complementarities, however, are crucial in the context of task scheduling in distributed computer systems, as already shown in section 1.4.4.

By meeting the challenges that arise from the fact that the resources have to be available either in parallel or sequentially, several heuristics have been developed and evaluated in the context of the PCRA scenario depicted in chapter 1.1. An ANN related approach has been combined either with a genetic algorithm or reinforcement learning. When comparing the RL method with a GA approach in terms of constructing good acceptance/rejection decision functions, RL clearly turns out to be the superior choice, since both methods are benchmarked by the results calculated using the results obtained employing classical SDP. While the application of RL shows good outcomes for a greater number of resources, the introduction of an appropriate state space compression seems to be recommendable in this context. The use of a *multi layer perceptron* as ANN representation might not be the best choice for the multidimensional resource allocation problem considered here. The following alternatives are therefore interesting for further evaluation:

- *Vector quantization methods* using a much finer granularity of the state space in those regions that are highly frequented at the expense of a coarse grain representation in other regions (Buhmann & Hofmann 1996),
- *Kohonen maps* making only those neurons which exhibit a certain proximity relation interact, unlike the multi layer perceptron that relates the entire neurons of a layer to all other neurons of the next layer (Kohonen 1995),
- *Neural gas algorithms (NGA)* that also restrict learning to the local neighborhood, but use an Euclidean-metric to determine the strength of interaction rather than the fixed neighborhood structure of the Kohonen maps (Fritzke 1995).

Figure 4.26 shows the three-dimensional representation of the value function learned by a classical TD(0) learning and an NGA approach. The diagram can be seen as a 3D interpretation of Figure 4.10, depicting the remaining requests N on the x-axis (in front), the residual number of requests i on the

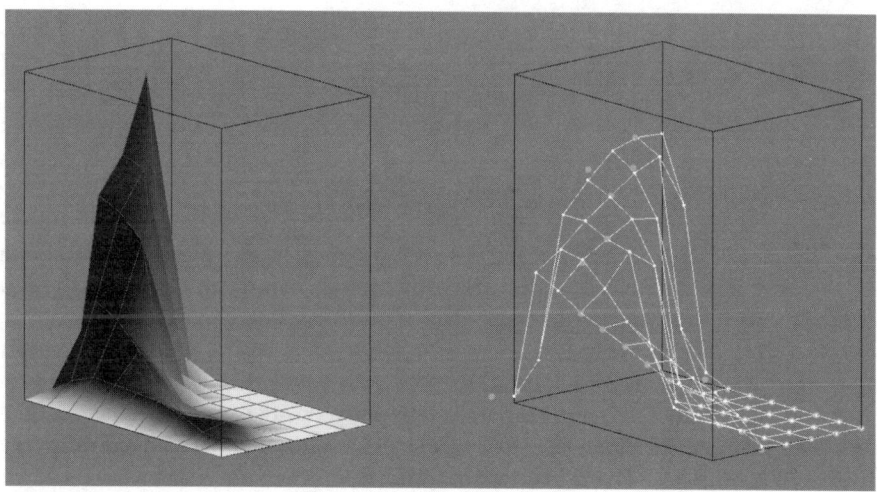

Fig. 4.26. Three-dimensional representation of the YM value function learned using plain TD(0) learning (left side) and a NGA (right side) for state space compression

y-axis (to the back) and the value of capacity $V^*(i)$ on the z-axis (at the top). The application of NGA shows a tendency to produce better function generalization and a faster learning behavior than YM-RL using classical ANN. The main problem, however, for resource allocation in ISIP environments using the RL-YM methods examined here lies in the fact that this method addresses the crucial complementarity issue only indirectly. RL methods therefore have to overcome the additional learning effort required to compensate for this constructive drawback. The additional learning effort results in a reduced ability to react to supply/demand fluctuations in the context of the RL applications presented here. Furthermore, the stochasticity of requests in the underlying SKP in combination with the enforcement of an immediate task acceptance decision leads to an extent of complexity that is not manageable for problem sizes relevant in real world ISIP environments. For this reason, the remaining part of the thesis will be dedicated to the evaluation of combinatorial auctions that are basically constructed to address the complementarity issue in PCRA systems as outlined in section 2.3.2. CAs also circumvent the stochasticity problem by collecting the requests and making the acceptance decision periodically at the end of the bid submission period, when the auction actually takes place.

Combinatorial Auctions for Resource Allocation

The application of *combinatorial auctions (CA)* for procurement and resource allocation processes has been intensively discussed in the last ten years. Motivated by the economic success of online auctions (see section 2.1.2) and accompanied by the development of sophisticated mechanisms for the auctioning of radio frequencies to telecommunication companies in Great Britain and the United States (McMillan 1995), CAs came into the focus of *electronic market engineering* for other economic sectors. CAs promise to achieve higher efficiency and fairness, as well as a reduction in transaction costs in industrial procurement and allocation processes (Cramton et al. 2005, Bichler et al. 2002). This thesis primarily discusses the application of CAs for dynamic pricing and automated resource allocation in ISIP processes. However, it is important to consider the broad spectrum of advantages, problems and implications that arise in connection with CA applications in a wider context. For this reason, this chapter will discuss CAs in a more general framework, whereas the following chapter regards their application more particularly in the ISIP process domain.

The first part of this chapter tries to give a comprehensive overview of the problems and challenges that arise with the use of CAs. It also includes fundamental issues like the construction of *incentive compatible bidding* schemes, *bidding languages*, and *benchmarks problems* for CAs. The second part of this chapter is solely dedicated to algorithms for the solution of the *combinatorial auction problem*, the most elaborate topic in CA literature next to incentive compatibility. Here, deterministic procedures such as integer programming, heuristics and equilibrium approaches that can be applied to solve the CAP, will be discussed in detail. To be able to approach the complexity of CAs adequately, the third part of this chapter tries to sketch a comprehensive *decision making framework* for the design of combinatorial auctions. The decision framework should allow one to handle various types of CAs by taking the broad range of criteria that have been discussed in the previous sections into account. The framework concentrates mainly on illustrating schematically the interdependencies between the design decisions. Measures for the realization of *complexity reduction* in CAs are also shortly addressed in this context. Finally a process model for the *simulative and experimental evaluation* of CA mechanisms will be presented and the application of CAs in ISIP provisioning will be discussed briefly in the light of the chapter's previous findings.

5.1 Basics of Combinatorial Auctions

Auctions are defined as a market situation where the price of the goods, services, or resources is determined in connection with their allocation according to predefined rules based on the bids of the auction participants (McAfee & McMillan 1987). Combinatorial auctions allow bidders to bid for bundles of goods (services or resources) while the valuation of the bundles depends on synergies between the individual goods, services or resources (Cramton 2005). Two types of valuation effects are of interest in this context (see Figure 5.1):

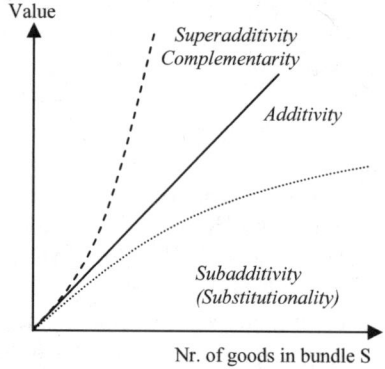

Fig. 5.1. Synergies in goods' bundle valuation

- *Subadditivity*: If substitutionalities between the goods in the package lead to a lower utility of the entire bundle compared to the sum of the utilities of the individual goods, subadditivity has to be assumed. The reduced utility results in a diminished W2P among the bidders for the package. Discriminatory prices (rebates) for the procurement of higher quantities of different goods in industry could be considered as an example of this phenomenon.[1]
- *Superadditivity*: If the valuation of a bid bundle is higher than the valuation of the individual goods the effect is described as superadditivity. This

---

[1] A set function f is locally subadditive in the disjoint sets S and T, if $f(S \cup T) \le f(S) + f(T)$ is true. A set function is subadditive, if it is locally subadditive for all disjoint S and T.

results from complementarities in the bidder's utility function for the single goods. People inquiring about a trip, for example, have a higher W2P for the acquisition of the whole travel package than for the separate purchase of a flight ticket, a hotel stay, and an evening activity at a special location.[2]

To solve the return-maximizing (cost-minimizing) *combinatorial auction problem)(CAP)*, the auctioneer has to calculate the result of the *winner determination problem (WDP)*. It is mathematically based on the weighted *set packing problem (SPP)* in the case of an *allocation auction* and on the weighted *set covering problem* in the case of a *procurement auction* (Vries & Vohra 2001).

5.1.1 Variants of the Combinatorial Auction

The literature names several auction versions associated with the combinatorial auction.[3] CAs can be regarded as a subset of *multidimensional auctions*, which are characterized by their ability to exchange complex preference information between bidder and auctioneer. In this context the term "multidimensional" denotes the various dimensions of the bids (besides the price, these dimensions include: quality of the goods or services, delivery time, quantity of different items, etc.), which play an important role generally, but especially in procurement processes (Bichler et al. 2002). Whereas in multidimensional auctions the impact of the goods' attributes on the bid price can be defined by various types of valuation functions (this kind of auction is often called a *"multi-attributive auction"* (Kalagnanam & Parkes 2003, p. 50)), in pure combinatorial auctions the existence of different predefined property constellations is the prerequisite for bid selection by the auctioneer. The classical variant of the combinatorial auction is the *multi-item auction*, where single items of different good types are combined in bundle bids. If, in addition, multiple items of a good type can be bid for in the bundles, a *multi-unit auction with combinatorial bids* occurs (Kalagnanam & Parkes 2003, p. 48). For the most part, both types of auctions are simply denoted as combinatorial auctions. According to the classification of classical auction types, combinatorial auctions can be divided into *procurement auctions (reverse auctions)* and *sales auctions (forward auctions)*.[4] With the former, suppliers offer goods (services, resources) to the auctioneer, who selects the cost minimizing bids on behalf of the buyer; with the latter, the auctioneer tries to sell the goods to the bidders on behalf of a seller, while maximizing the total return. If multiple buyers and sellers are involved in a combinatorial matching process the procedure is termed a *combinatorial exchange* (Tanner & Mühl 2003). In this case,

---

[2] A set function f is locally superadditive in the disjoint sets S and T, if $f(S \cap T) \geq f(S) + f(T)$ is true. A set function is superadditive, if it is locally superadditive for all disjoint S and T.

[3] The following argumentation can be found in König & Schwind (2005).

[4] Sales auctions are called allocation auctions in the context of this thesis.

bid-ask spread minimization or turnover maximization could be defined as the auctioneer's objective function. In the context of procurement auctions, there are a few special cases of multidimensional auctions, like the *discount auction*, which operates with falling, quantity-dependent price functions (Davenport & Kalagnanam 2001). A further decision of increased importance in combinatorial allocation and procurement process design concerns the choice of either a *one-shot* or an *iterative winner determination process*. Iterative combinatorial auctions allow the bidders to interactively adjust their bids if they are not awarded the desired bid combination. Unlike the one-shot combinatorial auction, which is normally carried out as a *sealed-bid auction* (Varian 1995), the iterative combinatorial auction is very sensitive to strategic bidder behavior, leading to sophisticated bid mechanisms (Kwasnica et al. 2005, Banks et al. 1989, Parkes 1999). In the following we will concentrate on the classical combinatorial multi-item allocation auction.

5.1.2 Advantages of the Combinatorial Auction

The ability to introduce the valuation of substitutionalities and complementarities into the bids for bundles of goods (or services, or resources) enables the auctioneer to achieve a *higher efficiency for the final allocation* than would be the case using classical auctions (Vries & Vohra 2001).[5] The relevance of this property for the PCRA environments regarded in this thesis has already been discussed in section 2.3. Besides higher allocation efficiency, CAs provide the following advantages in complex allocation and procurement processes:

- *Lower transaction costs*: A well-chosen auction format ensures a time-saving and efficient negotiation process when dealing with complex goods; without such an auction format this process would be subject to protracted decentralized negotiations. This is especially the case if web-based protocols are used.[6]
- *Higher transparency and fairness*: While defining a priori rules for bid acceptance by the auctioneer, a high level of market transparency and fairness is ensured for all participants (Hohner et al. 2003).

5.1.3 Problems with the Combinatorial Auction

The advantages of the CA over classical auctions are partially offset by several problems that confront auction users and operators. The first problem is the $\mathcal{NP}$-*hardness* of the WDP. As already discussed in section 1.5.4, $\mathcal{NP}$-hardness means that the WDP can not be solved algorithmically exactly in polynomial time for larger problem sizes. Solving the winner determination

---

[5] This requires that bidders to be able to express their preferences adequately.

[6] This has been shown e.g. by employing experimental game theory (Porter et al. 2003).

problem thus requires time-consuming calculations even for small problems, and is not tractable for a higher number of bids and goods. Numerous solution methods both *exact* and *approximate*, have been proposed for the WDP. Besides *integer programming* methods, often combined with *branch-and-bound search* methods (Sandholm 2002*a*), heuristics like *simulated annealing* or *genetic algorithms* have been applied to solving the WDP (Schwind, Stockheim & Rothlauf 2003). Section 5.2 will discuss this issue in detail. A further problem of the CA, that is closely related to the WDP, is the determination of *linear equilibrium prices*. The final allocation awarded by the auctioneer should maximize (or minimize) the bid prices by taking the bid properties into account. Due to the combinatorial character of the bids and their *nonlinear valuation*, it is not always possible to name *linear prices* for the bundle goods.[7] The introduction of *nonlinear-prices* $b^j(S) \neq \sum_{i \in S} b^j(i)(j \in \mathbb{N})$ does not solve the entire problem, because it is often impossible to determine consistent non-linear prices (Xia et al. 2004). For this reason Bichler et al. (2005) distinguish between *anonymous linear prices* (unit prices are equal for all bidders) and *personalized non-linear prices* $b^j(S) \neq b^k(S)(k \in \mathbb{N})$ (unit prices are individually different for the bidders). The auctioneers' *pricing problem* is always solvable by using personalized nonlinear prices. Bikhchandani & Ostroy (2002) define three types of prices in this context, "first-order" prices are anonymous and linear; "second-order prices" are anonymous and non-linear; and "third-order prices" are discriminatory (individual, personalized respectively) and non-linear. Users of the combinatorial auction are confronted with the *preference elicitation problem*. It is hardly possible for the bidders to enumerate all valid bid combinations $(2^j - 1)$ associated with their valuation (at least without auxiliary means); yet this is necessary to ensure an efficient outcome of the WDP process (Nisan 2005, Conen & Sandholm 2001). Another critical point lies in the design of an *incentive compatible and stable auction mechanism*, that is adequate to the allocation or procurement problem. Although there are a number of game theoretical approaches in the literature, most of them are not transferable to practicable auction processes.[8] The most prominent mechanism is the *Generalized Vickrey Auction* by Varian & MacKie-Mason (1994), which is based on the *Vickrey-Clarke-Groves (VCG)* mechanism (Vickrey 1963), and has the auction participants reveal their real valuation function. This mechanism will be illustrated in section 5.1.8. Another problem sometimes associated with the iterative combinatorial auction is the *threshold problem* (Banks et al. 1989, Milgrom 2000, Kwasnica et al. 2005). A bidder who wants to obtain only a single item and not a whole bundle has problems maintaining his bid in an iterative ascending auction, especially if a *beat-the-quote rule* is employed, which means that the partici-

---

[7] In this context the term linear price signifies that the goods prices determined by auctioneer are equal for all bidders, and add up linearly to the bundle prices.

[8] In many settings it is already a problem to make the complex rules comprehensible to the users of the auction.

pation in the next round requires a minimal bid increment over the temporary winning bid.[9]

5.1.4 Formalization of the Combinatorial Auction Problem

The CAP belongs to the class of *combinatorial optimization problems (COP)*. In general a COP is a maximization (minimization) problem with the associated instance bundles (Stützle 1999, p. 8). An instance is a tuple (X, f) where X is a finite solution set and $f : X \rightarrow \mathbb{R}$ an objective function assigning a value $f(x)$ to each $x \in X$. The target of a COP is to obtain a value maximizing $x^* \in X$, such that $f(x^*) \geq f(x) \forall x \in X$ is true. x^* is defined as *global optimum*.[10]

Figure 5.2 demonstrates the *combinatorial allocation problem*.[11] Nine objects can be bundled into arbitrary bids by the bidders in a combinatorial auction. *Bidder 1* wants to obtain bundle A consisting of four objects $(4, 5, 7, 8)$ for a price of 8 MU or a bundle B including the items $(7, 8, 9)$ for a price of 4 MU.[12] *Bidder 2* is interested in bundle C $(1, 4, 7)$ or in bundle D $(2, 3, 5, 6)$ both for a price of 8 MU, whereas *bidder 3* bids omly for bundle E $(1, 2, 4, 5, 7, 8)$ having a W2P of 5 MU. In a standard combinatorial auction, the auctioneer has the task of assigning the items to the bidders in such a way that the total return maximizes and each object is allocated only once. The optimal allocation in this example is to allocate bundle B to bidder 1 and bundle D to bidder 2, with no allocation to bidder 3, for a total value of 12 MU.

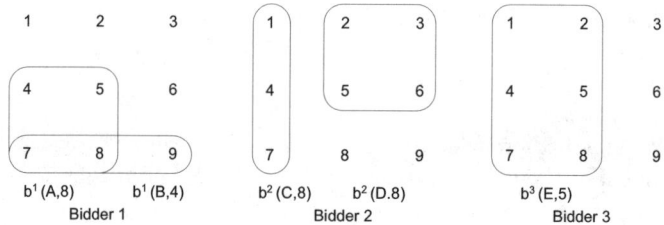

b¹ (A,8)   b¹ (B,4)   b² (C,8)   b² (D.8)   b³ (E,5)
         Bidder 1   Bidder 2   Bidder 3

Fig. 5.2. Example of the combinatorial allocation problem

---

[9] If for example a "big bidder" receives the temporary acceptance for a bundle with items A, B, C a "small bidder" bidding only for item A has difficulties obtaining the single good unless there are other participants bidding separately for the items B, C.

[10] Parts of the following sections follow the argumentation of a diploma thesis worked out in the context of the *PRISE* project (Halblau 2002).

[11] Figure in allusion to Parkes (2001 *a*).

[12] The term "or" denotes in this context that bidder 1 wants to receive either bundle A or bundle B, or both A and B (see also section 5.1.7).

The optimization problem of a combinatorial allocation auction can be expressed as shown in equations (5.1) to (5.3).[13]

Let N be the set of bidders j, M the set of goods i (items respectively) ($N, M \in \mathbb{N}$), while m is the number of items in set M ($|M| = m$). Each subset $S \subseteq M$ represents an arbitrary combination of goods and $b^j(S)$ is the bid of bidder $j \in N$ for the requested subset S. Let $b(S) = \max_{j \in N} b^j(S)$:

$$\max \sum_{S \subseteq M} b(S)x_s \tag{5.1}$$

$$\text{s.t.}$$

$$\sum_{i \in S} x_s \leq 1 \, \forall \, i \in M \tag{5.2}$$

$$x_s \in \{0,1\} \, \forall \, S \subseteq M \tag{5.3}$$

The auctioneer's optimization objective is to select bids $b^j(S)$ such that the return is maximized, where $x_s = 1$ denotes the acceptance of a bid and $x_s = 0$ indicates rejection. This form is only valid for purely superadditive bid valuation functions $b^j(A) + b^j(B) < b^j(A \cup B) \forall \{j \in N \wedge A, B \subseteq M \wedge A \cap B = \emptyset\}$. In the case of subadditive bid valuation functions $b^j(A) + b^j(B) < b^j(A \cup B)$ an alternative formulation must be given (Vries & Vohra 2001, p. 291):

$$\max \sum_{j \in N} \sum_{S \subseteq M} b^j(S)y(S,j) \tag{5.4}$$

$$\text{s.t.}$$

$$\sum_{i \in S} \sum_{j \in N} y(S,j) \leq 1 \, \forall \, i \in M \cup g \tag{5.5}$$

$$\sum_{S \subset M \cup g} y(S,j) \leq 1 \, \forall \, j \in N \tag{5.6}$$

$$y(S,j) \in \{0,1\} \, \forall \, S \subseteq M \cup g \wedge j \in N \tag{5.7}$$

For the CAP formulation in equations (5.1, 5.2, 5.3) it can not be ruled out that a bid $b^j(A \cup B)$ is assigned to a bidder j instead of the bids $b^j(A) + b^j(B)$ which would lead to a higher income for the auctioneer in the case of subadditivity. This problem can be resolved by introducing dummy goods g into the formulation.[14] In the resulting formalization (5.4) to (5.7), bids of bidder j have been replaced with the expressions $b^j(A \cup g) + b^j(B \cup g)$ and $b^j(A \cup B)$ while M has been replaced with $M \cup g$. Equation (5.5) ensures the acceptance of overlapping bids by using $y(S,j) = 1$ to indicate that bundle $S \subseteq M \cup g$ is allocated to bidder j and equation (5.6) guarantees that a bidder does not receive more than one subset as in the previous CAP formulation.

---

[13] The formalization in this chapter follows mainly the terminology used in Vries & Vohra (2001, p. 284 seq.).

[14] The introduction of dummy goods enhances the complexity of the CAP problem, see Nisan (2005).

5.1.5 Formalization of the Weighted Set Packing Problem

As already mentioned at the beginning of this chapter, the CAP can also be expressed in the form of a weighted SPP in the case of a forward auction. The formalization of the weighted *set packing problem(SPP)* used in this thesis follows Vries & Vohra (2001, p. 293):

$$\max \sum_{j \in V} d_j x_j \tag{5.8}$$

$$\text{s.t.}$$

$$\sum_{j \in V} e_{ij} x_j \leq 1 \; \forall i \in M \tag{5.9}$$

$$x_j \in \{0, 1\} \; \forall j \in V \tag{5.10}$$

Here d_j is an element in the weight vector and e_{ij} denotes an element in the $m \times n$ incidence matrix of a hypergraph $\mathcal{A}$ spanned by the set $M = (1, \ldots, m)$ and the node weights d_j. The optimization objective is to find a combination of subsets V that maximizes the total of weights, where each edge of $\mathcal{A}$ is pairwise disjoint and must not be cut more than once. The binary variable $x_j = 1$ indicates the inclusion of subset j with weight d_j in the set of selected elements, otherwise $x_j = 0$. $e_{ij} = 1$ if the j-th set in V describes an element $i \in M$. If M denotes the set of all goods in a CA, $V = S \subseteq M$ is a subset of M and x_j describes a binary value, it can be seen that the CAP is an instance of the SPP. Due to the fact that the SPP has been thoroughly examined in operations research, a multitude of algorithms exist that can be transfered into CAP solution methods (Andersson et al. 2000).

5.1.6 Complexity of the Combinatorial Auction Problem

The calculation of the return maximizing allocation in *non-combinatorial auctions* is a simple task requiring the selection of the highest bids. The upper bound of computationally complexity for single item sealed bid auctions is in $\mathcal{O}(|N||M|)$. In combinatorial auctions the calculation of the optimal bids requires the testing of all bid combinations and combination of their assignment to all bidders. Figure 5.3 illustrates the feasible combinations of potential bids for four items $\{1, 2, 3, 4\}$.[15] In the case of bidding for all feasible combinations, the CAP search space can be characterized as follows. Each node of the search space represents a subset $S \subseteq M$. The value of the objective function has to be calculated in each single node and must be compared to the values of the other bids. The number of feasible combinations increases according to:

$$\sum_{q=1}^{m} \binom{m}{q} \text{ for } 0 \leq q \leq m \tag{5.11}$$

---

[15] Figure and argumentation in allusion to Sandholm (2002a, p. 5 seq.).

where $\binom{m}{q}$ is the number of feasible combinations on the level q (potential subsets $S \subseteq M$, set of feasible bids respectively) and m the number of items.

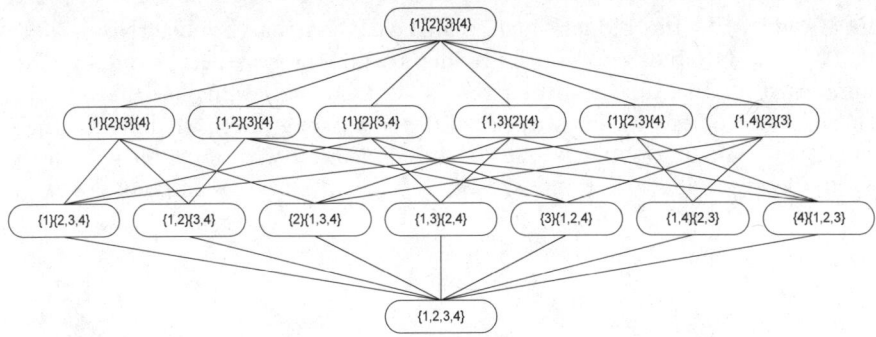

Fig. 5.3. Example of the combinatorial allocation problem.

Exact CAP solution calculation requires total enumeration with a resulting *computational complexity* that can be determined by using the *Stirling equation*. Asymptotic complexity of the CAP is then settled between $o(m^{m/2})$ and $\mathcal{O}(m^m)$ (Sandholm 2002a, p. 6).

5.1.7 Bidding Languages for Combinatorial Auctions

To enable the bidders to express the synergies in the valuation of bundled goods valuation a suitable bidding language has to be integrated into the auction protocol. If a bidder wants to bid for good $A, B, A \cup B$ or either for A or B exclusively, this must be expressible in the bidding language. An important criterion for the usability of a bidding language is its *expressiveness* and *understandability* for the users. Closely related to the usability is the communication complexity of a bidding language. It defines how many expression are needed to formulate the users' preferences in the bidding language. Eight types of bidding languages are common in literature (Nisan 2005, p. 5 seq.):

Atomic Bids

Atomic bids have the notation (S, v^j), while S denotes an arbitrary bundle of items in the bid associated with W2P v^j. They often represent the simplest instance of a bid by denoting only one item in a complex bid structure.[16]

---

[16] Atomic bids are also called "single minded bids" in literature (Lehmann et al. 1999).

OR Bids

OR bids consist of an arbitrary number s of atomic bids, e.g. a collection of (disjoint) pairs (S_i, v_i^j), where each S_i is a subset of the items available in M and v_i^j is the bidder's maximum W2P for that bundle (Nisan 2000, p. 11). The OR bids express the bidder's willingness to accept one or more bundles from the collection of pairs. This kind of bidding language is not suitable for goods with substitutionalities; e.g. using plain OR bids for three bundles $b^j(A)$, $b^j(B)$ and $b^j(A \cup B)$ in connection with subadditivity leads to an incorrect allocation unless *dummy goods* are used (Sandholm 2002a, p. 13).

XOR Bids

They have the form (S_i, v_i^j) and express the willingness of a bidder to receive the acceptance of exactly one bundle from the bid set (Nisan 2000, p. 11).[17] XOR bids allow the formulation of all synergy effects (complementarities and substitutionalities). This implies a higher communication effort because the formulation of an additive valuation for m items requires 2^m XOR bids.

OR-of-XOR Bids

OR-of-XOR bids provide a combination of OR and XOR bids that express the bidders' willingness to accept of one or more bundles from a set of XOR bids. The XOR bid restrictions lead to a reduction of the worst case communication complexity to m^2 if the bids' valuation has a symmetric downward slopping structure. A *symmetric downward slopping bid structure* is given if each player values all items as if they were identical, and has a sequence of valuations $v_1^j \geq v_1^j \geq, \ldots, v_m^j$ (Nisan 2000, p. 26).[18] Otherwise the complexity of the bids, increases exponentially with 2^m. As is the case with XOR bids all types of synergies can by expressed by using the OR-of-XOR bidding language (Sandholm 2002a, p. 41).

XOR-of-OR Bids

Another variant of combined OR and XOR bids are XOR-of-OR bids. With XOR-of-OR bids each bidder can submit an arbitrary number of OR bids,

---

[17] XOR bids are also called "boolean bids" in literature (Steinberg 2000, p. 14,17)

[18] Nisan defines bid valuations as *general symmetric valuation* if $v(S) = \sum_{i=1}^{|S|} v_i$ while v_i is the W2P for the additional i^{th} item awarded by the auctioneer. In this notation the *additive valuation* is specified by $v_1 = 1$ for all i. The *single item valuation* is specified by $v_1 = 1$ for all $i > 1$. The *K-budget valuation* is specified by $v_i = 1$ for $i \leq K$ and $v_i = 0$ for $i > K$. The *majority valuation* is specified by $v_{m/2} = 1$ and $v_i = 0$ for $i \neq m/2$ (Nisan 2000, p. 10).

while willing to obtain just one of these bids (Nisan 2000, p. 14). However the XOR-of-OR language is less intuitive than the OR-of-XOR formulation.[19]

OR/XOR Bids

With OR/XOR bidding language bidders can submit XOR-of-OR bids as well as OR-of-XOR bids. This language can be considered as a top level class including both bidding types. Except in some special cases the OR/XOR formula requires exponential communication complexity (Nisan 2000, p. 14).

Applet Bids

Applet bids reflect a special algorithmic view of bid formation. According to the applet bid definition, a bid is a polynomial time computer program b that accepts two inputs, a subset S and a string w, and outputs a number. The value of such an "applet bid" (according to JAVA) or "universal touring machine bid" b on a subset S is $\max_w b(S, w)$ (Nisan 2000, p. 17). There is no practical use for this kind of bid, because no reasonable allocation can be attained by using the applet bid language. However, for the sake of complexity considerations, the definition of such a bidding language is useful.

OR* Bids

OR* bids differ from OR bids by introducing *dummy goods* with a zero value. The introduction of dummy goods avoids the multiple allocation of goods by retaining the expressiveness of the OR bids (Fujishima et al. 1999, p. 3). OR* bids can emulate all bidding languages listed above and are at least as strong as XOR-of-OR and OR-of-XOR bids as well (Nisan 2000, p. 18). The emulation is performed by replacing a bid $(S_1, v_1^j) \, XOR \, (S_2, v_2^j)$ with $(S_1 \cup \{g\}, v_1^j) \, OR \, (S_2 \cup \{g\}, v_2^j)$ where g is a *phantom (dummy)* bid. Nisan shows that any valuation that can be represented by OR-of-XOR (XOR-of-OR, OR/XOR) bids or by using s atomic bids can be represented by OR* bids employing s atomic bids and at most s^2 dummy bids.[20]

Expressing the synergies with auction languages encourages the auction designer to make a trade-off between expressiveness and simplicity. In special cases of restricted bidding languages the CAP can by solved in polynomial time by simply using linear and dynamic programming (see section 5.2.1). This, however, reduces the bidders' ability to express their utility function appropriately and therefore eventually deters an efficient allocation process.

---

[19] Additionally, Nisan (2000, p. 14) shows that for a K-budget valuation a minimal communication complexity of $m^{1/4}$ can be reached.

[20] Later Nisan (2000, p. 19) shows that for a *majority valuation* the minimal communication complexity of the OR* bidding language is $\binom{m}{m/2}$ denoting the limitations of the dummy bid representation

5.1.8 Incentive Compatibility for Combinatorial Auctions

As in the general auction literature, incentive compatibility also plays an important role in the CA domain. The *Vickrey-Clarke-Groves (VCG)* (Vickrey 1963, Clarke 1971, Groves 1973) mechanism achieves incentive compatibility by charging the social-cost generated by the participation of bids that are not based on the true W2P of the bidder to the corresponding agent. Varian & MacKie-Mason (1994) therefore proposed a *Generalized Vickrey Auction (GVA)* procedure as an extension of the classic sealed-bid, second-price Vickrey auction.

Agents a^j $(j = 0, \ldots, n)$ in the GVA bid for consumption bundles b^j including goods $i = 0, \ldots, m$ where good $i = 0$ denotes money the bidder has to pay for the agents' bid bundle $b^j = (b_1^j, \ldots b_m^j)$ of goods i. Each agent has an initial allocation $\bar{b}^j$ and an initial amount of money $\bar{b}_0^j$.

In this context Varian (1995, p. 6) states: The set of all bids except the bid of agent a^j is referred as $b^{\sim j}$. An allocation $b = (b^1, \ldots, b^n)$ is *feasible* if the total amount of each good held (including money) equals the total amount available

$$\sum_{j=1}^{n} b_i^j = \sum_{j=1}^{n} \bar{b}_i^j \tag{5.12}$$

for each good $i = 0, \ldots, m$.

Each consumer a^j has a utility function $u^j(b) + b_0^j$ which is known as quasilinear utility function.

Pareto efficiency is achieved by allocating the goods among the bidders in a way that maximizes the sum of utilities. The Pareto-efficient frontier is described by (Varian & MacKie-Mason 1994, p. 2):

$$\max_{(b_i^j)} \sum_{j=1}^{n} u^j(b) + b_0^j \tag{5.13}$$

$$\sum_{j=1}^{n} b_i^j = \sum_{j=1}^{n} \bar{b}_i^j \tag{5.14}$$

for all bidders $i = 0, \ldots, m$

In the classic Vickrey auction the utility functions denote the difference between the bundle value v^j and the payment made by the bidder. In this resource allocation process, participants are not guaranteed to be willing to reveal their true valuation for the bids for strategic reasons. Therefore the allocation mechanism must be designed in such a way that it motivates the bidders to give a truthful revelation of their private value for the bids. According to Varian (1995, p. 7) the *GVA* mechanism is defined as follows:

1. Each consumer j reports a utility function $r^j(\cdot)$ (which may or may not be the truth) to the center (auctioneer).
2. The center calculates the allocation (b^{*j}) that maximizes the sum of the reported utilities subject to the resource constraints.
3. The center also calculates the allocation $(\hat{b}^{\sim j})$ that maximizes the sum of the utilities other than that of consumer j subject to the constraint that the allocation may not use any of consumer j's resources.
4. Agent j receives the bundle b^{*j} and a payment of $\sum_{i \neq j} \left[r_i(b^*) - r_i(\hat{b}^{\sim j}) \right]$ from the center.

The final payoff of the VCG mechanism results in:

$$u^j(b^*) + \sum_{j \neq i} r_i(b^*) - \sum_{j \neq i} r_i(\hat{b}^{\sim j}) \qquad (5.15)$$

The GVA calculates the utility maximizing allocation, subject to the resource constraints using the valuation function reported by the bidders $r^j(\cdot)$ (which can deviate from true utility function) with (b^{*j}) and without the bidder $(\hat{b}^{\sim j})$. The bidder j is then charged with the costs of welfare-loss caused in the auction process by his possibly not truthfully reported valuation. This is done by calculating the value difference for an awarded bid bundle while using the utility maximizing allocations calculated before. Varian (1995, p. 8) shows that this mechanism makes truth revelation a dominant strategy.

The third term (named K in the following) in equation 5.15 is irrelevant to the bidder j's decision, because he has no influence on this by definition. It is useful for bidder j to reduce the magnitude of the side payment resulting from the GVA mechanism but this has no effect on the mechanism.

The auctioneer wants to maximize his payoff, subject to the resource constraints:

$$r^j(b) + \sum_{j \neq i} r_i(b) \qquad (5.16)$$

The consumer wants to maximize his payoff:

$$u^j(b) + \sum_{j \neq i} r_i(b) - K \qquad (5.17)$$

This means that both parties have the same optimization objective $r^j(\cdot) = u^j(\cdot)$ and it is therefore optimal for the bidders to report their true bundle valuation function v^j to minimize the magnitude of the side payments. Unfortunately, the calculation of the side payments in the GVA requires the $j+1$ fold solution of the $\mathcal{NP}$-hard WDP. For this reason the GVA is computationally intractable for larger problem sizes.

GVA mechanisms play a prominent role in recent CA literature. Many variations of the GVA have been proposed because the GVA seems to be the only mechanism that is capable of effectively addressing the incentive compatibility issues in CAs (Parkes 2001b, Sandholm 1996, Conen & Sandholm 2002, Ronen & Nisan 2000, Nisan et al. 2003, Ausubel & Milgrom 2005). The focus of this thesis, however, is not on incentive compatibility issues due to the fact that automated bidding strategies are used to avoid the incentive problems.

5.1.9 Benchmarking Combinatorial Auctions

Due to the $\mathcal{NP}$-hardness of the CAP, benchmarking is an important issue in CA literature. The trade-off between the solution quality and the execution time of a CAP algorithm can be best measured and compared by solving specifically designed benchmark problems. Additionally, criteria such as worst case and mean execution time play a role for algorithm rating. The average performance of a combinatorial auction algorithm is strongly determined by three factors (Fujishima et al. 1999, p. 551):

  (i) variation in the number of bids
 (ii) variation in the number of goods
(iii) distribution of bids during the auction

While artificially generated CAP benchmarks are easy to design by simulating the variation in the number of bids submitted by the agents and the number goods comprised in those bids, there is almost no real data deriving from real CA applications. The rare combinatorial auction deployment cases described, for instance, by Ledyard et al. (2002), Epstein et al. (2002) or Hohner et al. (2003) are usually not sufficient to model a satisfactory statistical distribution of incoming bids in a real world CA. Three major artificial benchmark problems for the CAP are known from literature. Two of these three benchmarks will be presented in the following list in connection with their underlying distributions:[21]

- *Random:* For each bid a randomly selected number of items $1 \ldots l$ is drawn from the set of m goods $i \in M$ without replacing. The associated W2P is picked randomly from interval $[0, 1]$ (Sandholm 2002a, p. 28).
- *Weighted random:* Like random distribution, but W2P for the bundles is drawn from a $[1, l]$ distribution (Sandholm 2002a, p. 28).
- *Uniform:* Generates a constant number of randomly chosen items in all bids. W2P is picked randomly from a $[0, 1]$ distribution (Sandholm 2002a, p. 28).
- *Decay:* The construction of a new bid is started using a randomly drawn item. Then new items are repeatedly added with probability α. The process is continued either until it stops with probability $(1 - \alpha)$ or all l items have been added to the bid. W2P is drawn from a $[1, l]$ distribution (Sandholm 2002a, p. 28).

---

[21] The relevant sources in literature are Vries & Vohra (2001, p. 54), Sandholm (2002a, p. 28), Fujishima et al. (1999, p. 551) and Andersson et al. (2000, p. 11).

- *Binomial:* In this distribution inclusion probability p for l items out of m goods in a given bid is kept independent of other goods' inclusion: $f_{bin}(l) = p^l(1-p)^{(m-l)}\binom{m}{l}$. W2P for l goods associated with the bids is uniformly distributed in the interval $[l(1-d), l(1+d)]$ with $d = 0.5$ (Fujishima et al. 1999, p. 551) (Andersson et al. 2000, p. 11).
- *Exponential:* Defines the probability distribution $f_{exp}(l) = Ce^{-l/p}$ with $C = \sum_l^m f_{exp}(l) = 1$ that represents the case where a bid for $l+1$ goods appears $e^{-1/p}$ times less than a bid for l goods. W2P for l items in these bids is uniformly distributed between $[l(1-d), l(1+d)]$ with $d = 0.5$ (Fujishima et al. 1999, p. 551) (Andersson et al. 2000, p. 11).

The third benchmark is the *combinatorial auction test suite(CATS)* which has been conceptualized to simulate combinatorial bids that are more appropriate to realistic applications scenarios, such as the combinatorial allocation of routes to truck lanes and railway tracks, bandwidth in networks, as well as real estate, spectrum, and scheduling auctions (Leyton-Brown et al. 2000). It also includes the CA benchmarks described above in so-called 'legacy distributions'.

Leyton-Brown et al. (2000) present benchmark problems based on five real-world situations. As in the previous benchmarks the CATS bid formation consists of two phases: the construction of the bundles out of the relevant goods or services according to the underlying allocation problem and the calculation of the associated W2P coupled with a separately defined distribution function.

- The first group of combinatorial bids that can be generated using CATS describes *paths in space*. Examples of application cases in this domain are CAs for the assignment of routes to truck lanes and railway tracks (Sandholm 1993, Caplice 1996, Ledyard et al. 2002, Cantillon & Pesendorfer 2006). Goods are represented as edges in a nearly planar graph with agents submitting XOR-bids for paths connecting two nodes. The W2P associated with the bundles depends on a utility-cost structure for the edges and their combinations into routes.
- As a second real-world scenario for CA benchmark problems *proximity in space* is considered. This kind of problem structure is found in real estate and spectrum auctions (McMillan 1994, Ausubel et al. 2005). CATS constructs adjacency graphs that are close to the proximity distribution of related real estate objects in cities or cellular telephony frequency band distribution over regions in order to capture this scenario. The valuation of the bundles is calculated using the common values of the goods and the private values of the bidders. Private values are treated as bidders' offsets to the common values. The bundles preferably include goods that are adjacent in the graph structure. Based on the common and private values their valuation is calculated using a superadditive function.
- The CATS function *arbitrary relationships* covers bid structures that cannot be expressed with the more specific scenarios. Bid structure is constructed by using a fully connected graph with weighted edges. Bundle

valuations are computed by using the likelihood of a good's being added to an existing bid structure and the bidders' private valuation.

- The case of *temporal matching* could be regarded as an instance of constructing scheduling problem bids in CATS. Due to the popularity of the related airport take-off and landing slot allocation problem in the CA literature Leyton-Brown et al. (2000) created a special CATS benchmark instance (Rassenti et al. 1982, Ball et al. 2006). The bid packages for the acquisition of starting and landing slots matching the flight schedule of the airlines are formed using real airport data of the US. Again a common value for the slots and a time dependent utility function for the flights that is specific for each airline is used to determine the W2P for the bundles.

- The case of *temporal scheduling* problems that are expressed by using a CAP formulation is most appropriate to the ISIP scenario considered in this thesis. A number of approaches try to address real world scheduling problems in supply chains using CA approaches (Collins et al. 1998, Ygge et al. 2000, Wellman et al. 2001, Elendner 2004, Stockheim & Schwind 2004). CATS defines a benchmark problem for XOR-bids in a single machine setting, where bidders submit resource requests with a decreasing valuation for later job completion. Due to the inflexibility of the CATS benchmark instance with respect to the starting time of the jobs Stockheim & Schwind (2004), extend the CATS benchmark at this point to test an economic scheduling problem in a supply chain environment. Unfortunately, CATS can not deal with multiple resources in the scheduling context and an extension of the existing bid construction model seems to be too costly. For this reason, the author of this thesis decided to create his own benchmark problem for the ISIP scenario. The benchmark will be introduced in section 6.1.1.

Unfortunately, these benchmarks are not widely used to measure CAP algorithm performance, especially in case of CAP heuristics. For this reason a broad evaluation of which circumstances a specific winner determination method might perform best under a benchmark problem has not been conducted yet (Vries & Vohra 2001, p. 24). In any case, the following criteria should be considered when it comes to CAP algorithm evaluation:

(i) Solution quality compared to the exact solution in case of heuristics (Barr et al. 1995, p. 13).
(ii) Calculation time of diverse CAP mechanisms corresponding to the problem size.
(iii) Robustness of the algorithm with respect to the ability to solve the different problem classes.
(iv) Manageability in implementation and application issues.
(v) Information efficiency vs. total return especially for iterative methods.

For deterministic methods the average computation time performance of an algorithm is of special interest, whereas heuristics should also be rated according to their ability to reach solutions near the global optimum. For the use of iterative auctions (v) the number of rounds that are necessary to reach a satisfying equilibrium state should be regarded as an important evaluation criterion. In this case, a trade-off must be made with respect to achievable efficiency improvement associated with an increasing number of rounds (Kwasnica et al. 2005, p. 430).

5.2 Solving the Combinatorial Auction Problem

Seeking an optimal, or at least a near-optimal, solution for the $\mathcal{NP}$-hard CAP (WDP respectively) is one of the most frequently discussed problems in CA literature. This section tries to summarize and shortly explain almost all prominent approaches that deal with efficient WDP solutions. Three main directions can be categorized according to the solution quality and information efficiency of the algorithms: *deterministic procedures*, *heuristic approaches* and *equilibrium methods*.

5.2.1 Deterministic Procedures

Deterministic methods deliver the optimal value $f : X \rightarrow \mathbb{R}$ of the objective function by searching (enumerating) the complete search space explicitly or implicitly. In many cases a-priori knowledge about the problem and constraints is employed to reduce the search space. Well-known examples for these algorithms are *branch-and-bound*, *dynamic programming*, and *linear optimization*. Normally, these methods are only viable for small sizes of combinatorial optimization problems.

Explicit Full Enumeration

Explicit full enumeration calculates the objective function value for the set of all polyhedron vertices that spans the search space of the combinatorial optimization problem and must then determine the optimal vertex (Papadimitriou & Steiglitz 1998, p. 34). The maximum number of vertices in an n-dimensional bounded polyhedron (polytope) grows exponentially with the number of non-negativity and less-or-equal constraints introduced and reaches *intractable computational effort* for advanced problem sizes (see also section 5.1.6).

Integer Programming

The objective of *linear programming (LP)* is to search the extrema of a linear objective function with a finite number of variables regarding a finite number of constraints (Bronstein et al. 2001, p. 870). *Integer programming (IP)* adds the additional constraint that x_j could either be 0 or 1. As seen before, the

CAP is an IP optimization problem due to the indivisibility condition for the bundles in the final allocation. A general formulation of the IP approach following the SPP formalization of the CAP is given in equation (5.18) where $x_j \leq c_i$ is the allocation constraint and d_j denotes the weight vector:[22]

$$\max \sum_{j=1}^{n} d_j x_j \qquad (5.18)$$

$$\text{s.t.}$$

$$\sum_{i=1}^{m} e_{ij} x_j \leq c_i \ \forall \ i \in \mathbb{N}$$

$$x_j \in \{0, 1\} \ \forall \ j \in \mathbb{N}$$

One of the first approaches capable of solving IP problems was the *cutting plane method (CPM)* proposed by Gomory (1958). This procedure initially calculates the result of the LP problem by omitting the $x_j \in \{0, 1\}$ condition (*linear relaxation*). The result naturally does not satisfy the integer condition in most cases. The search space is therefore refined by using cutting planes. A cutting plane is constructed by introducing a new inequality $e_{(m+1)j} x_j \leq c_{m+1}$ that is satisfied by all integral solutions of the original system $e_{ij} x_j \leq c_i$. The simplex method can be used to produce such a new inequality while the new introduced constraint c_{m+1} does not satisfy the integer condition. In a next step c_{m+1} is rounded down to the nearest integer. The resulting inequality $e_{(m+1)j} x_j \leq c_{m+1}$ then is the cutting plane. This cutting plane is added to the LP system and the simplex method is the applied again. Iterating the entire procedure leads to an LP system in which the simplex method will give an optimum that satisfies the integer condition.

A more sophisticated method of solving the CAP has been proposed by Andersson et al. (2000) based on the Garfinkel & Nemhauser (1969) algorithm. Table 5.1 sketches the *Garfinkel-Nemhauser algorithm*:

Table 5.1. Pseudo code of the Garfinkel-Nemhauser algorithm

```
Garfinkel-Nemhauser Procedure
 BEGIN
      PERFORM reduction to obtain Q, P, P₁
          CHOOSE next list
          BUILD the partial solution
          BACKTRACK
      TEST for optimal solution
 END
```

---

[22] Regarding the SPP notation in equation (5.10) the single unit constraint $x_j \leq 1$ is replaced with a multi unit formulation $x_j \leq c_i$.

Before starting the main algorithm the incidence matrix $E = (e_{ij})$ consisting of the column partitions $E = (C_1, \ldots, C_n)$ and the row partitions $R = (R_1, \ldots, R_m)^T$ (see Table 5.2) can generally be reduced.[23] Garfinkel & Nemhauser describe five reduction rules. Only two of them are described here, because they are crucial for solving the CAP effectively.

In the following, an example is used to illustrate the Garfinkel & Nemhauser algorithm. In an allocation auction seven bundle bids have been submitted for four single items: $b^1(1, 2, 4) = 20, b^2(2, 4) = 15, b^3(1, 4) = 10, b^4(2, 4) = 17, b^5(1, 2) = 25, b^6(1) = 1$. For these given bids the value of the objective function results as $Z = 20x_1 + 15x_2 + 10x_3 + 17x_4 + 25x_5 + x_6$ and the corresponding constraint matrix is shown in Table 5.2.

Table 5.2. Constraint matrix for Garfinkel-Nemhauser algorithm

$R_i \cdot C_j$	C_1	C_2	C_3	C_4	C_5	C_6
R_1	1	0	1	0	1	1
R_2	1	1	0	1	1	0
R_3	0	0	0	0	0	0
R_4	1	1	1	1	0	0
Z	20	15	10	17	25	1

Step one of the Garfinkel & Nemhauser (1969) algorithm applies the reduction rules. The first reduction rule (*reduction 1* in the Garfinkel & Nemhauser (1969) algorithm) deletes the 'zero' row vectors R_i. In the example given here this means that row vector R_3 is removed because no bid was submitted for item 3. The second reduction rule (*reduction 4*) searches for linear combinations that can be found in the columns of the constraint matrix. This implies that a column C_t can be deleted if $C_t = \sum_{j \in \Xi} C_j$ (Ξ is an arbitrary subset of $\{1, \ldots, n\}$ with $t \notin \Xi$) and $e_t = \sum_{j \in \Xi} e_j$ (e_j are weights). In our example C_4 and C_6 are a linear combination of C_1 and have to be deleted because they provide a lower joint return ($17 + 1 = 18$) than C_1 (20). Let $Q = (i | i \in S$ and R_i not deleted by reduction), $P = (j | C_j$ and not deleted by reduction) and $P_1 = (j | C_j = 1$ deleted by reduction) be the results of step one of the Garfinkel & Nemhauser algorithm (see Table 5.1 line 2).

We now consider step two of the algorithm. A list is generated for each item (rows of Table 5.3) and each bid (columns of Table 5.3) is stored exactly in one list. Given an ordering of the rows, each bid is stored in the list corresponding to the first row that occurs. Within these lists, the bids (columns in Table 5.3) are sorted in descending order according to their return value (Andersson et al. 2000, p. 3-4).

---

[23] The incidence matrix is also called constraint matrix in the following because the integer incidence condition imposes constraints on the optimization problem.

The search for the optimal solution (line 3-5 in Table 5.1) works as follows:

1. CHOOSE the first list as active list and select the first bid from the list.
2. Add the first disjoint bid (rows not yet selected) from the active list into the current solution. Repeat the BUILD process until either *(i)* the current return can not exceed the best solution found so far or *(ii)* no further bid can be added: check if this is a valid solution and the best so far.
3. BACKTRACK the current solution by replacing the latest chosen bid with the next valid bid in the active list and goto the BUILD step 2. If all bids have been selected from the active list back up recursively. If no backtracking can be done terminate and TEST for optimal solution.

Table 5.3. Sorted constraint matrix after reduction

$R_i \cdots C_j$	C_5	C_1	C_2	C_3
R_1	1	1	0	1
R_2	1	1	1	0
R_4	0	1	1	1
Z	25	20	15	10

Applied to our example this means: Select C_5 and check for further disjoint bundles. Because no column is disjoint to C_5, condition *(ii)* is satisfied and the solution can be tested for optimality without employing the BACKTRACK procedure. The solution is $x_5 = 1$ and $x_1 = x_2 = x_3 = x_4 = x_6 = 0$ with $Z^* = 25$ feasible and optimal. A variant of the Garfinkel & Nemhauser algorithm is used in the IP solver package CPLEX 6.5[24]. Andersson et al. (2000) tested the algorithm with benchmarks formulated by Fujishima et al. (1999) and Sandholm (2002*a*). The bid distributions for the CAP were *random, weighted random, uniform, decay* all from Sandholm (2002*a*) and *binomial, exponential* used by Fujishima et al. (1999).[25] In most of the cases the application of CPLEX 6.5 performed better than the algorithms presented by Sandholm (2002*a*) and Fujishima et al. (1999).

Linear Relaxation

A common way to solve an IP problem is to employ *linear relaxation (LR)* to transform the integer problem into an LP problem. The integer optimization problem can then be solved using a *cutting plane (CP see previous section)* or a *branch-and-bound (B&B see following section)* method. Additionally several heuristics operate with LR (see section 5.2.2). When LR is used the CAP

---

[24] www.cplex.com
[25] See section 5.1.9 for the benchmark definitions.

allocation is allowed to include partitions $\wp$ of the bids $b^j(S)$ while $\wp \cdot (S)$ denotes the resulting proportion of the bundle and $\wp \cdot p$ the corresponding fractional bundle price.

The relaxed CAP can be expressed as follows:[26]

$$\max \sum_{S \subset M} b(S)\, x_S \qquad (5.19)$$

$$\text{s.t.}$$

$$\sum_{i \in S} x_S \le 1 \;\forall\, i \in M$$

$$x_S \ge 0 \;\forall\, S \subset M$$

Table 5.4 shows the pseudo code for a LR-based CAP algorithm:

Table 5.4. Pseudo code of a linear relaxation algorithm

```
Linear Relaxation Procedure
  BEGIN
      compute an optimal basic solution Z_LP for LP(·)
      satisfy all bids j for which x_j = 1
  RETURN
```

The result of the linear relaxation for the CAP can be used to calculate the upper bound in B&B procedures (Gonen & Lehmann 2001, p. 2). In general LR on its own is not capable of providing a feasible CAP solution. Some CAP formulations, however, can be solved directly by employing LR. This is on the one hand the *fractional combinatorial auction* which requires only parts $\wp$ of the desired bundles to be filled $x_s \in [\wp, 1]$ (Nisan 2000, p. 21).[27] On the other hand the auctioneer can specify rules which only allow constraint subsets $C \subseteq M$ as permitted bid combinations (Rothkopf et al. 1998, p. 1136). The family of all permitted combinations is denoted by $\mathcal{P}$, defined by $\mathcal{P} = \{C \subseteq M | C \text{ is a permitted combination}\}$ with $|\mathcal{P}| \le 2^m$.

Nisan (2000, p. 24) describes three cases of bid restrictions that directly lead to an optimal solution while employing LR:

- *Linear ordered*: The sets of items i can be linearly ordered $M = \{1 \ldots m\}$ such that all bids C form a consecutive subrange $C = \{k \ldots l\}$ of items. A special case of linearly ordered bids is *hierarchical nested bids* that form a tree structure. A family of sets $\mathcal{P}$ forms a tree structure if for all $C, C' \in$

---

[26] See Nisan (2000, p.21) and Vries & Vohra (2001, p.19).

[27] This kind of CA is often used in financial applications, where the filling of a minimum portfolio share is required in the final allocation (Polk & Schulman 2000, Bossaerts et al. 2002).

P, $(C \cap C')$ is one of $\emptyset, C, C'$. That means that every two sets in the family are disjoint or one is a subset of the others (Rothkopf et al. 1998, p. 1138).[28]

- *Mutual exclusion*: The use of XOR-of-OR bidding language with *single item bids* (singletons), where a bidder can win at most a single item always leads to an optimal allocation in connection with LP winner determination. A special case of mutual exclusion is *downward slopping symmetric bids* (see section 5.1.7).
- *Substructure*: Auctions that have integer LP solutions can be combined and modified while retaining this property. E.g., it can be shown that the sum of two auctions, each having integer LP solutions, yields integer solutions as well (Nisan 2000, p. 26).

The applicability of bid constraints to assure the integrality of LP solutions for the WDP is very limited because in most practical environments such restrictions could not be fulfilled.

Primal-Dual Algorithms

Primal-dual algorithms are often employed to solve combinatorial optimization problems. The optimization problem is formulated as a primal and a dual optimization problem that could be described as follows.[29]

The initial primal problem is:

$$\max Q(x) = \sum_{j=1}^{n} d_j x_j \tag{5.20}$$

$$\text{s.t. } \sum_{j=1}^{n} e_{ij} x_j \leq c_i \ \forall \ i \in \mathbb{N}$$

$$x_j \geq 0 \ \forall \ j \in \mathbb{N}$$

The complementary dual problem is:

$$\min G(w) = \sum_{i=1}^{m} c_i w_i \tag{5.21}$$

$$\text{s.t. } \sum_{i=1}^{m} e_{ij} w_i \leq d_j \ \forall \ j \in \mathbb{N}$$

$$w_i \geq 0 \ \forall \ i \in \mathbb{N}$$

---

[28] The *tree structure* allows the separate calculation of the winners in the subproblems that can be used to determine the winning bids for the whole problem in dynamic programming (see also section 5.2.1).

[29] See Dantzig & Thapa (1997, p. 135), Papadimitriou & Steiglitz (1998, p. 104), Sandholm et al. (2005, p. 375) and Vries & Vohra (2001, p. 29).

Equation (5.20) denotes the relaxed CAP as described above. A common interpretation of the primal LP reads as follows: select activities from $\sum_{j=1}^{n} d_j x_j$ with respect to resource constraints $\sum_{j=1}^{n} e_{ij} x_j \leq c_i$ such that the total return maximizes.

Within the dual problem (5.22) a non-negative vector $w = (w_1 ... w_m)$ is searched minimizing the prices of the individual resources while satisfying the minimal program $\sum_{i=1}^{m} e_{ij} w_i \leq d_j$. For this reason the values of w_i are called *shadow prices* (Dantzig & Thapa 1997, p. 129 seq., p. 171 seq.).

The computational efficiency of the primal-dual method is based on the fact that the algorithm searches iteratively for the solution of a sequence of restricted primal and the dual problems while using the *complementary slackness condition (CSC)*, instead of searching for an optimal primal (dual) solution directly. The restricted primal or dual problem is often much simpler to solve than the full primal (dual) problem (Parkes 2001a, p. 89). The CSC expresses the relationship between the primal and the dual values of the LP formulation that are necessary and sufficient for the optimality of the LP solution. According to the CSC a feasible primal (dual) solution is only optimal if $G(w) \leq G_{max} = Q_{min} \leq Q(x)$ is satisfied. If $G(w) = Q(x)$ vectors x and w represent the optimal solution for both problem formulations. The use of primal-dual methods for the search of optimal solutions is therefore reduced to a simple test of the CSC.

In CA settings bid prices represent a feasible dual solution and the agents' bid bundle structure denotes the integrality conditions for the primal problem, which is sufficient to test the CSC and adjust the solution towards optimality. Primal-dual algorithms are therefore consistent with the decentralized information inherent in distributed agent-based systems (Parkes 2001a, p. 89). Normally a primal-dual algorithm maintains a feasible solution $G(w)$ of the dual problem and tries to find a feasible solution of the restricted primal problem $Q(x)$. Because this is not possible in general unless the dual solution is optimal, a relaxed version of the primal problem is formulated:

"Compute a feasible primal solution $Q(x)'$ that minimizes the violation of CSC with dual solution $G(w)$" (Parkes 2001a, p. 90).

A primal-dual-based auction method therefore comprises the following steps (Parkes 2001a, p. 91):

(i) Maintain a feasible solution (prices).
(ii) Compute a feasible primal solution (provisional allocation) to minimize violations with CSC given agents' bids.
(iii) Terminate if all CSC are satisfied ("are the allocation and prices in competitive equilibrium?").
(iv) Adjust the dual solution towards an optimal solution, based on CSC and the current primal solution ("increase prices based on agent bids").

This approach is often used in *iterative combinatorial auctions*.

Dynamic Programming

A different method of solving the CAP in an exact manner is to refine simple search strategies which would normally lead to a full enumeration of the problem in the search space (Papadimitriou & Steiglitz 1998, p. 448). One way to produce such a refinement is to employ *dynamic programming (DP)* which represents a multiple stage search strategy (Bellman 1957). The solution is obtained in a step-by-step process while calculating subproblems of the original optimization task following a *bottom-up* principle. Bellman's optimality principle thereby ensures that the calculation of the higher level optimization problems can be performed without recalculating the results of the subproblems. This leads to a significant reduction in computational effort. An important prerequisite for a WDP calculation in polynomial time by employing DP is a *unimodular structure* of the constraint matrix in the CAP formulation. This implies that only WDPs with a special bid structure can be solved by using DP techniques. A quadratic matrix $(e_{ij}) \in \mathbb{R}$ is unimodular if its determinant $det(e_{ij}) = \pm 1$. In this case, all vertices of the corresponding bounded polyhedron will be integer values (Vries & Vohra 2001, p. 298).

- **Rothkopf's Algorithm**

  The solution approach proposed by Rothkopf et al. (1998) uses the unimodularity property to identify feasible bid structures that have a polynomial runtime for the CAP. The bidders may formulate bundles only according to a specific *bid tree structure*. Three variants of bid trees are distinguished in this context:

  (i) *Nested structures* can be formed if the set of allowed bid subsets $\mathcal{P} = \{C \subseteq M | C \text{ is a permitted combination}\}$ has the property that $(C \cap C') \, \forall \, C, C' \in \mathcal{P}$ is a subset of C, C' or $\emptyset$ (see linear relaxation in previous section). The resulting CAP constraint matrix is unimodular and the problem can be solved employing a DP algorithm.

  (ii) *Cardinality-based structures* are bid combinations where multiple bids for a single good stand in competition with *large bid combinations*. Rothkopf et al. show that there is a threshold for the minimal number of bids in a large bid combination that guarantees that the resulting constraint matrix of the CAP is unimodular. The underlying CAP problem can then be solved in polynomial time by using DP.

  (iii) *Geometric structures* can be used for bid construction if the underlying valuation of the items in the bundles has a well-defined geometrical structure. E.g., in bidding for spectrum licenses (McMillan 1994) only adjacent regions may be included in a bundle. The application of such geometrical construction principles likewise leads to a unimodular structure of the CAP constraint matrix and can therefore also be solved using DP algorithms.

  Rothkopf et al. present a DP formulation to solve these types. As already discussed in section 5.1.7, such constraints reduce the expressiveness of the bidding language significantly and lead to a decline in allocation efficiency.

- **CASS Algorithm**
  Fujishima et al. (1999) introduce an allocation mechanism named *combinatorial auction structured search (CASS)*. The algorithm enhances a *depth-first search* strategy by the reduction of the search space while pruning the search path whenever it is clear that a better combination can not be found in the corresponding branch of the search tree (see also next section). The basic structure of the algorithm is depicted in Table 5.5 where α denotes a pointer to the first element in a collection $\mathcal{A}$ of bids and ω stands for a pointer to the last. The set of bids in the *recent* allocation is described as $\mathcal{R}$. The CASS algorithm can be regarded as a modified version of the Garfinkel-Nemhauser algorithm with respect to the use of *bins*[30] to identify conflicting bids and *caching* to avoid repeated calculation of sub-problems. The caching process is organized as a DP approach, leading to the categorization of CASS in this section. However, the CASS algorithm tends to be a hybrid of the DP and the B&B technique addressed in the next section.[31]

Table 5.5. Pseudo code of CASS algorithm

```
CASS Procedure
 INITIALIZE the collection A:= (α,...,ω)
 ADD bid α to the allocation if it does not conflict with R
     WHILE more disjoint bids can be added to R DO
         INCREMENT α
         UPDATE best revenue and allocation R found so far
         WHILE ω is NOT in current allocation AND NOT first bid DO
             DECREMENT ω
         ELSE REMOVE bid ω, SET α := ω + 1, RETURN
     ELSE RETURN
 END
```

The performance of the CASS algorithm has been tested in comparison to CPLEX 6.5 for the bid distributions described in the first part of section 5.1.9. CASS turns out to provide a superior performance for binomial and exponential distributions in terms of computation time (Andersson et al. 2000). The CASS algorithm has the *anytime* property, meaning that the computation can be stopped at any time yielding a suboptimal approximation result. The distance to the optimal solution can not be determined in such cases.

---

[30] The bins contain the bids in such a way that the goods included in the bids are disjoint. By including only one single bid from a bin into the set of allocated bids the number of unfeasible allocations is reduced (Fujishima et al. 1999, p. 3).

[31] The distinction between DP and B&B methods is often difficult due to their parallels (Lawler & Wood 1966, p. 714).

Branch-and-Bound

Another way to attain exact solutions of the CAP is to use *branch-and-bound (B&B)* procedures in connection with a search tree. The basic idea of a B&B algorithm, that can be considered as a *top-down* search, is to divide a hard problem into subproblems (nodes of the search tree) that also represent results of the main problem (Lawler & Wood 1966, p. 699, 703). In the worst case, B&B methods lead to a full enumeration of all partial problems (nodes ν) (Vries & Vohra 2001, p. 304). B&B methods can be characterized by five elements $(\mathcal{T}, \mathcal{B}, \mathcal{L}, \mathcal{U}, \mathcal{S})$.[32] $\mathcal{T}$ stands for *termination rules*, that determine whether a further search from a node in the search tree can lead to an optimal solution. $\mathcal{B}$ denotes the *branching rules* in the subdivision process of a partial problem if there is neither optimality nor termination for the current solution. Principles of *exclusion* and *inclusion* thereby guarantee the formation of a tree structure of the search paths. *Lower* and *upper bounding rules* $(\mathcal{L}, \mathcal{U})$ are used to decide whether a node and its adjunct subtree should be included in or excluded from the search path. A node will be excluded if there is no way to find a better solution than the current one by searching the subtree. Lower and upper bounds are calculated by using the evaluation function $f(\nu)$ which should have the monotonicity property to guarantee the correctness of the bounds. Lower and upper bounds can be generated using various methods. One of these methods is to use LR of the CAP as described in section 5.2.1. *Search strategies* $\mathcal{S}$ finally decide which order is used to enumerate the nodes in the search process. Table 5.6 shows the general scheme of a B&B process.

Table 5.6. Pseudo code of a branch-and-bound algorithm

```
B&B Procedure
 BEGIN
      INITIALIZE the collection A:=(X)
      WHILE |∪(A)| ≠ 1 DO
         A:= BRANCH(A) if the sub-problem can not be solved directly
         A:= PRUNE by using cost/return function and lower/upper bound
 END
```

The B&B procedure starts with the definition of subsets of the problem set $\mathcal{X}$ in a collection $\mathcal{A}$. This collection $\mathcal{A}$ is split by branching in the search tree, and elements (nodes ν) that are dominated by the evaluation function $f(\nu)$ are pruned (excluded from further examination). Dominance is identified using the evaluation function $f(\nu)$. The evaluation function operates on costs for minimization problems and on returns for maximization problems while running through the search path. Such an evaluation function can for instance be the composed term $f(\nu) = g(\nu) + h(\nu)$, where $g(\nu)$ denotes the cost/return of the search path from the starting node in the search tree to the current node

---

[32] The argumentation follows Zhang (1999, p. 10 seq.).

and $h(\nu)$ identifies an estimator for the cost/return from the current node to the target node. This procedure is denoted as A^*-*algorithm* and guarantees to find the optimum, as long as $f(\nu)$ never overestimates the cost of the path $f(\nu) \leq f^*(\nu)$ (lower bound) in a minimization problem (Hart et al. 1968). In the case of maximization problems, as is the case for the COP in a forward CA, the estimated revenue is regarded as an objective function and the A^*-algorithm must not underestimate the maximum achievable return $f(\nu) \geq f^*(\nu)$ to attain the exact solution (Russel & Norvig 2003, p. 97).

Three major search strategies S are common in B&B methods: *best-first search, breadth-first search*, and *depth-first search*. Extensions and combinations of these methods are feasible. *Best-first search (BFS)* generates a tree by further expanding that node from the set of uninvestigated nodes that has the most promising value $f(\nu)$. The A^*-algorithm can be regarded as a BFS method because it expands the node with the minimal estimated cost (or maximum revenue for a maximization problem). *Breadth-first search (BrFS)* is similar to best-first search and generates a tree by further expanding the node with the minimal distance to the root node (Zhang 1999, p. 11). Both algorithms have memory requirements that grow exponentially with the depth of the search tree (Russel & Norvig 2003, p. 74). In contrast to this, depth-first search algorithms have a memory requirement that has only linear growth with increasing depth of the search tree. For this reason, and due to the particular relevance for efficient CAP solutions, the *depth-first search (DFS)* method will now be illustrated in more detail. In the following outline we will regard return optimization only, because the COP of a forward CA is a maximization problem.

Table 5.7. Pseudo code of the depth-first search B&B algorithm

```
DFS-B&B Procedure
   DFS(ν)
          GENERATE all κ children of ν : ν₁, ν₂, ..., νκ
          EVALUATE and SORT the children of ν in decreasing order of return
          FOR i = 1 to κ DO
                IF (return(νᵢ) > z)
                      IF (νᵢ is a goal node)
                            z ← return(νᵢ)
                      ELSE DFS(νᵢ)
                ELSE RETURN
          RETURN
   END
```

With the start of the DFS-B&B algorithm (see Table 5.7 for a recursively formulated variant) the value z is set to zero for the root node ν and new child nodes are generated. Then the algorithm investigates the node that has the maximum depth in the search tree (or that has been generated last according to the LIFO method). If a node ν is found that has higher return than the current value of z according to $f(\nu)$ the value of z will be replaced with the current value of $f(\nu)$ and will be used as a lower bound in the further evaluation. If a node is found that has a return less than or equal to z, the subsequent branch of the tree is cut. The computational complexity of the DFS algorithm grows exponentially with the depth δ and the branching factor ρ of the search tree (number of possible child nodes per node) $\mathcal{O}(\rho^\delta)$.

- **Sandholm's IDA* Algorithm**
  Sandholm (2002a) presents a *branch-on-items* search technique for CAs that uses a graph representation depicted in Figure 5.4. While employing branch-on-items search, the branching rule $\mathcal{B}$ is formulated as follows: "What bids should this item be assigned to?" (Sandholm 2005, p. 338). Sandholm uses a modified A^*-algorithm called *iterative deepening A^* (IDA*)* that is a hybrid of a DFS-B&B algorithm and an A^*-algorithm (Korf 1985). The IDA* algorithm performs a DFS while using the evaluation function $f(\nu) = g(\nu) + h(\nu)$ of the A^*-algorithm. A lower bound *f-limit* for the maximum obtainable return $f(\nu)$, – that is the sum of bid prices in the context of the forward CA –, is estimated for the search process. Whenever the current estimated return is below this bound $(g(\nu) + h(\nu) < \text{f-limit})$ the subsequent branch of the search tree is pruned. If no solution is found, the lower bound of the IDA* process was too optimistic, *f-limit* is increased, and the BFS algorithm is repeated (Sandholm 2005, p. 344).

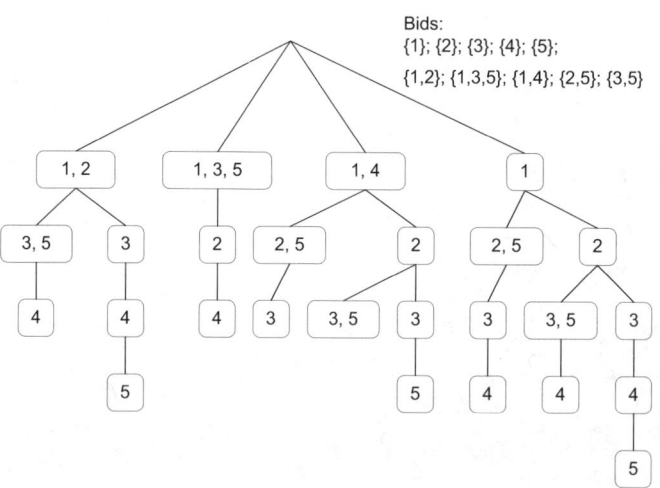

Fig. 5.4. Search tree for Sandholm's IDA* algorithm

The Sandholm algorithm uses the IDA* mechanism to build a search tree where the bids included in the path do not include items that have already been allocated on this path (see Figure 5.4 for illustration from Sandholm (2002a, p. 17)).[33] The search makes use of the fact that the order of the bids in the branches of the tree does not matter to achieve a significant reduction of the branching factor ρ in the search process. A DFS algorithm is used in connection with the IDA* mechanism to guarantee fast generation of child nodes in lexicographic order of the bid's set of items. Moreover the Sandholm algorithm performs a preprocessing phase that assures that: (i) only the highest bids for a combination are kept in the bidtree, (ii) provably noncompetitive bids are removed from the tree[34], (iii) bids are decomposed into connected components such that no item is shared by bids from different sets and (iv) non-competitive tuples of bids are marked as not to be considered in the same search path. The Sandholm algorithm has been benchmarked by using the bid distributions that have been described in section 5.1.9 for a varying number of items and bids.[35]

- **CABOB Algorithm**
  Sandholm et al. (2005) propose another algorithm called *combinatorial auction branch on bids (CABOB)* which is faster than the IDA* strategy proposed in Sandholm (2002a). The algorithm uses a *branch-on-bids* search instead of the branch-on-items strategy.[36] With this strategy the question for the branching rule $\mathcal{B}$ is: "Should this bid be accepted or rejected". This $\mathcal{B}$ rule constructs a binary search tree whose size is polynomial in bids N and exponential in items M. According to Sandholm (2005, p. 341) this is a desirable property because the auctioneer can usually control the numbers of items in an auction, but does not want to restrict the number of bids submitted. In addition to the search tree a *bid graph* Γ represents the bids that remain to be allocated.
  Table 5.8 gives a pseudo code representation of CABOB. Before each search process, the graph Γ is investigated for the special cases COMPLETE and NO_EDGES. In the COMPLETE case the graph Γ is complete and only remaining bids can be accepted. The highest remaining bid is accepted, $f(\nu)$ is updated and the branch is cut off. In the NO_EDGES case, none of the nodes is connected to the others. All bids can be accepted, $f(\nu)$ is updated and the corresponding branch is cut off. A further improvement of CABOB is achieved by using LR to calculate upper and lower bounds for the use in the branching rules $\mathcal{B}$. The upper bound $\mathcal{U}$ is calculated by

---

[33] The branch-on-items structure has the drawback that dummy bids have to be introduced for those items that have received no one-item bids (Sandholm 2002a).

[34] This means that the combinations of bids with a summarized revenue that is dominated by other bid combinations are removed from the tree.

[35] Unfortunately no comparison to other CAP algorithms has been performed.

[36] This avoids the drawback of the branch-on-items strategy that dummy bids must be introduced for items with no one-item bids.

Table 5.8. Pseudo code of CABOB algorithm

```
CABOB Procedure
  CABOB(ν_i)
    apply special cases COMPLETE and NO_EDGES
    GENERATE all κ children of ν : ν_1, ν_2, ..., ν_κ in graph Γ
    apply special case INTEGER
    EVALUATE κ's and sort them according to SORTING_RULE
    FOR i = 1 to κ DO
        IF (U > z > L)
            IF (ν_i is a goal node)
                z ← return(ν_i)
            ELSE CABOB(ν_i)
        ELSE RETURN
  END
```

employing the LP relaxation Z_{LP} of the remaining WDP. In the special case of an INTEGER value of Z_{LP} the remaining solution is also an integer, $f(\nu)$ can be updated if it is greater than the solutions previously found and no further branches are investigated below this node. In addition, a lower bound $\mathcal{L}$ is computed at each search node based on the revenue that the remaining items can contribute to the total return. This procedure drastically reduces branching in the search tree at low extra computational cost. For the identification of candidate nodes in the branching phase of the algorithm, Sandholm et al. (2005, p. 380) propose six principles (SORTING_RULES). Two of these sorting methods should be discussed here in more detail: The first principle is called *normalized shadow surplus (NSS)* and uses the shadow prices w_i of the items i from the dual formulation of the remaining subproblem as a proxy for the price of a bundle $p^j = \frac{v_j - \sum_{i \in S} w_i}{\log(\sum_{i \in S} w_i)}$. The algorithm branches on the bids with the highest p^j. The second principle is defined as *one-bids* and describes a branching behavior that prefers bids with x_j close to 1. The basic idea behind this rule is that the more of a bid is accepted in the LP the more likely it is to be competitive. Sandholm et al. (2005, p. 381) show empirically that a combination of NSS and *one-bids* rules which is dynamically changed during runtime performs best for the node selection in the branching process.

CABOB was tested against the IP solver CPLEX 7.0 employing instances of the benchmarks presented in section 5.1.9 in terms of computing time for different numbers of bids. For the random distribution CABOB performs slightly better than CPLEX, whereas for the weighted random distribution both algorithms are roughly equal. In the case of the decay distribution CABOB clearly outperforms CPLEX, while for the uniform distribution CPLEX has a computing time that is significantly shorter. Additionally two problem types from CATS (paths and temporal matching) were used to test CABOB, both indicating the predominance of CPLEX.

5.2.2 Heuristic Approaches

A common way to deal with $\mathcal{NP}$-hard problems like the CAP is to accept a trade-off between the solution quality and the calculation effort by using *heuristic approaches*.

"*A heuristic method (also called an approximation algorithm, an inexact procedure, or simply, a heuristic) is a well-defined set of steps for quickly identifying a high-quality solution for a given problem, where a solution is a set of values for the problem unknowns and 'quality' is defined by a stated evaluation metric or criterion. Solutions are usually assumed to be feasible, meeting all problem constraints. The purpose of heuristic methods is to iden-tify problem solutions where time is more important than solution quality, or the knowledge of quality.*"(Barr et al. 1995, p. 2)

Heuristics are defined as global optimization methods that either use stochastic variations of the method itself or randomization while modeling the objective function $f : X \to \mathbb{R}$ (Barr et al. 1995, p. 11). Heuristics are not guaranteed to find the global optimum, they should however provide a solution of high quality (local optima) by using a-priori knowledge about $f : X \to \mathbb{R}$.

Local Search

Local search methods are a well known heuristic and could be used for the most combinatorial optimization problems (Papadimitriou & Steiglitz 1998, p. 454 seq.). Formally expressed, such a method works by using a series of solutions $\{\beta_0, \beta_1, \ldots, \beta_N | \beta_i \in \mathcal{M}(x)\}, N \in \mathbb{N}_0^+$ beginning with an arbitrarily created solution $\beta_0 \in \mathcal{M}(x)$ while $\mathcal{M}(x)$ is the set of randomly created feasible solutions. In the i-th solution β_i is developed by small modifications of β_{i-1}, where the solution quality should improve over the iteration steps. The method stops if no further improvement of quality is attainable or a time limit is reached. The result is a feasible solution $\beta' \in \mathcal{M}(x)$ which represents a local optimum for x with respect to the proximity $f : \mathcal{M}(x) \to f(\mathcal{M}(x))$ if the value of the objective function $f(n)$ is at least as high as the neighboring target function values.

For $f(n)$ the following must apply:

(i) $\beta \in f(\beta)$ for each $\beta \in \mathcal{M}(x)$,
(ii) if $\beta_j \in f(\beta_i) \, \forall \, \beta \in \mathcal{M}(x)$ then is $\beta_i \in f(\beta_j)$ and
(iii) $\forall \, \beta \in \mathcal{M}(x)$ exists a positive k and $\gamma_1, \gamma_2, \ldots \gamma_k \in \mathcal{M}(x)$, such that $\gamma_i \in f(n_i), \gamma_{i+1} \in f(\gamma_i)$ for $i = 1, \ldots, k - 1$ and $\beta_i \in f(\gamma_k)$.

According to (iii) it is possible to reach β_j from β_i $(i \neq j)$ for all feasible solutions $\beta \in \mathcal{M}(x)$ by employing the *proximity relationship*. With respect to the CAP this means that β_j can be reached from β_i (or S_1, S_0) by associating a new bid b with β_i, while the old bid b' can be replaced. It is therefore true that $\beta_j = \beta_i \cup \{b\} \setminus \{b' \in \beta_i : b' \cap b \neq \emptyset\}$ for an arbitrary $b \notin \beta_i$ (Hoos

& Boutilier 2000, p. 3). Local search methods can be characterized by the state transition method employed in the search process. Several types of local search methods that are most relevant for the solution of the CAP will be described in the following.

Gradient Descent Method

The *gradient descent method*[37] is the most simple strategy for local search (Zhang 1999, p. 27). Search starts with a randomly selected starting point and proceeds with the point in search space that has the lowest cost (or highest reward). The search process is repeated until a local minimum (maximum) is found. The search result can be improved by restarting the search with other starting points and by selecting the best of the local solutions *(multiple restart schema)*.

Table 5.9. Pseudo code of the gradient descent algorithm

```
Gradient Descent Procedure
 BEGIN
     SET the best solution n* ← ∅
         WHILE (no stopping criterion is met) DO
             GENERATE initial solution n'
                 WHILE (f(n') < f(n)) DO
                     n ← n'
                     GENERATE the neighboring solutions of n : N(n)
                     SELECT the best neighbor n' in N(n)
                     IF (f(n') < f(n*))
                         n ← n*
                 RETURN
             RETURN
 END
```

Table 5.9 shows the pseudo code of the gradient search method. The optimal solution n^* is empty at starting time. The starting solution is initiated n' and its neighborhood N determined. The value $f(n)$ of the objective function is calculated for each neighbor $N(n)$. The search transition follows the minimum (maximum) value (or steepest descent, steepest ascent) and n^* is replaced by the improved value of the objective function until a stopping criterion is met. This can, for example, be a calculation time limit or an upper (lower) bound for the value of the objective function. Gradient search is mostly used as an auxiliary optimization technique in CAP heuristics (Likhodedov & Sandholm 2005, Zurel & Nisan 2001). It also plays a role in CAP optimization methods that are based on Lagrangean relaxation (Vries & Vohra 2001).[38]

---

[37] For maximization problems the method is called *gradient ascent method*.
[38] See section 5.2.3 for further discussion.

Greedy Heuristic

Another simple approach to finding a CAP solution by local search is to employ a *greedy procedure*. Greedy algorithms sort the search points (nodes ν) of an optimization problem according to an optimization function. Then they construct the problem solution by greedily selecting the nodes with the highest objective function values into the solution set (Papadimitriou & Steiglitz 1998, p. 282).

Table 5.10. Pseudo code of the greedy algorithm

```
Greedy Procedure
 Greedy(π)
      ORDER the nodes νᵢ by decreasing value of π
      SET R← ∅, R∈A
          WHILE no solutions AND A≠ ∅ DO
              greedy choice νᵢ ∈R
              IF νᵢ∪R feasible allocation
                  REMOVE νᵢ from A
                  INSERT νᵢ in R
          RETURN
 END
```

Table 5.10 shows the algorithmic formulation of a greedy algorithm. The greedy strategy starts with an empty solution set $\mathcal{R}$ and extends it stepwise by highest ranking element of the candidate list $\mathcal{A}$. The ordering in the candidate list functions as follows: The higher the value z_i of node ν_i, the higher is the ranking π in the ordered list $\mathcal{A}$ and thus the probability of being selected first by the greedy algorithm. Some important rankings π are (Mu'alem & Nisan 2002): (i) *value ranking* $\pi(\nu_i) = z_i$ and (ii) *density ranking* $\pi(\nu_i) = z_i/\varpi_i$ (Sandholm et al. 2005, p. 380). In the first case (i) the costs (return) is the only deciding factor. The second case (ii) is useful if the number of items or quantity of goods ϖ_i is different per node[39] and performs a value standardization (value weighted by number of items or quantity of goods). Greedy algorithms normally provide suboptimal solutions. The allocation quality for the CAP achieved with such a greedy procedure strongly depends on the sorting criteria for the bids and the bidders' utility function. The computational complexity of greedy procedures is usually in the range of $\mathcal{O}(n^2 \log n)$ (Zhang 1999, p. 107), the asymptotically best complexity for greedy CA algorithms is $\mathcal{O}(n/(\log n)^2)$ where n denotes the number of bids (Lehmann et al. 2005, p. 310).[40]

---

[39] This is usually the case for most combinatorial auctions.

[40] For further complexity considerations for SPP heuristics see Fisher & Wolsey (1982), Vries & Vohra (2001, p. 24), and Lehmann et al. (2005, p. 309 seq.).

Zurel & Nisan (2001) present a combined greedy and an approximate LP procedure for the CAP. The method is divided into two phases:

In the *first part* of the algorithm a near optimal allocation is calculated using a *primal-dual formulation* of the CAP. The algorithm stops if the primal solution of the LP is at a distance of ε from the dual solution of the LP to save computation time, meaning that the *CSC* must only be partially fulfilled. The result of the first part of the algorithm is a vector of estimated prices[41] $\overline{p_i} = (\overline{p}_1 \ldots \overline{p}_N)$ for the goods i in the allocated bundles and a vector of their fractional allocation $\overline{\wp_j} = (\overline{\wp}_1, \ldots \overline{\wp}_M)$ to each bundle $b^j(S)$.

Table 5.11. Pseudo code of the Zurel-Nisan greedy algorithm

```
Zurel-Nisan Greedy Procedure
  INITIALIZE A(π) ← InitialOrder(p̄₁,...,p̄ₙ,℘̄₁,...℘̄ₘ)
     REPEAT UNTIL no improvement is made
        LocallyImprove(π)
        Greedy(π)
     RETURN A(π)
  END
```

The estimated prices of the items and their fractional allocation in the bundles are the input for the *second part* of the Zurel & Nisan algorithm which is depicted in Table 5.11. In a first step the bids are sorted into a list according to the ranking π by ordering them primarily by descending values of $\overline{\wp}_j$. If $\overline{\wp}_j = 0$ for a bid, the ordering criterion π is the ratio of the bidders' valuation for a bundle divided by the estimated prices of the goods in the bundles $v^j \setminus \sum_{i \in S} \overline{p}_i$. The rationale behind π is similar to the one-bids ordering strategy in CABOB. The higher the fractional allocation $0 \leq \wp_j \leq 1$ of goods to a bundle the more likely it is to be competitive. The resulting allocation is further ameliorated by a *local improvement* method in a second step, which means that bids are exchanged randomly in the resulting π and maintained in π' if the overall return increases. This process accords with the multiple restart schema of the gradient descent method. The local improvement process is iterated until no further improvement is attainable. The results of the Zurel & Nisan algorithm has been tested with nearly all bid distribution benchmarks formulated in section 5.1.9. The solution quality was usually within 1% of the optimal result, and in no case more than 4% away from optimal. The algorithm is fast enough to run in less than a minute on problems with thousands of items and tens of thousands of bids (Zurel & Nisan 2001, p. 135).

---

[41] The estimated prices in the Zurel & Nisan algorithm refer to the concept of shadow prices. Shadow prices can be considered as the price of an added item, which is the upper bound to how much that item will actually contribute to the auction's revenue. See sections "primal-dual algorithms", "Lagrange relaxation and subgradient method" and 5.2.3 for further explanation.

Simulated Annealing

A more elaborate technique for local search that is still one of the most important *stochastic search* methods of solving $\mathcal{NP}$-hard problems (Catoni 1998)
is *simulated annealing (SA)*. It was introduced by Kirkpatrick et al. (1983)
who established an analogy between the minimization of a cost function (or
maximization of a return function) for an optimization problem and the slow
cooling process of solid matter.

Table 5.12. Pseudo code of a simulated annealing algorithm

```
Simulated Annealing Procedure
 INITIALIZE the collection A:= (ν₁,...,νₖ)
     SET R← ∅, R∈A
         FOR Θ = Θstart TO Θstop DO
             RANDOMLY SELECT νᵢ ∈A and νⱼ ∈R
             ΔE = return(R\νⱼ ∪ νᵢ) - return(R)
             IF rand(0,1) ≤ pacc(e^{ΔE/Θ})
                 IF R ∪ νᵢ feasible allocation
                     INSERT νᵢ in R
                 ELSE IF (R\νⱼ ∪ νᵢ) feasible allocation
                     REMOVE νⱼ from A and INSERT νᵢ in R
         DECREASE Θ by ΔΘ
 END
```

Table 5.12 shows the pseudo code of an SA algorithm. The algorithm starts
with the initialization of a set of nodes $\mathcal{R}$ that will carry the bids included in
the final allocation and the collection of bids $\mathcal{A}$ available to the auctioneer.
The algorithm adds new bids to the allocation as long as the constraints are
not violated. If the allocation reaches the constraints, bids are exchanged with
probability $p_{acc} \sim (e^{\frac{\Delta E}{\Theta}})$ where ΔE is the differential increase (decrease) in
the solution resulting from the bid swap and Θ is a temperature factor that
controls the stochastic optimization. The temperature is decreased stepwise
(annealed) to enforce a stabilization of the process. Allowing for temporary
deterioration in the SA process makes it possible to leave local optima and to
find the global maximum (Russel & Norvig 2003, p. 115).

Hoos & Boutilier (2000) present a *combinatorial auction search algorithm
(CASANOVA)* which substitutes bids sorted by revenue in a greedy allocation, controlled by stochastic exchange probabilities that are modified by time
factors denoting the age of the bid in the current allocation. The CASANOVA
algorithm has been tested for the uniform, decay, and exponential distribution described in section 5.1.9 in a comparison with the CASS algorithm.
CASANOVA outperforms CASS on large problem instances with fixed cutoff
times over all distributions and generally finds optimal solutions for smaller
problem instances. On uniform distributions, CASANOVA finds optimal solutions faster than CASS, whereas for the others improvement is small.

Genetic Algorithm

Genetic algorithms (GA) and related hybrids are a common variant of stochastic search capable of dealing with COPs (Goldberg 1989a, Wendt 1995). GAs start exploration at a multitude of points in the search space and proceed with the evolutionary process by mutation, reproduction, and selection (see also section 4.5.2). Figure 5.13 depicts the pseudo code of a GA that can be employed to solve the CAP. The GA starts by constructing an *initial population* $\Pi(0)$ of candidate solutions (set of bids in the resulting allocation $\mathcal{R}$ for the CAP). Traditionally, this population is generated randomly and covers the entire range of the search space. The way an individual is represented in the GA is very important for the manner in which a solution can be manipulated by the GA and is therefore relevant for the performance and quality of the algorithm. A solution can be, for instance, represented by a string of bits that indicates the inclusion of a bid in the allocation set for the CAP. However, sophisticated problem codings have been developed (Rothlauf et al. 2002).

Table 5.13. Pseudo code of a genetic algorithm

```
Genetic Algorithm
 INITIALIZE population Π(0)
        REPEAT
                Evaluate(Π(i))
                Π' ← SelectForVariation(Π(i))
                RecombineAndMutate(Π')
                Evaluate(Π(i), Π')
                Π'' ← SelectForReproduction(Π(i), Π')
                Π(i + 1) ← Reproduce(Π''); i:=i+1
        UNTIL termination condition satisfied
 END
```

In the optimization phase the algorithm enters a loop which aims at the stepwise improvement of the population's fitness. The iterations are called *generations* in analogy to biological evolution. Each individual is *evaluated* in the stochastic optimization process by employing an associated *fitness function* that is formulated according to the optimization criteria (this is the auctioneer's return for an allocation in the CAP). At each generation a proportion Π' of the existing population $\Pi(i)$ is randomly *selected for variation* based on the results of the evaluation process. New child individuals are generated by *recombining* parent individuals (mating the information about the representation of different individuals) and *mutation* randomly changing information about the genes (see also Figure 4.20 in section 4.5.2 for illustration). In the next step of the GA the individuals in $(\Pi(i), \Pi')$ are evaluated and selected for *reproduction and survival*. Better fitness leads to a higher probability of reproduction and survival in this stochastic selection process. This process is often designed in such a way that a small proportion of less fit solutions is selected to avoid premature convergence on poor solutions. Due to the process of selecting the fittest individuals the population size of the current generation has diminished. The generation is replenished by reproducing the fittest individuals Π'' in the current generation and goes on to the next iteration step. The GA *terminates* if a predetermined numbers of generations (operations) has been reached or no further fitness improvement has been achieved over the last generations.

Easwaran & Pitt (2000) propose a GA to match ISIP tasks to service providers in a CA setting. The algorithm tries to find a cost minimizing matching of the task requests to the service providers, while the providers either cooperate or not, and therefore generate various cost profiles depending on the combination. For this purpose, services, providers, and task allocation are coded in the GA representation. The fitness function of the optimization problem is based on selection criteria for the services like e.g. required bandwidth and server capacity. By applying the genetic optimization technique to the combinatorial matching problem, Easwaran & Pitt achieve a good performance for small problem sizes in terms of computation time and solution quality. For larger problem sizes (number of services and providers), however, the GA performed poorly in terms of solution quality compared with the CABOB algorithm that has also been applied to the minimization problem while the GA's computation time remained on a low level.

5.2.3 Equilibrium Methods

So far, only one-shot CAs have been considered in the survey on CAP solution techniques that is provided in this chapter. The introduction of *simultaneous iterative auctions (SIA)* where bidders are allowed to bid simultaneously on bundles of items in multiple rounds offers the possibility of reducing the algorithmic and communication complexity of the auction process by making use of the temporal price and allocation information gained in the equilibrium tâtonnement process. The advantage of such auctions is based on the fact that an allocation which is formed in one round may not be optimal in the Pareto-efficiency sense unless all alternative preferences have been expressed by the bidders,[42] because bidders can revaluate the allocation that has been attributed by the auctioneer in the previous round and reformulate their bids appropriately for the next round (Parkes 2005, p. 42). The main problem of the SIA is the tendency to encourage bidders to engage in strategic bidding behavior, e.g. an agent could exaggerate its W2P to deter other bidders from further bidding as early as in the first round and then be able to submit a lower offer in the second round that corresponds to its true W2P (see section 5.1.8). Many SIA designs therefore integrate the GVA mechanism into the auction protocol to reduce the *incentive compatibility problem* (Parkes & Shneidman 2004, Parkes & Kalagnanam 2005). Another problem of the SIA

---

[42] This is known as the preference elicitation problem (Hudson & Sandholm 2004).

is the provision of appropriate *pricing information* to auction participants to enable them to adjust the W2P associated with their bids. This problem is often addressed by using a relaxed version of the IP approach and employing LP duality to calculate the corresponding shadow prices (Xia et al. 2004, Bichler et al. 2005). SIAs are often called *decentralized auctions* because the price building process takes place in interaction with the distributed participants (Parkes 2005, p. 43). The following sections portray prominent representatives of this auction type.

VSA Allocation Design

The *virtual simultaneous auction (VSA)* proposed by Fujishima et al. (1999) is the first decentralized auction protocol that will be discussed in this survey. The VSA protocol starts with the submission of the bid bundles to the auctioneer by the participants. For each bid a virtual bidder is created that is trying to win all the goods comprised in a bundle b^j for a price p based on its W2P and a fixed monetary budget. The VSA starts with an empty allocation and prices of zero and is repeated until an optimal allocation is attained or a time limit is reached. The VSA has three phases per auction round:

- In the *virtual auction phase* the available items are assigned to the bids according to the bidders' W2P. If a bidder receives all desired goods, he is declared temporary winner. Otherwise the bidder has to return the goods to the auctioneer.
- In the *refinement phase* each of the losers is examined in random order to see whether making the agent a temporary winner and consequently another agent a loser would increase the total revenue. If this is the case, the list of current winners will be updated.
- In the *update phase* the current highest prices of the goods are updated to the level given by the price of the current winning bundle. The prices for unallocated goods are set to zero.

The strategy of the virtual bidder is not to bid in a current round, if it has been the temporary winner in the previous round. The losing bidders, however, use the current highest good prices to calculate the W2P that is required to obtain the desired bundle. If the calculated price for a bundle exceeds the agent's budget, the bid is not submitted. If the bundle price is below the agent's budget, the remaining budget is equally divided into shares and added to the prices of the goods contained in the bid. Under certain conditions the VSA is able to find an optimal allocation. This is the case if no virtual bidder bids have been submitted in a round. Otherwise, the VSA ends after a predefined number of rounds. The performance of the CASS algorithm has been tested in comparison to the CASS mechanism (see section 5.2.1) in connection with the binomial and exponential distribution described in section 5.1.9. In terms of computation time and solution quality the performance of CASS is superior to that of VSA except for some anytime solutions.

Lagrange Relaxation and Subgradient Method

The method of *Lagrange relaxation (LAR)* of an integer problem is often used instead of the linear relaxation. The solution of an integer problem relaxed such a way provides an upper and a lower bound for the optimal value of the original problem and can be used for branch methods. The Lagrange method is based on the idea that the underlying integer problem can be solved by transforming one or more side conditions into the objective function by using *Lagrange multipliers*. Non-feasible solutions are penalized in the objective function in proportion to the level of non-feasibility. The precondition for finding an optimum is given by the partial derivations being zero. Formally expressed the SPP in equation (5.10) is transformed into the relaxed problem,

$$Z(\lambda) = \max \sum_{j \in V} d_j x_j + \sum_{i \in M} \lambda_i (1 - \sum_{j \in V} e_{ij} x_j) \tag{5.22}$$

$$\text{s.t. } 1 \le x_j \le \forall j \in V \tag{5.23}$$

where $\lambda = (\lambda_1, \ldots, \lambda_M)$ is the vector of Lagrange multipliers with $\lambda_i \ge 1$ and $Z(\lambda)$ is the solution of the relaxed SPP with $Z \le Z_{LP}$.[43]

The calculation of $Z(\lambda)$ for a given λ can be accomplished easily by using the following transformation:

$$\sum_{j \in V} d_j x_j + \sum_{i \in M} \lambda_i (1 - \sum_{i \in V} e_{ij} x_j) = \sum_{j \in V} (d_j - \sum_{i \in M} \lambda_i e_{ij}) x_j + \sum_{i \in M} \lambda_i \tag{5.24}$$

To determine $Z(\lambda)$ it is necessary to set $x_j = 1$ if $(d_j - \sum_{i \in M} \lambda_i e_{ij}) > 0$ and to zero otherwise. If $Z(\lambda)$ is piecewise linear and convex, the following holds (Vries & Vohra 2001, p. 31):

$$Z_{LP} = \min_{\lambda \ge 0} Z(\lambda) \tag{5.25}$$

The determination of the optimal vector $\lambda^* = (\lambda_1^*, \ldots, \lambda_M^*)$ by using equation (5.25) and the optimal value Z_{LP} of relaxed LP formulation is defined as the *dual problem*. $Z(\lambda)$ can be used instead of Z_{LP} to calculate the upper bound for the next branch decision in a B&B algorithm (see section 5.2.1) or as an instrument for the worst-case behavior analysis in the development of heuristics (Fisher 1981, p. 2).

If we find an optimal vector λ^* that solves equation (5.25), the calculation of Z_{LP} is easy. It is therefore necessary to determine a vector λ^* that is capable of solving $\min_{\lambda \ge 0} Z(\lambda)$. This can be accomplished by using a *subgradient algorithm*. Let λ^t the value of the Lagrange multiplier in the tth iteration of the subgradient algorithm and the vector $\gamma^t = (E x^t - \mathbf{1})$ the subgradient at point

---

[43] The description of this section is based essentially on Vries & Vohra (2001).

λ^t. Let σ^t further be the step width in the algorithm. The following iteration rule defines a sequence of Lagrange multipliers λ^t with $\lambda^{t+1} = \lambda^t + \sigma^t \gamma^t$ that converges to the optimal solution x^t associated with $Z(\lambda^t)$:

$$\lambda^{t+1} = \lambda^t + \sigma^t (E\,x^t - 1) \tag{5.26}$$

It is important that $\lambda^{t+1} > \lambda^t$ for each i such that $\sum_j e_{ij} x_j^t > 1$. For any constraint that is violated the penalty term has to be increased at each round (Vries & Vohra 2001, p. 32). The subgradient method can be derived from the gradient descent method introduced in section 5.2.2 because at each stage it adjusts the multiplier λ to reduce the function value. Due to the fact that the ordinary gradient method is only capable of dealing with steady and monotonous functions and $Z(\lambda)$ is only partially differentiable the subgradient has to be employed for local search. For an appropriate step size the procedure can be shown to converge: $\sigma^t \rightarrow 0$ while $\sigma^t \rightarrow \infty$ and $\sum_{t=0}^{\infty} \sigma^t$ diverges. However, it is not assured that the algorithm converges in a finite number of steps for all instances. An additional stopping rule has to be implemented and the step width σ has to be adopted for fast convergence behavior. LAR is not guaranteed to find the optimal solution, but it provides the optimum of the relaxed problem.[44]

The subgradient method can be used for price adjustment in the WDP process. The expression $(d_j - \sum_{i \in M} \lambda_i e_{ij})$ is of special interest in this context, because it can be interpreted as the W2P d_j for a bid $b^j(S)$ on the one hand and a *Walrasian price vector* λ on the other hand. If the highest W2P for a bid in d_j exceeds the summarized prices of the goods in a bundle the equation

$$\sum_{j \in V} (d_j - \sum_{i \in M} \lambda_i e_{ij}) > 0 \tag{5.27}$$

is true and the bundle is temporarily allocated to the bidder. For this purpose the auctioneer does not need to know d_j. It is sufficient if the bidders communicate whether they accept the bids at the announced price vector λ. If the auctioneer identifies conflicting bids he can adjust the prices of the goods in the competing bids by using the subgradient algorithm and repeat the auction process (Vries & Vohra 2001, p. 33).

RSB Allocation Design

One of the first CA designs was created by Rassenti et al. (1982) for the procurement of airport take-off and landing time slots by airlines. The allocation process is split into a primary market using the classical sealed bid one-shot CA and a second market organized as an open-outcry exchange to give the

---

[44] The percentage of convergence to the optimal solution can be calculated by using the difference of the bounds and the value of the dual problem: $(Z - Z(\lambda^*))/Z(\lambda^*)$ (Fisher 1981, p. 4).

agents the opportunity to reallocate time slots that do not fit after the single round auction process. The *Rassenti-Smith-Bulfin mechanism (RSB)* for the combinatorial allocation and pricing process is based on the LAR. The procedure employs *shadow prices* to solve the pricing problem, whereas the original integer problem is decomposed into two *pseudo-dual* programs to construct a bid acceptance/rejection rule. A traditional IP relaxation delivers non-integer x_j values that do not indicate bid acceptance/rejection ($x_j = 0, x_j = 1$) clearly (see LR in section 5.2.1). The pseudo-dual decomposition transforms the IP side conditions using Lagrange-relaxation (Sandholm et al. 2005, p. 375). The Lagrange multipliers λ (shadow prices) are used to determine the upper and lower bounds for the acceptance/rejection decision. The first dual formulation for the bid acceptance condition provides the acceptance shadow price λ_A and a set of upper bound prices for the single resources (goods) i. The second dual formulation for the bid rejection condition provides the rejection shadow price λ_R and a set of lower bound prices for the resources i. Figure 5.5 following Rassenti et al. (1982, p. 407) depicts the acceptance shadow price λ_A and the rejection shadow price λ_R. In this way the abscissa a denotes the resources constraints given by e_{ij} and the ordinate b represents the bid prices (W2P) for the bundles $b^j(S)$. If the sum of the single upper bound resource prices exceeds the acceptance shadow price the bid is accepted for sure $b^j(S) \geq \sum_{i \in M} \lambda_A e_{ij}$. By analogy, if the sum of the single lower bound resource prices falls below the rejection shadow price ($b^j(S) \leq \sum_{i \in M} \lambda_R e_{ij}$) the bid is also definitely rejected. The space between the two slopes corresponds to the core of the IP problem.

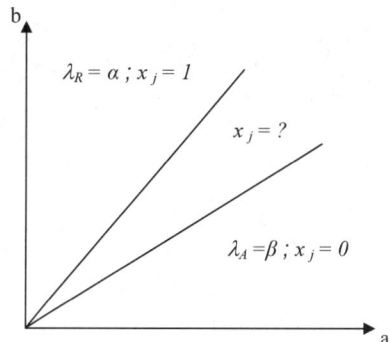

Fig. 5.5. Categorizing of combinatorial bids with RSB

The RSB reduces the complexity by enforcing a price adjustment only in case of $b^j(S) \neq \sum_{i \in M} \lambda_i e_{ij}$ (Steinberg 2000, p. 11). The resulting allocation calculated with the LAR, however, is not exactly the same as that which the IP formulation yields. The RSB was only tested in settings with human bidders and is therefore not comparable to using the benchmarks of section 5.1.9.

AUSM Allocation Design

The *adaptive user selection mechanism (AUSM)* was designed to solve a priced scheduling problem for Space Shuttle payloads, where non-convexities in the task valuations occur due to mission execution uncertainties (Banks et al. 1989, p. 6). AUSM is conceived as an ascending iterative auction mechanism, where bidders can continuously bid for bundles of resources (X, Y) on two markets (shuttle A, B). Highest bids leading to a feasible allocation are posted on a bulletin board. In the case of a new bid submission a reallocation takes place, if the resource constraints are exceeded and the acceptance of the new bid bundle leads to a higher total return. Formally expressed, this definition of the WDP optimization objective can be written as: "Accept bid $\hat{b}^j(S)$ if $\hat{b}^j(S) \geq \max_{j \in N} b^j(S)$ and $S \subseteq M$". The winning bids on the bulletin board (analogous to the price indicators in a *clock auction* which is an auction type where the current prices of the single goods are permanently visible to all participants during the auction phase) denote the lower bounds the bidders have to surpass to reach their desired allocation. The targeted *simple equilibrium* is reached if no more bundle S^* exists in the bid set $\mathcal{A}$ for any participant $j \in N$ such that he could further maximize his utility v^j. That means that $\xi^j(S^*) \cap \{S | v^j(S) > v^j(S^*)\} = \emptyset \ \forall j \in N, S \subseteq M$ is true, where $\xi^j(S^*)$ denotes the feasible allocations according to the budget restrictions. The AUSM mechanism faces the *threshold problem* that is solved by a modification called *adaptive user selection mechanism with queue (AUSMQ)*: "smaller" bidders can post their bids in a queue which is employed to maintain their bids until complementary bids of other "smaller" bidders can be used to fill the missing items and eventually outbid the package bidder if the summarized W2P of the "small" bidders exceeds the packet price. On the one hand AUSM rules are more intuitive for the bidders to handle than the mechanism of a one-shot CA where the bidders face the *preference elicitation* problem.[45] On the other hand, AUSM distributes the WDP across the bidders, while the auctioneer is simply required to verify that a new provisional allocation is better than the current allocation that is formed from the bids in the queue (Parkes 2005, p. 67). Though *decentralized auctions* reduce the CAP calculation load for the auctioneer, they might favor technologically advanced bidders that can handle larger COPs to optimize their bids. However, empirical experiments using the AUSM (or AUSMQ) mechanism in the two-resources two-markets example showed an efficiency of 78% (81% for AUSMQ) compared to 78% with the implementation of an iterative VCG mechanism. Banks et al. define the 100% benchmark as an *ex ante Pareto optimal allocation* that could be achieved by exactly solving the CAP associated with the scheduling problem without knowledge about later mission disturbances. Other authors report even higher levels of allocation efficiency for AUSM and AUSMQ (see Kwasnica et al. (2005, p. 430,431), Bykowsky et al. (1995, p. 21) and next section).

---

[45] This is the case for the most iterative combinatorial auctions.

RAD Allocation Design

The *resource allocation design (RAD)* mechanism proposed by Kwasnica et al. (2005) is an iterative auction that uses the AUSM approach and adds some elements from the *simultaneous multiple round (SMR)* spectrum auction initiated by Milgrom (2000). Bids are submitted synchronously in each round. The auctioneer determines the winners at each round and transmits the temporary allocation to the participants. The winner determination is done according to the return maximization objective of the CAP and remains valid as long as the current winning bid is not outbid. If a bid was valid in period t and is not overbid in period $t+1$ it remains unchanged. This implies $\mathcal{W}_t^+ \subseteq \mathcal{A}_{t+1}^j \forall\, j \in N$ with $\mathcal{W}^!$ being the set of winning bids, $\mathcal{W}^-$ the set of losing bids, and $\mathcal{A}^j$ the set of all temporary available bids. Unlike AUSM, RAD employs a modified version of the *eligibility stopping rule* for the bids which can be described as $\mathcal{E}_t^j \leq \mathcal{E}_{t-1}^j$ with $\mathcal{E}_0 = |M|$ and $\mathcal{E}_t^j = |\{M|b^j(S)x_s \in \mathcal{A}_t^j \wedge x_s = 1, S \subseteq M\}|$. This eligibility stopping rule ascertains that bidders cannot bid for more goods $\mathcal{E}_t^j$ in the current round than in the previous round $\mathcal{E}_{t-1}^j$ and ensures that participants only bid for goods they are willing to buy right from the start *(use-it-or-lose-it rule)*. The auction stops at the end of t if $\sum_{j \in N} \mathcal{E}_t^j \leq |M|$. In combination with a *minimum bid increment* ι this rule leads to a faster convergence of the auction process. However, it is important to find the appropriate bid increment ι to guarantee an efficient allocation in iterative auctions. RAD uses a flexible adaption of bundle prices in the next rounds based on the pricing of the single goods in the bundles. The minimal bundle prices p^j for the next auction round are determined based on *shadow prices*. These are calculated by using LAR of the dual CAP (Kwasnica et al. 2005, p. 426)[46]:

$$\min_{\lambda_i^t, Z, \varsigma_i} Z \tag{5.28}$$

$$\text{s.t.} \sum_{i \in S} \lambda_i^t x_s = v^j \;\forall\; b^j(S) \in \mathcal{W}^+, \lambda_i^t \geq 0 \tag{5.29}$$

$$\sum_{i \in S} \lambda_i^t x_s + \varsigma_i = v^j \;\forall\; b^j(S) \in \mathcal{W}^-, 0 \leq \varsigma_i \leq Z \tag{5.30}$$

The price vector λ_i^t for the goods i at time t is adjusted iteratively until equations (5.29 and 5.30) fit approximately. ς_i is employed as a slack variable in this process and often described as reduced price (Bjørndal & Jørnsten 2001).

$$p^j \geq \sum_{i \in S} x_s(\lambda_i^t + \iota) \tag{5.31}$$

The RAD performance was evaluated with respect to efficiency, total return and process time. Five persons participated in the RAD experiments with ten

---

[46] The procedure includes further steps to eliminate the discrepancy between the individual item prices p_i and bundle prices p^j in the bid (Xia et al. 2004).

objects to bid on 25 combinations of goods. RAD's mean efficiency measured against the utility maximizing allocation was 90.42% compared to 94.0% with AUSM and 66.95% in the non-combinatorial SMR. The average return was 79.4% of the maximum achievable return in RAD compared to 96.25% in SMR and 71% in AUSM.[47] The auction time was shortest for RAD compared to AUSM and SMR (Kwasnica et al. 2005, p. 432).

AkBA Allocation Design

The ascending k-bundle auction *(AkBA)* mechanism designed by Wurman & Wellman (2000) concentrates on a dynamic price tâtonnement process during the auction to attain a general market equilibrium and addresses especially the *combinatorial auction pricing problem*. The procedure is based on a linear relaxation of the integer programming formulation where elements of the optimal solution from the dual LP problem are used to determine equilibrium prices. AkBA is an iterative auction with *nonlinear anonymous prices* that can be categorized as a *progressive combinatorial auction (PCA)*. In a PCA the bidding rules require that each iteration's bid be an improvement over the previous. This enforces a convergence of the auction towards an equilibrium.

AkBA employs the pricing principle of the *k-double auction* proposed by Satterthwaite & Williams (1989). The k-double auction is a generalization of the classical first price auction which is especially used in connection with multi-unit auctions. To determine the winning bids in a k-double auction the bids are sorted according to the W2P of the bidders. Then the first L bids are selected from the sorted list while K is the number of selling bids. The price range between the Lth winning bid and the subsequent $(L + 1)^{th}$ bid represents the range where supply meets demand in the double auction. The k-double auction computes a clearing price that is a ratio of two boundary prices. The upper bound p_L is the Lth price, the lower bound p_{L+1} is the $L+1^{th}$ price. The k-double auction sets the clearing price to the linear combination $p_c = k\,p_{L+1} + (1-k)p_L$ with $0 \le k \le 1$. The pricing rule of the k-double auction is used to construct a *price lattice* containing the price ranges for the temporary winning bids calculated from the dual LP problem. The properties of the k-double auction mechanism guarantee that none of settled prices within the lattice bounds produce excess supply (or demand) (Bao & Wurman 2003).

The AkBA allocation and pricing mechanism works as follows:

(i) After bid submission temporary winners of the auction are determined by using an arbitrary winner determination algorithm.

---

[47] Due to the small number of experiments (42 in total), however, the results might not be very significant (Kwasnica et al. 2005, p. 428).

(ii) In a second step equilibrium price bounds $\overline{p_i}$ and $\underline{p_i}$ and are calculated for the temporary allocated goods using the dual values of the WDP's linear relaxation.

(iii) In the third step a price lattice is calculated using the price bounds from step two $p_c = k\,\underline{p_i} + (1-k)\,\overline{p_i}$.

The auction begins with a quote setting all lattice prices to zero. After the bid submission the tentative allocation is calculated and the price lattice updated. After each round the bidders are told their allocation and the associated prices. The bidders must increase their bids to win the bundles they have not been awarded yet in the temporary allocation (*ascending price rule, PCA*); in addition participation in the next round requires a minimal bid increment (*beat-the-quote rule*). The auction clears if a period of bidding inactivity occurs. A*k*BA has been tested in an environment with myopically bidding agents. These agents increase bids on bundles that are not currently winning, if they can further improve a real surplus derived from their utility function at prices. Compared with the outcome of an optimal allocation calculated by using an A*-algorithm for the WDP the agents reached 99.8% allocation efficiency in the simulations (Wurman & Wellman 2000, p. 28).

*i*Bundle Allocation Design

With the *iterative bundle auction (iBundle)* Parkes (1999) presents an SIA using the primal-dual optimization principle. *i*Bundle is designed to handle *nonlinear anonymous prices* as well as *nonlinear personalized prices* in several variants. The basic *i*Bundle algorithm is formulated in Table 5.14.

Table 5.14. Pseudo code of the *i*Bundle algorithm

```
iBundle Procedure
  BEGIN
      p_ask(b_i^j) := 0
      REPEAT for each round t
          collect best-response bids with p_bid(b_i^j)
          calculate (provisional) return maximizing allocation
          update (personalized) ask prices p_ask(b_i^j)
      UNTIL termination condition
  END
```

In each round t bidder agent j submits the *best response set* of desired good bundles, which denotes the set of all bundles and corresponding personalized prices that maximize the bidders' utility, either as OR or XOR formulation. The bids are collected and reformulated by the auctioneer in a first step, so that a closed XOR term exists for each bidder j. The provisional return maximizing allocation is then determined iteratively starting with price $p_{ask}(S)$

set to zero for the bundles in round one.[48] The auctioneer uses these prices $p_{bid}(S)$ (solution $G(u)$ of the dual problem) to calculate the provisional allocation in a second step (solution of the primal problem $Q(x)$). Based on this provisional allocation the prices of the bundles are recalculated to adjust the dual solution $G(u)$ towards an optimal solution that can be idetified by the CSC. Figure 5.6 illustrates this tâtonnement process for the primal-dual algorithm in allusion to Parkes (2001a, p. 93) (see also section 5.2.1).

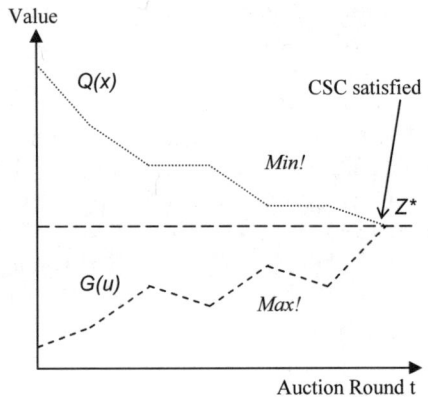

Fig. 5.6. Price tâtonnement process with primal-dual algorithms

Three different variations of iBundle exist, depending on the pricing mechanism applied: iBundle(2) uses anonymous and non-linear prices *(second order prices)*; iBundle(3) uses individual and non-linear prices *(third order prices)* and iBundle(d) switches *dynamically* between anonymous and individual non-linear prices (second and third order prices) to avoid discriminatory bundle prices whenever possible (Parkes 1999, p. 149). The bundle prices $p_{ask}(S)$ are communicated to the bidders and new bids are only accepted in the next round if they are positioned above this price. The auction repeats until at least one of two **termination conditions** is reached: All agents are *happy* or all agents submit the same bids in successive rounds. An agent is defined as happy ($j \in happy$) at the end of round t if it received exactly one bundle placed in its XOR bid.

The price update in iBundle(2) which employs anonymous prices $p_{ask}(S)$ depends on all bids of the agents that are unhappy. $p_{ask}(S)$ is increased by the auctioneer if the maximum rejected price for the bundle from an unhappy agent ($j \in unhappy$) is within a minimal bid increment ι of the current ask

---

[48] The prices submitted with the bundles by the bidders are called *bid prices* $p_{bid}(S)$ in the iBundle context and the minimum prices set by the auctioneer at each round are consequently called *ask prices* $p_{ask}(S)$.

price (Parkes 1999, p. 151). With $p^j_{bid}(S)$ being the bid of agent j for bundle S the new ask price $p_{ask,\,t+1}(S)$ can be calculated by using equation (5.32):

$$p_{ask,\,t+1}(S) = \max\left(p_{ask,\,t}(S),\ \max_{j \in unhappy} p^j_{bid}(S) + \iota \right) \qquad (5.32)$$

If the individual pricing rule (5.33) of iBundle(3) is employed, a set of ask prices is calculated for each agent. The individual prices $p^j_{ask,\,t+1}(S)$ are increased by the auctioneer based only on the bids $p^j_{bid}(S)$ of agent j if this agent is unhappy and its bids exceed the current $p^j_{ask,\,t}$ price by ι:

$$p^j_{ask,\,t+1}(S) = \max\left(p^j_{ask,\,t}(S),\ p^j_{bid}(S) + \iota \atop j' \in unhappy \right) \qquad (5.33)$$

Using this pricing and a sufficiently small ι, iBundle finds an equilibrium that fulfills the CSC $(Q(x) = G(u) = Z^*)$ in a finite number of steps.

iBundle has been tested with respect to allocation efficiency and auctioneer's return for the benchmark sets (weighted) random, uniform, and decay that have been described in section 5.1.9. Additionally Parkes (1999) tested the auction against the problem sets used in the AUSM and RAD cases (Kwasnica et al. 2005, Banks et al. 1989, Ledyard et al. 1997). The comparability, however, is poor due to a lack of experiments with other CA mechanisms for the benchmark sets mentioned first, and due to the fact that the original AUSM and RAD designs use human bidders in contrast to proxy-agents used as bidders in the Parkes (1999) setting.

5.3 Design of Combinatorial Auctions

The previous two sections of this chapter introduced a broad range of problems and proposed solution mechanisms accompanying the application of CAs for ARA and dynamic pricing. Motivated by the complexity of these problems and approaches, the rest of this chapter tries to outline a design framework for CAs after sketching various measures to reduce CA complexity. In a further step a validation approach will be presented for the auction mechanisms developed using the criteria presented in the framework. The argumentation of this section follows mainly Schwind (2005a) and König & Schwind (2005).

5.3.1 Complexity Reduction in Combinatorial Auctions

A very prominent problem in CAs is the *incentive compatibility* issue (see section 5.1.8). The GVA provides a mechanism that is capable of ensuring truthful bidding by the auction participants, the Vickrey-Clarke-Grooves process. Yet it also requires a computation of the $\mathcal{NP}$-hard WDP for each bidder and is therefore intractable for most applications unless heuristics are used

to solve the WDP (see section 5.2.2). A further difficulty when applying the
GVA in a one-shot auction is the *pricing problem* described in section 5.1.3.
For these reasons iterative auctions are increasingly used to reduce complexity
in auction design. Iterative auctions diminish the *preference elicitation* effort
by making sure that bidders only need to evaluate optimal strategies along the
equilibrium pricing path. Moreover, the *pricing problem* is reduced by using
an appropriate sequential bidding and winner feedback process. Kalagnanam
& Parkes (2003, p. 42) classify two types of complexity reduction in SIAs:

- *Bid space restrictions* using bidding languages: The auction designs RAD
  (Kwasnica et al. 2005) and AUSM (Banks et al. 1989) only allow OR-bids,
  while participants in other combinatorial auctions operate with XOR-bids
  (Parkes 1999, Parkes & Ungar 2000, Ausubel & Milgrom 2002). The basics
  of this approach have been sketched in section 5.1.7.
- *Feedback of current prices* in the bidding rounds to guarantee consistent
  equilibrium prices. While AUSM uses linear prices in the provisional allocations, *i*Bundle (Parkes 1999) and A*k*BA (Wurman & Wellman 2000)
  employ nonlinear prices for the bundled goods. The RAD (Kwasnica et al.
  2005) auction and the ascending proxy auction (Ausubel & Milgrom 2002)
  work without explicit price feedback to the bidders (see section 5.2.3).

Before we turn to the implications of these auction mechanism simplifica-
tions we should look briefly at existing *electronic auction* frameworks.

5.3.2 Framework for Combinatorial Auction Design

At the moment only a few designs and realizations of auction frameworks explicitly dedicated to the realization of combinatorial allocation processes exist.
Most of them are derived from simpler concepts for electronic markets, like e.g.
Kasbah (Chavez & Maes 1996). The *Michigan Internet AuctionBot* created
by Wurman et al. (1998) is such an auction platform. The associated system
design is universally configurable and allows the detailed modeling of the auction flow control. *eMediator* is a versatile e-commerce framework proposed by
Sandholm (2002*b*) that is capable of dealing with multiple variants of combinatorial auctions and exchanges and allows the definition of price-quantity valuation functions (discount auction) as well as the use of proxy-bidding agents.
Besides these features, eMediator permits the withdrawal of bids by paying
penalty charges *(leveled commitment)* (Andersson & Sandholm 1998, Kaelbling et al. 2000). Special attention is paid to the suitable selection of the
bidding language and pricing mechanism dependent on the auction scenario
(Sandholm 2002*b*, p.664). No universal *decision support tool* for the design
of combinatorial auctions however has yet been developed. The development
of such a decision support tool is very demanding. A number of allocation
processes must be analyzed in order to select applicable auction variants in
the context of the design problems outlined in the previous sections: *allocation efficiency, pricing problem, preference elicitation and stability of the*

mechanism. The following factors play a major role in the design process for Web-based combinatorial auctions:

- *Modeling of the pre and post auction phase:* The organization of the auction preparation and post processing phase, e.g. the publication of auction participation rules or the transaction management, are naturally part of the auction design process.
- *Design of the main auction phase:* This is the most complicated part of the design process, because it has a major impact on the auction outcome and a large number of mechanism design variants.
- *Modeling the auction process flow control:* The process flow control of most auctions is structured in phases. The bidding sequence, the closing, and the clearing time are important parameters for the auction outcome. For this reason process flow control plays a major role especially in online auction design.
- *Legal, security, and system stability issues:* They are of increased interest when implementing bidding and transaction management protocols.

Some of the design parameters of auctions are easy to deduce from a given allocation task. E.g. *seller-buyer cardinalities* are clearly defined within procurement and allocation processes as well as in exchanges (see Figure 5.7).

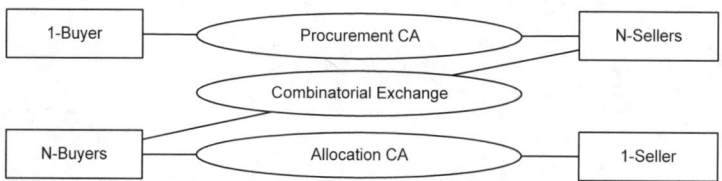

Fig. 5.7. Seller-buyer cardinalities in auction variants

Other design decisions are very complex. An impression is given by the possible choices raised by the following question: "When does the application of the following action variant does make sense?"

- Use of *one-shot* or *iterative* auction design
- Organization as *open-outcry* or *sealed-bid* auction
- Operating with *ascending* or *descending* bids
- Option of *bid withdrawal* possibly with penalty fees *(leveled commitment)*
- *Multi-attributive* or pure *price valuation* of bundles
- Use of encapsulated bidding agents *(proxy-agent sealed-bidding)*
- Implementation of the *Vickrey-Clarke-Groves* mechanism
- Ability to restrict the final allocation by *winner determination constraints*
- Introduction of *bidding language constraints*
- Use of a *bid-valuation module* for automated bundle pricing

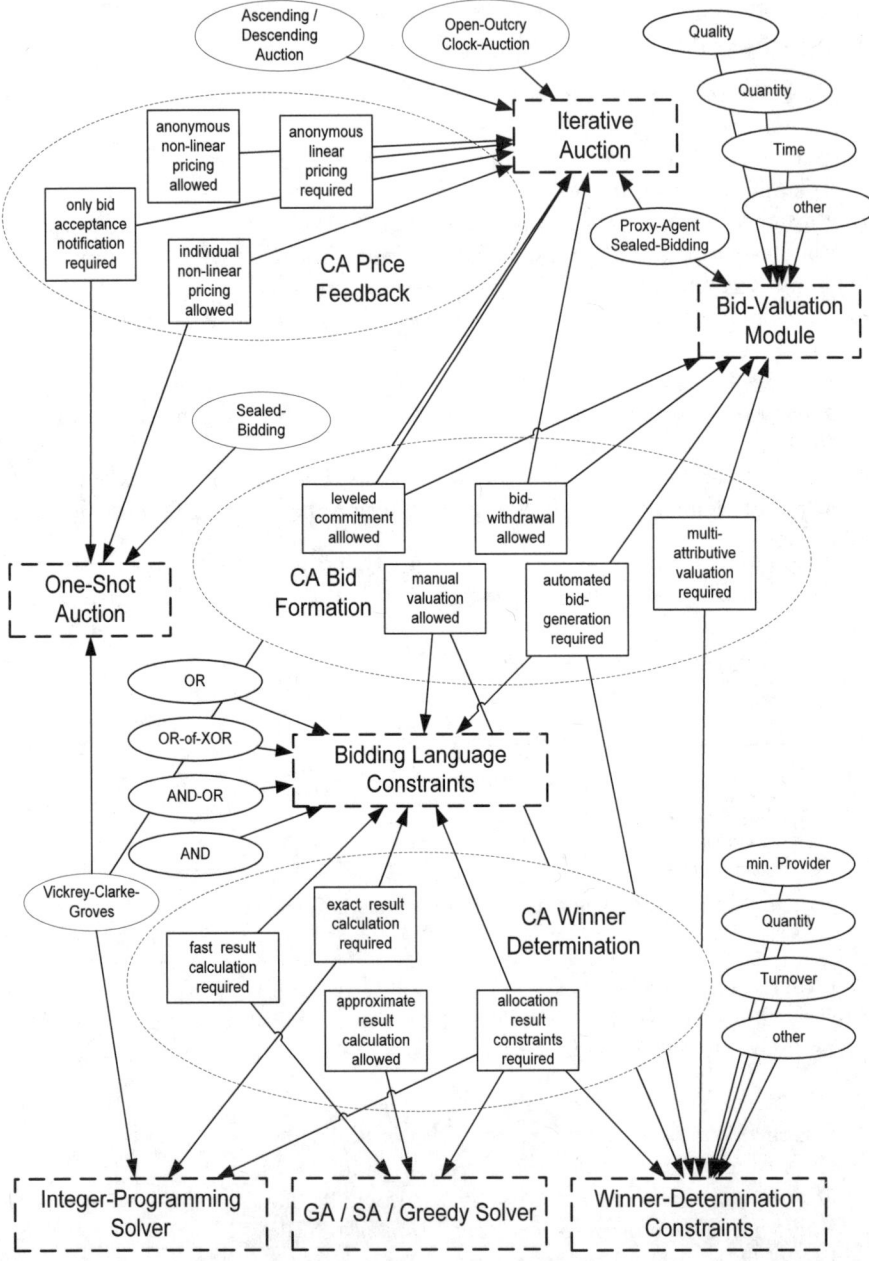

Fig. 5.8. Decision diagram for the design of combinatorial auctions dependent on required system properties

A number of design decision accompany these variants:

- Organization of the auctioneers *price feedback process*
- Selection of an appropriate *winner determination mechanism*
- Choice of suitable *bid formation*

Figure 5.8 depicts the design options in a survey diagram. Some core decisions *(one-shot / iterative auction, bid valuation, bidding language / winner determination constraints, WDP-solver)* are singled out as central instances (boxes in dotted lines), while further parameters that could act as properties of these central instances are depicted as attributes of these decisions (small ellipses). In addition, design requirements that directly derive from the practical requirements of allocation processes are arranged as decision variables in categories *(winner determination, bid formation, price feedback)* (boxes in big ellipses). The arrows in the diagram specify the effect of the design requirement on the core decisions, the mapping of the attributes on the auction variants, and system components.

The most fundamental decision in auction design is whether to carry out the auction as a one-shot process or to go iteratively over multiple rounds. Besides the various game theoretical implications (Parkes & Ungar 2000) this decision is of great relevance to the visibility of equilibrium prices for the participants during the auction process (see section 5.2.3). If only bid acceptance notification is demanded by the bidders, a *one-shot sealed-bid* auction with VCG option can be implemented.[49] For all other price feedback mechanisms the *iterative auction design* offers better design possibilities to provide *anonymous linear or non-linear prices*. In addition there are various extensions for the incentive compatible and transparent design of iterative auctions, like the use of encapsulated *sealed bidding proxy agents* or the realization as a *clock auction*, where the current prices of the individual goods are permanently visible to all participants during the auction phase (Ausubel et al. 2005).

Besides price feedback, the desired manner of *bid construction* has an important role in auction design. If an automated evaluation of the bidders W2P for the bundles is required, the application of a *bid-valuation module* is necessary. After putting some selected questions to the bidders, such a module can calculate (or interpolate) the valuation function, which is then deposited for further use in automated bid construction and consequently reduces the *preference elicitation problem*. On the other hand it is possible to integrate *multi-attributive valuation functions* depending on bundle properties (e.g. quality of service and execution time in ISIP process scheduling) into the bid formation process. In addition, technically determined bundle construction, like the combination and valuation of tasks bundles in a PCRA framework can be processed by a bid valuation module. The question whether *bid withdrawal*, possibly combined with *leveled commitment*, is allowed in an iterative auction

---

[49] The same applies if the auctioneer has to provide only individual non-linear prices.

setting has to be regarded when implementing the bid valuation module, especially if *proxy-agents* are involved in the bidding process. The bid generation is tightly coupled with the selection of the bidding language and the *bidding language constraints*. Normally the semantic range of a bidding language that consists of simple AND/OR or XOR terms is sufficient (see section 5.1.7). By restricting the valid bid combinations (e.g. the number of different items connected with AND terms) the *preference elicitation problem* can be simplified considerably, especially if *manual valuation* of bids is necessary. WDP complexity also can be reduced by restricting bidding languages to yield *fast result calculation* (see section 5.2.1). This, however, implies the loss of bidding language expressiveness, and with it diminished allocation efficiency of the CA (*expressiveness vs. simplicity* (Nisan 2005) and *computational speed vs. economic efficiency* (Sandholm 2002a, p.14). Particularly if an exact result calculation of the WDP by an *IP solver* is essential, as in the case of the VCG mechanism application, bidding constraints can achieve polynomial runtime for some problem instances (see section 5.2.1). If an *approximate result calculation* is sufficient in the WDP process and a *fast result calculation* is desired, heuristics like GAs, SAs or greedy methods can be used to implement the *WDP solver (GA/SA/greedy solver)* (see section 5.2.2). Approximate and deterministically working *WDP solvers* should also provide the possibility of defining *winner determination constraints*. In allocation processes it is often requested to restrict the *turnover* or *quantity* share that is delivered by a single provider to avoid strong interdependence. The same goal can be reached by defining a *minimal* number of *providers* for a specific service in the resulting allocation. Such *winner determination constraints* should be directly considered in the *WDP solver* as *multi attributive valuations* of the bids.

5.3.3 Validation of Combinatorial Auction Design

Game theoretical factors play a crucial role in auction design and dominate the entire design process. This view of auction design is covered by the mathematically grounded discipline of *mechanism design*, which analytically evaluates all variants of a previously designed game. In the conceptualization process of practicable CAs the game theoretical requirements are generally not fulfilled by applying a straight-forward VCG mechanism.[50] In addition, aspects like computational complexity are not considered in the classic mechanism design process. Kalagnanam & Parkes (2003) thus propose the use of *automated mechanism design*. With this, the auction mechanism performing the analysis is prototypically implemented and the performance evaluated by simulating auctions (see Figure 5.9 lower part). Bidding agents track different strategies and thus cover the entire strategy space. After a series of simulation runs, the parameterization of the auction is changed stepwise to search successively for an efficient auction design.[51] Figure 5.9 presents an approach that goes beyond

---

[50] To ensure comprehensibility, VCG mechanisms are often implemented indirectly.
[51] GAs can be valuable in searching for an optimal parameterization.

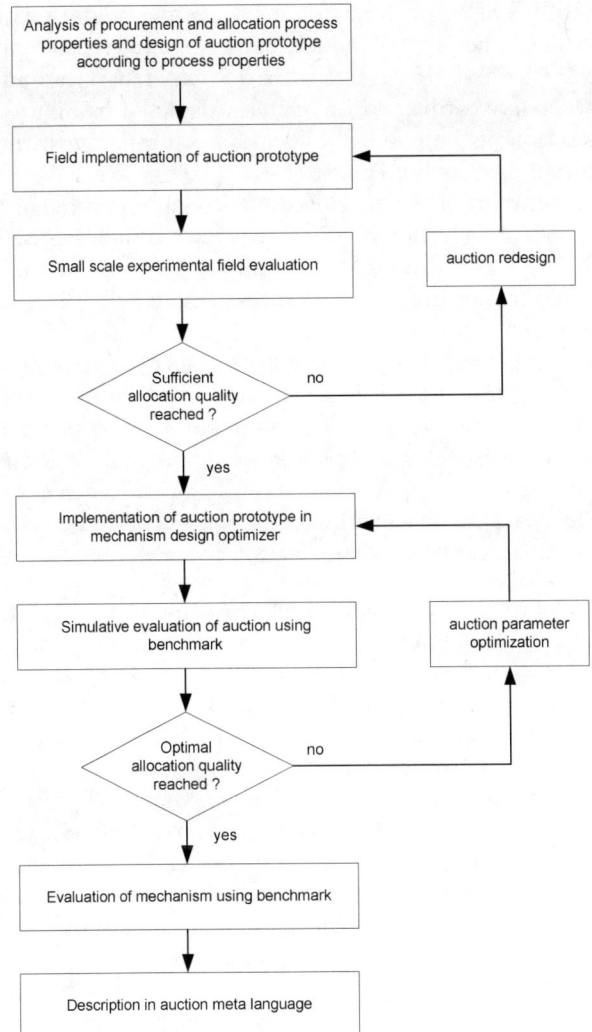

Fig. 5.9. Adjustment and verification of a proposed combinatorial auction by automated mechanism design and field tests

the Kalagnanam & Parkes proposal for the economic validation of *combinatorial auction design*. The process model comprises two steps of validation and optimization. In the first step (see Figure 5.8 upper part) an auction design draft is constructed according to the criteria depicted in Figure 5.8 after an extensive evaluation of the planned allocation or procurement process. Then the auction design draft is implemented as a prototype model.

The implementation of the auction draft is elaborated in such a way that observation of bidder behavior is made possible through experimental game theory (Smith 1994) by carrying out a limited field test in the future environment. By redesigning the auction process, the results of such a field test could then be taken into consideration according to the allocation quality. The redesigned process is then subjected to further evaluation. After having reached a satisfying mechanism design in step one, the auction is optimized in parameters in step two (Figure 5.8 lower part) by employing automated mechanism design. Using an iterative auction approach, for example, the appropriate bid increment is of particular interest for the efficiency of a CA (see section 5.2.3 last part). The same holds for the maximum number of rounds and the termination condition of the auction process.

This chapter gave a comprehensive survey of almost all CA related problems and related solution approaches that have been proposed in the last two decades. As an essential result of this analysis a CA design framework has been proposed which should capable of capturing all system design options that are relevant for implementing a CA for ISIP resource allocation, which will be presented in the context of our PCRA scenario in the next chapter. The main design consequences that result from the insights gained from the issues in this chapter, are: (i) the use of an *iterative CA* with *anonymous prices* communicated to the market participants at the beginning of each auction round, (ii) the introduction of *proxy-agents* that bid on behalf of their principal with a fixed set of rules which is equal for all market participants to avoid the incentive compatibility problem, and (iii) the development of a set of various algorithms (deterministic and approximate) that are capable of managing the CAP allocation process for different problem sizes by making a trade-off between solution quality and computation time.

Dynamic Pricing and Automated Resource Allocation Using Combinatorial Auctions

This chapter addresses the application of CAs for the allocation of resources needed for ISIP task execution in the PCRA scenario described in section 1.4.3. For this purpose, the first part of this chapter is dedicated to the development and benchmarking of algorithms for CAP solutions in ISIP environments. The second part of this chapter is concerned with the description of an agent-based simulation environment that employs a combinatorial scheduling auction for ISIP task allocation. Within this simulation environment dynamic pricing mechanisms based on the resources' shadow prices as well as resource scarcity information are developed. The system behavior is then tested in the context of varying resource availability. Additionally, different bidding strategies used by proxy-bidding agents that submit bids for purpose of ISIP resource acquisition are evaluated in the context of a utility function that allows the simultaneous valuation of two impact factors by making a trade-off between the amount of resources acquired and the speed at which the resources can be acquired.

6.1 Solving the Combinatorial Auction Problem in ISIP Environments

As outlined in section 2.3.2, CAs can be considered a suitable tool to allocate resources in a distributed computing environment. For this reason, the following section will present a formalization of the CAP in the context of ISIP resource allocation and some suitable solution heuristics for the WDP. Performing CAs in a PCRA context means that a bidder agent a_i submits bid bundles $b_{i,j}$ which include the requests $q_{i,j}(o,t)$ for the resources o in a specific quantity q at a particular point of time t; i denotes the number of the agent, and j the number of the agent's bid.

Due to the bidders time-dependent nonlinear valuation of the resource allocation and especially the complementary property of the bidders utility function $v_i(\{b_{i,j}\})$, the CAP is $\mathcal{NP}$-hard (Parkes & Ungar 2000, Fujishima et al. 1999) (see also section 5.1.6). The valuation of two bids ($j = 1, 2$) submitted by an agent i is superadditive if technological complementarities exist:[1]

$$v_i(\{b_{i,1}\}) + v_i(\{b_{i,2}\}) < v_i(\{b_{i,1}\} \cup \{b_{i,2}\}) \tag{6.1}$$

The formal description of the CAP could be considered as a special variant of the *weighted set packing problem* (Vries & Vohra 2001) and is formulated as:[2]

$$\max \sum_{i=1}^{I} \sum_{j=1}^{J} p_{i,j}\, x_{i,j}$$

subject to

$$\sum_{i=1}^{I} \sum_{j=1}^{J} q_{i,j}(o,t)\, x_{i,j} \leq q_{max}(o,t), \tag{6.2}$$

where $o \in \{1, \ldots, O\}, t \in \{1, \ldots, T\}$ and

$$\sum_{j=1}^{J} x_{i,j} \leq 1, \text{ where } i \in \{1, \ldots, I\} \text{ and } j \in \{1, \ldots, J\}.$$

Resources:	o	$\in \mathbb{N}$
Time slots:	t	$\in \mathbb{N}$
Number of bid bundles:	I	$\in \mathbb{N}$
Number of bids in bundle:	J	$\in \mathbb{N}$
Resource requests:	$q_{i,j}(o,t)$	$\in \mathbb{N}$
Price for XOR-bundle $b_{i,j}$:	$p_{i,j}$	$\in \mathbb{R}^+$
Acceptance variable:	$x_{i,j}$	$\in \{0,1\}$
Bid j of agent i:	$b_{i,j}$	$\in B$

The goal is to maximize the auctioneer's income. $q_{max}(o,t)$ is the maximum capacity of resources at time t available to the auctioneer and B is the set of all bids $b_{i,j}$. Furthermore, the set of accepted bids is expressed by B^+ whereas the set of rejected bids is defined as B^- (with $B^- \cup B^+ = B \wedge B^- \cap B^+ = \emptyset$).

The search for an optimal solution to the CAP is mainly performed by using approaches like IP (Chandru & Rao 1998, Andersson et al. 2000) or B&B (Sandholm 2002a, Sandholm et al. 2005) as shown in section 5.2.1. Due to

---

[1] The bidder (producer) applies for a particular combination of resources at a specific point of time t. If not all requested resources are assigned to him, the partial acquisition of the resources has much less value because of production delays or even production failures. This results in the higher than linear (superadditive) valuation of the bundled goods compared to the single items valuation.

[2] See section 5.1.5: however, the version of the SPP employed in this chapter is especially formulated to fit the bid matrix formulation of the PCRA scenario.

the high computational effort of such approaches, heuristics based on greedy allocation strategies are employed to solve the CAP, accepting a trade-off between solution quality and computational effort. A *simple greedy CA algorithm (SG-CAA)* normally consists of two steps:[3]

- According to revenue oriented criteria (e.g. average price per bid-bundle or single-item) submitted bids are sorted in an ordered list.
- CA allocation is made by adding ordered bids from the list as long as they are not ruled out by bids which are already included.

The allocation quality achieved using the SG-CAA usually depends on the sorting criteria and the bidder's utility function (see section 5.2.2). Many approaches combine greedy allocation with more sophisticated heuristics like SA (Hoos & Boutilier 2000). Collins et al. (2002) presented a SA-based approach to solve the CAP in a supply chain setting, where contracts for task allocation are negotiated based on temporal and precedence constraints. The use of a GA as a metaheuristic to solve the CAP has been proposed by Easwaran & Pitt (2000). In this work tasks are matched to a service provider using a fitness function based on criteria such as bandwidth and server capacity. Other GA approaches addressed the weighted set packing problem to find an optimal solution for real-world problems like flight crew scheduling (Chu & Beasley 1995). Levine (1994) implemented a parallel GA, which uses up to 128 subpopulations. Although this approach can handle a large number of bundles, the algorithm has difficulties when dealing with highly constrained set partitioning problems.

6.1.1 Price-Controlled Resource Allocation Scenario and Benchmark Problems

In the following section a benchmark problem is proposed which fits the PCRA scenario depicted in section 1.4.3 for evaluating and comparing the performance of three heuristics presented in Schwind, Stockheim & Rothlauf (2003).

A request $b_{i,j}$ is formulated as a 2-dimensional BM describing which resource[4] o, where $o \in \{1, \ldots, O\}$, is requested at time t, where $t \in \{1, \ldots, T\}$. A simple example of a PCRAS request is presented in Table 6.1. Each entry in the BM denotes the amount $q_{i,j}(o,t)$ of resource units needed. To every bid matrix a bid price p belongs which indicates the bidders W2P.

After describing the load request on the demand side, the resource allocation on the supply side should be modeled. In our scenario this is done by employing an AM, which has the same structure as the BM. Integrating a task into the schedule of the resource provider is done simply by aggregating the current BM into the AM (according to the definition in Table 1.1). A

---

[3] See also section 5.2.2 for an algorithmic formalization of the greedy heuristic.

[4] Computing power (CPU), volatile computer memory (MEM), non-volatile storage capacity (DSK), and data transfer bandwidth (NET).

violation of a resource load constraint can be detected by comparing the AM with the constraint q_{max} (which is equal for all resource types over all time slots). The agents' W2P $p_{i,j}$ for a bid depends on the overall resource load

$$q_{i,j}^{ovl} = \sum_{o=1}^{O} \sum_{t=1}^{T} q_{i,j}(o, t) \tag{6.3}$$

and is calculated as

$$p_{i,j} = q_{i,j}^{ovl} \, p_{fact}. \tag{6.4}$$

p_{fact} is a random decimal number from the interval $[p_{min}, p_{max}]$ and q_{bmax} denotes the maximum resource load that can be requested by a bidder for a single BM-element $q_{i,j}(o, t)$. In the test instances each entry in the BM is occupied with probability p_{tso}. This means, $q_{i,j}(o, t) = 0$ with probability $1 - p_{tso}$.

In the following work three different test instances of BMs are used. The thesis does not make use of the benchmark problems presented in section 5.1.9. This is due to the fact that the existing benchmarks are not capable of expressing resource requests that are representative instances of PRCA scenario here. The *temporal scheduling* benchmark in CATS might seem to be especially suitable; however, it does not support multi resource problems. For this reason this work employs customized BMs that are composed as follows:

- **Unstructured Bids:** The first type of BMs contains single resource requests with a maximum bid length of one time slot associated with a specific resource load. A BM depicting this type is shown in Table 6.1.
- **Substructured Bids:** In a more realistic environment, bidder agents require resources with the same intensity for a longer period of time (up to l_{max} slots). This results in continuous bids of varying length, called substructured bids. An example is shown in Table 6.2.
- **Structured Bids:** Some application cases require resources to be allocated synchronously within shifts. The coherent bids must not cross the shift limits, but may have different constant load intensities for the particular resource types during the shift. A sample of such a structured BM is displayed in Table 6.3.

Table 6.1. Example of an unstructured PCRAS bid matrix

resource	time slot t											
	1	2	3	4	5	6	7	8	9	10	11	12
o_1	1		2		3		1			2		2
o_2		3			1	2	3		1		3	1
o_3			1	1			1				2	
o_4	1				1					1	3	1

Table 6.2. Example of a substructured PCRAS bid matrix

resource	time slot t											
	1	2	3	4	5	6	7	8	9	10	11	12
o_1	3	3	3			2	2	2				
o_2		2	2	2				1	1	1		
o_3			1	1					2	2	2	
o_4	1	1	1							3	3	3

Table 6.3. Example of a structured PCRAS bid matrix

resource	time slot t											
	1	2	3	4	5	6	7	8	9	10	11	12
o_1	3	3	3			2	2		2			
o_2	2	2	2			3	3		1			
o_3	1	1	1			1	1		2			
o_4	1	1	1			2	2		3			

6.1.2 Three Heuristics for the Combinatorial Auction Problem

In the following section newly developed algorithms for the solution of the CAP in the ISIP resource bundle allocation context are presented. They are based on three common types of heuristics: the greedy method, the simulated annealing procedure and the genetic algorithm. The nature of these heuristic types has already been explained in section 5.2.2. The following outline therefore concentrates on the specific description of the three algorithms:

Simple Greedy CA Algorithm

The algorithm description starts with the presentation of the simple greedy CA algorithm (SG-CAA) that is depicted in Table 6.5. The SG-CAA implementation uses a max_f-function to sort the bids according to the ratio $f_{i,j} = p_{i,j}/q_{i,j}^{ovl}$ (price over the overall resource load). Table 6.4 provides an example of such a sorted list.

Table 6.4. List of bids

bid nr.	2	6	4	5	1	3
res. load	111	141	133	85	126	93
bid price	294	344	251	132	157	104
$p_{i,j}/q_{i,j}^{ovl}$	2.65	2.44	1.89	1.55	1.25	1.12

All bids $b_{i,j}$ are labeled from 1 to L, where $L = I * J$ is the total number of bids. The SG-CAA iteratively inserts a bid into the accepted bid set B^+ if no other agent's bid is already in B^+ (XOR-condition), and if the resource load constraint is not violated (compare step 4 of Table 6.5). The results of the SG-CAA are used as a benchmark for the algorithms presented in the next sections.

Table 6.5. Pseudo code of the simple greedy CA algorithm (SG-CAA)

1. Let $i = 0$, B^+ be an (initially) empty set of accepted bids, and r^g be a sorted bid list according to the criteria illustrated in Table 6.4.
2. Let v be the number at the ith position of the permutation r^g and a_v the corresponding bidder agent.
3. If B^+ contains a bid of a_v continue with step 6.
4. If the insertion of bid number v in B^+ would violate the resource load constraint ($\sum_{i,j} q_{i,j}(o,t)x_{i,j} < q_{max} \forall o \in \{1,..,O\}, t \in \{1,..,T\}$) continue with step 6.
5. Insert the bid with number v in B^+.
6. Stop, if $i > L$.
7. Increment i and continue with step 2.

Based on the GR-CAA an improved variant of the greedy algorithm, called *improved greedy CA algorithm (IG-CAA)* has been proposed by Schwind, Stockheim & Seibel (2003). In contrast to the procedure outlined in Table 6.5, the IG-CAA adds all the bids in the sorted list r^g to the set B^+ and subsequently removes the lowest ranked bids, causing a constraint violation in the CM. Coupled with this process the algorithm tries to re-fit the rejected bids to achieve a reduction in the *resource load to price ratio* of the resulting allocation using a stochastic replacement process that comes near to the *simulated annealing* procedure defined in the following section. The evaluation of the improved variant of the greedy algorithm will be made not until section 6.3.1.

Simulated Annealing CA Algorithm

Due to its simple functionality, the greedy strategy implemented in the SG-CAA is not very promising. One way to enhance the efficiency of the greedy strategy is to apply a stochastic improvement process to the initial allocation, trying to remove suboptimal bids from the AM and to replace them by bids which result in a higher reward for the auctioneer. The algorithm proposed here uses a simulated annealing approach, which is based on the original proposal of Kirkpatrick et al. (1983). The fitness function employed is the expected income of the auctioneer. The probability of acceptance of a worse solution is controlled by the temperature Θ.

The simulated annealing CA algorithm *(SA-CAA)* proceeds as follows (see also section 5.2.2 for another SA formalization):

Starting with an empty AM, the auctioneer tries to add a bid which is submitted by an agent that has had none of its bids accepted. If the resulting allocation AM violates the resource allocation constraint, the new allocation is discarded and another bid is tried. As well as adding and removing bids from the AM, the SA algorithm can also handle both operations simultaneously to obtain a new solution. The probability of accepting a new solution p_{acc} is determined by the *metropolis probability* (Metropolis et al. 1953), which depends on the temperature Θ.

$$p_{acc}(\Delta E) \sim \exp\left(\frac{\Delta E}{\Theta}\right). \tag{6.5}$$

ΔE denotes the change in the auctioneer's income due to the insertion/removal step. The SA-CAA algorithm used for the following experiments is summarized in Table 6.6.

Table 6.6. Pseudo code of the simulated annealing CA algorithm (SA-CAA)

1. Let $i = 0$, B^+ be an empty set of accepted bids, and b_{out}, b_{in} empty.
2. With a probability of 0.25 continue with step 4.
3. Select a bid b_{in} randomly from B^-. Continue with step 5 with a probability of 0.33.
4. Select a bid b_{out} randomly from B^+.
5. If the insertion of b_{in} and removal of b_{out} from B^+ would violate the resource load constraint, continue with step 9.
6. If the insertion of b_{in} and removal of b_{out} from B^+ would increase the auctioneer's revenue, or if $random(0, 1) \leq p_{acc}(\Delta E_i)$, continue with step 8.
7. Insert b_{in} into B^+ and remove b_{out} from B^+.
8. If $i > maxSteps$ stop optimization.
9. If $i \bmod N_e = 0$ and thermodynamic equilibrium reached, decrease Θ.
10. Increment i, let b_{out} and b_{in} be empty, and continue with step 2.

When using SA, the starting temperature and cooling rate are essential for the performance of the algorithm. However, optimizing these parameters manually seems to be unpromising. Therefore a temperature control technique proposed by Sundermann & Lemahieu (1995) is employed. This determines the starting temperature such that the algorithm accepts about 80% of exchange operations leading to a deterioration of the fitness function:

$$\sum_{i=1}^{N} \frac{1}{N} \exp\left(\frac{\Delta E_i}{\Theta}\right) = 0.8 \tag{6.6}$$

yielding $\Theta = 5/N * \sum_{i=1}^{N} \Delta E_i$ after *Taylor* expansion.

The annealing process is executed awaiting the occurrence of a thermodynamic equilibrium indicated by a constant $\sum_{i=1}^{N_e} \Delta E_i / N_e$, where N_e is the number of observed steps. The temperature between successive annealing stages is decreased by a cooling factor $\alpha = 0.995$, which is a suitable rule of thumb. The annealing process can be stopped if lowering the temperature does not significantly change the average energy difference.

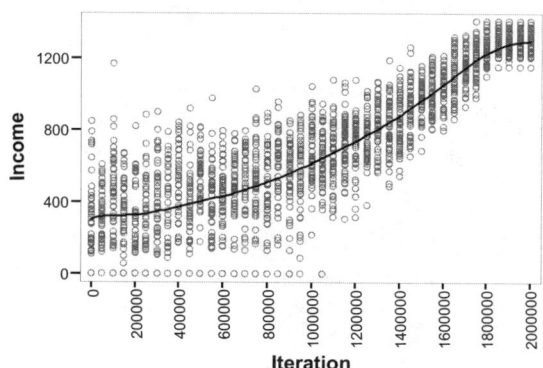

Fig. 6.1. Fitness (auctioneer's income) when using SA-CAA for unstructured bids

Figure 6.1 shows the increasing fitness of the SA-CAA during the annealing process averaged over 50 runs. The regression line has the typical shape of an annealing process. Initially, the temperature is high. Therefore, there is a random walk and no increase in the fitness of the solutions. This is followed by an accelerating, and later decelerating, increase in fitness. Finally, when the temperature is low enough the fitness is fixed in a (local) optimum.

Genetic CA Algorithm

The third approach is a standard generational GA, called *genetic CA algorithm (GA-CAA)*, using the *random key (RK)* encoding.[5] The RK encoding was introduced by Bean (1992) and can advantageously be used for ordering and scheduling problems if the relative ordering of tasks is important. Using the RKs for the combinatorial auction problem allows one to incorporate the resource load constraint into the random key encoding and to generate and process only valid solutions. Therefore, no additional penalties for invalid solutions are necessary.

Later, this encoding was also proposed for single and multiple machine scheduling, vehicle routing, resource allocation, quadratic assignments, and

---

[5] The basic principle of a GA for CAs has already been discussed in section 5.2.2.

traveling salesperson problems (Bean 1994). Norman & Bean (1994) refined
this approach (Norman & Bean 2000) and applied it to multiple machine
scheduling problems (Norman & Bean 1997). An overview of using random
keys for scheduling problems can be found in the (unpublished) PhD thesis of
Norman (1995). Norman et al. (1998) applied random keys to facility layout
problems. Rothlauf et al. (2002) used random keys for the representation of
trees.

The RK encoding uses random numbers for the encoding of a solution. A
key sequence of length L, where L is the number of bids $b_{i,j}$, is a sequence of
L distinct real numbers (keys). The values are initially chosen at random, are
floating numbers between zero and one, and are only subsequently modified
by mutation and crossover. An example of a key sequence of length $L = 4$
is $r = (0.07, 0.75, 0.56, 0.67)$. Of importance for the interpretation of the key
sequence is the position and value of the keys in the sequence. If we assume
that $Z_L = \{0, \ldots, L-1\}$ then a permutation σ can be defined as a surjective
function $\sigma : Z_L \rightarrow Z_L$. For any key sequence $r = r_0, \ldots, r_{L-1}$, the permutation
σr of r is defined as the sequence with elements $(\sigma r)_i = r_{\sigma(i)}$. The permutation
r^s corresponding to a key sequence r of length L is the permutation σ such
that σr is decreasing (i.e., $i < j \Rightarrow (\sigma r)_i > (\sigma r)_j$). The ordering corresponding
to a key sequence r of length L is the sequence $\sigma(0), \ldots, \sigma(L-1)$, where σ is
the permutation corresponding to r.

For better understanding, a brief example of the construction of the permutation list r^s of bids from the key sequence r is given. The positions of
the keys in the key sequence r must be ordered according to the values of the
keys in descending order. In the example the position of the highest value in
the key sequence (0.75 at position 2) has to be identified. The next highest
value is 0.67 at position 4. By continued ordering the complete sequence is
attained with the permutation $r^s = 2 \rightarrow 4 \rightarrow 3 \rightarrow 1$. In the context of the CA
problem this permutation can be interpreted as a list of bids that are con-
sidered sequentially by the auctioneer. (The auctioneer starts by accepting
bid 2, then accepts bid 4, bid 3, and bid 1). From a key sequence of length
L, one can always construct a permutation of L numbers (bids). Every bid
number between 1 and L (or 0 and $L-1$) appears in the permutation only
once as the position of each key is unique. In a next step the accepted bids are
calculated from the permutation r^s representing an ordered list of bids. With
this construction directive the GA-CAA can now be formulated as described
in Table 6.7.

For the experiments a simple generational GA (Goldberg 1989b) with two-
point crossover, no mutation, and tournament selection with replacement of
size two was chosen. The encoding of the bids was done by using the RK
encoding. The fitness of the individuals is the resulting income. Due to the
use of the RK encoding there are no invalid solutions and no penalties have
to be added to the fitness of the individuals.

Figure 6.2 shows an example of the increasing fitness of the GA-CAA for
unstructured bids averaged over 50 runs. The plots depict the averaged best

Table 6.7. Pseudo code of the genetic CA algorithm (GA-CAA)

1. Let $i = 0$, B^+ be an (initially) empty set of accepted bids, and r^s the permutation of length L that can be constructed from the key sequence r. All bids $b_{i,j}$ are labeled from 1 to L.
2. Let v be the number at the ith position of the permutation r^s and a_v the corresponding bidder agent.
3. If B^+ already contains a bid of a_v continue with step 6.
4. If the insertion of the bid with number v in B^+ would violate the resource load constraint, continue with step 6.
5. Insert the bid with number v in B^+.
6. Stop, if $i > L$.
7. Increment i and continue with step 2.

GA Unstructured Bids

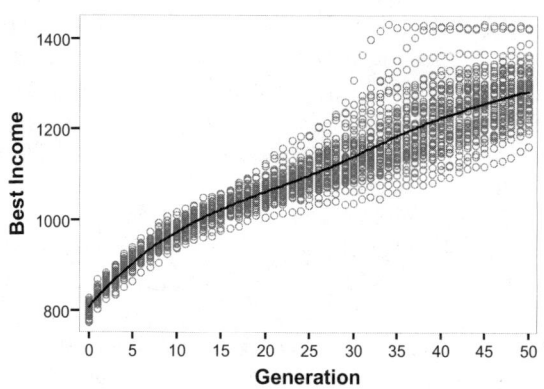

Fig. 6.2. Best fitness (auctioneer's income) in the process of the GA-CAA with 500 individuals for unstructured bids

income of the auctioneer over the number of generations and use a population size of $N = 500$. They show a steady improvement in the auctioneer's income.

6.1.3 Performance of the Three CAP Solution Heuristics

To evaluate the performance of the three CA solution heuristics simulations are performed using the JAVA Agent Development Environment (JADE)[6]. All bidders and the auctioneer are implemented as agents. After emitting a call for proposals the auctioneer collects the bids submitted by the agents and calculates the optimal allocation according to the chosen heuristic. The results of 50 simulation runs per parameter setting is logged and analyzed using SPSS[7].

---

[6] JADE: Java Agent DEvelopment Toolset http://jade.cselt.it.
[7] SPSS: Statistical Product and Service Solutions http://www.spss.com.

Table 6.8. Performance of the SG/SA/GA CA-algorithm for 10 agents submitting four bids measured in income and CPU time (sec) for 50 simulations

	bid type	income	time	perf.	stdev.
SG	unstr.	887.28	0.007	1.00	74.86
	substr.	991.82	0.011	1.12	127.89
	str.	1123.04	0.011	1.27	162.74
SA	unstr.	960.22	48.334	1.08	68.15
	substr.	1048.04	84.792	1.18	75.01
	str.	1200.78	84.768	1.35	110.35
GA	unstr.	961.70	9.051	1.08	54.98
	substr.	1082.50	15.067	1.22	63.77
	str.	1269.04	13.455	1.43	89.80

Table 6.9. Performance of the SG/SA/GA CA-algorithm for 50 agents submitting four bids measured in income and CPU time (sec) for 50 simulations

	bid type	income	time	perf.	stdev.
SG	unstr.	1084.98	0.049	1.22	92.18
	substr.	1291.08	0.092	1.46	99.94
	str.	1602.72	0.089	1.81	110.11
SA	unstr.	1251.70	55.046	1.41	48.67
	substr.	1447.08	92.369	1.63	62.78
	str.	1718.22	92.481	1.94	76.38
GA	unstr.	1241.42	81.453	1.40	44.57
	substr.	1474.16	139.999	1.66	50.08
	str.	1766.82	132.324	1.99	74.12

Table 6.10. Performance of the SG/SA/GA CA-algorithm for 100 agents submitting four bids measured in income and CPU time (sec) for 50 simulations

	bid type	income	time	perf.	stdev.
SG	unstr.	1114.36	0.114	1.26	78.93
	substr.	1393.32	0.207	1.57	79.35
	str.	1831.66.	0.208	2.06	116.46
SA	unstr.	1302.32	63.335	1.47	47.61
	substr.	1531.70	100.809	1.73	66.55
	str.	1876.12	100.911	2.11	117.12
GA	unstr.	1282.91	189.512	1.45	55.59
	substr.	1525.80	314.326	1.72	56.67
	str.	1929.76	302.544	2.17	87.44

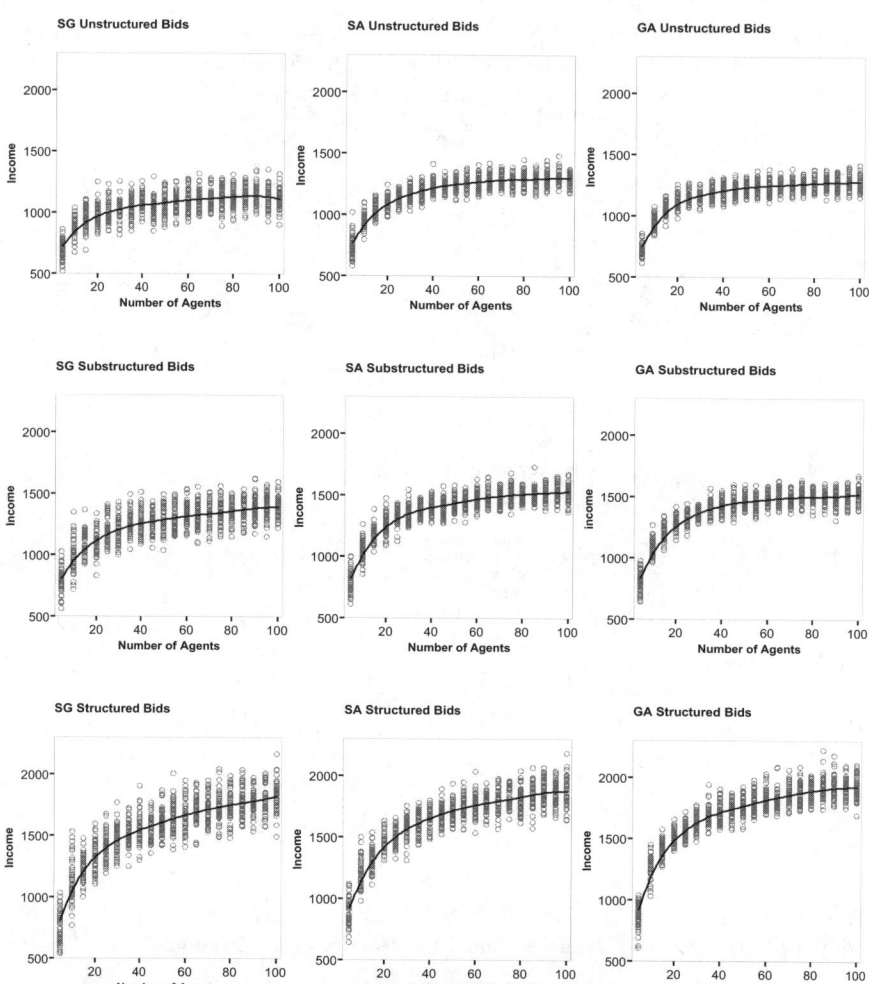

Fig. 6.3. Performance of the SG-CAA, SA-CAA and GA-CAA for unstructured, substructured and structured bids indicated by the auctioneer's income over the number of bidding agents

Our experimental setting tries to reach the optimal allocation of ISIP tasks in the AM ($q_{max} = 8$) for four basic resources $O = 4$ and $T = 24$ time slots. The maximum resource request per BM of the agents is limited to $q_{bmax} = 3$. Each agent is allowed to submit four bids $J = 4$ of which only one could be allocated (XOR-condition). Bids are generated with time slot occupancy probability $p_{tso} = 0.33$. For the calculation of the bid price a parametrization of $p_{min} = 1$ and $p_{max} = 3$ is used. The number of agents varies between $I = 5$ and $= 100$ in steps of 5 for each heuristics type as can be seen in Figure 6.8. The plots show that the auctioneer's income Inc^{acc} increases with a larger number of agents. This can be explained as the probability of finding a compact allocation increases with the number of bids the auctioneer can choose from. More interestingly, the auctioneer's maximal revenue depends on the structure of the bids and increases from unstructured over substructured to structured bids. A higher structure in the bids allows the auctioneer to allocate the bids more efficiently and to gain a higher revenue. Tables 6.8 to 6.10 summarize the results of Figure 6.3 and compare the three different optimization heuristics for 10 (Table 6.8), 50 (Table 6.9), and 100 (Table 6.10) bidding agents. Comparing the maximal auctioneer's revenue for different optimization heuristics shows that the SG-CAA is outperformed by the SA-CAA and the GA-CAA. When using the SA-CAA the auctioneer's revenue is about 20% higher in comparison to the SG-CAA. However, computing time increases to a factor of 10,000, the GA-CAA performs slightly better than the SA-CAA although the increase in performance comes at the expense of a higher computational effort.

6.2 An Agent-Based Simulation Environment for Combinatorial Scheduling

Based on the WDP algorithms described in the previous section an entire simulation environment for the allocation of ISIP tasks in the PCRA scenario (see Figure 1.2) has been developed. The implementation of the simulation environment presented in this thesis is based on the FIPA[8] conform JAVA MAS platform JADE 3.3[9] (Poggi & Bergenti 2001). The *task agents* that bid for the use of the resources needed to complete the tasks they intend to provide to the ISIP system users the combinatorial auctioneer (*market mediator*) performing the scheduling auction, as well as the *resource agents* that manage the resources are represented as software agents (see section 1.4.4). These agents can be located on distributed computer hosts in JADE, enabling the transferability of the entire setting into real world B2B or B2C environments without great effort (see section 2.4.1). The combinatorial task scheduling environment, see Figure 6.4 for a visualization of the corresponding GUI of the system, goes beyond the recent approaches discussed in section 2.3.2 in several respects:

- The system makes it possible to use several winner determination algorithms like SA, GA, and IP methods according to the users' requirements in terms of allocation quality and computation time.
- The simulator provides tools to investigate various bidding behaviors of the proxy agents in the resources acquisition process.
- The framework can simulate changing resource capacities to test the combinatorial Grid scheduler's reaction with respect to allocation efficiency and system stability.

---

[8] FIPA: Foundation for Intelligent Physical Agents `http://fipa.org`
[9] JADE: Java Agent DEvelopment Toolset `http://jade.cselt.it`

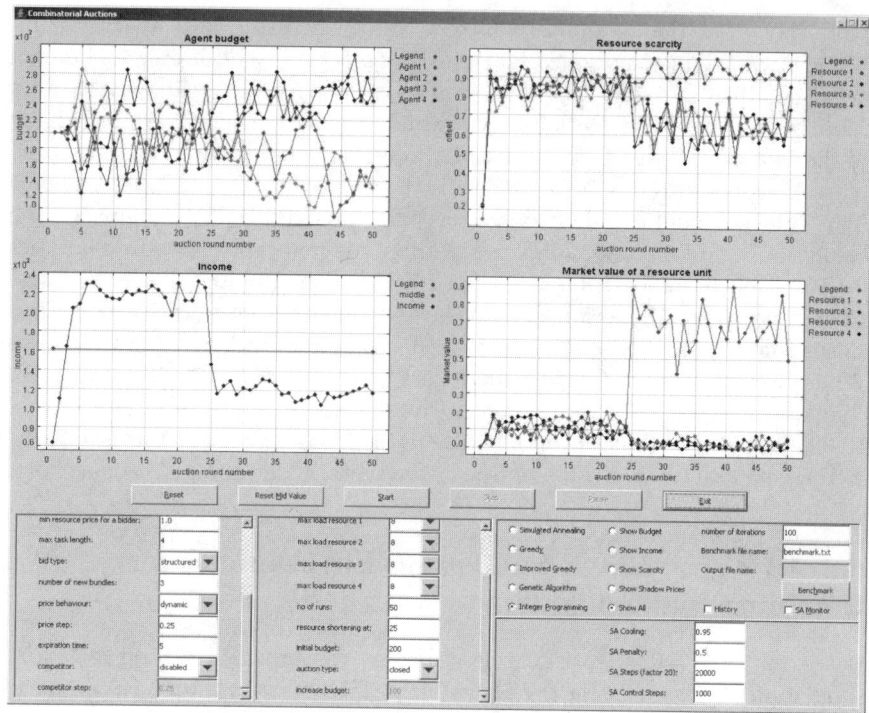

Fig. 6.4. The combinatorial Grid simulation environment

A critical point in MAS is usually communication design. Even auctions generate a non-negligible communication overhead between agents. Therefore communication acts have been designed according to the FIPA Agent Communication Language (ACL) definition.[10] A helpful approach in this context is to deploy the communication structure by designing an FIPA Agent-UML (AUML) sequence diagram (Bergenti & Poggi 2000).[11] Additionally, as shown in section 2.4.2, ACL transmission requires the formulation of an ontology to define communication content (e.g. communication predicate submitBid has slot classes: Pinc, nrBundle, bundleId, agentId, budget, nrRound). This was achieved by using the Protégé 3.0 ontology editor[12], which is able to produce ready-for-use JAVA code. The class diagram of the communication ontology employed in the combinatorial Grid simulation environment[13] is depicted in Figure 6.5.

---

[10] FIPA ACL Message Structure Specification (Document number: XC00061E)

[11] FIPA Interaction Protocol Library Specification (Document number: XC00025E)

[12] http://protege.stanford.edu

[13] The entire simulation software is available at www.combinatorial-auction.de.

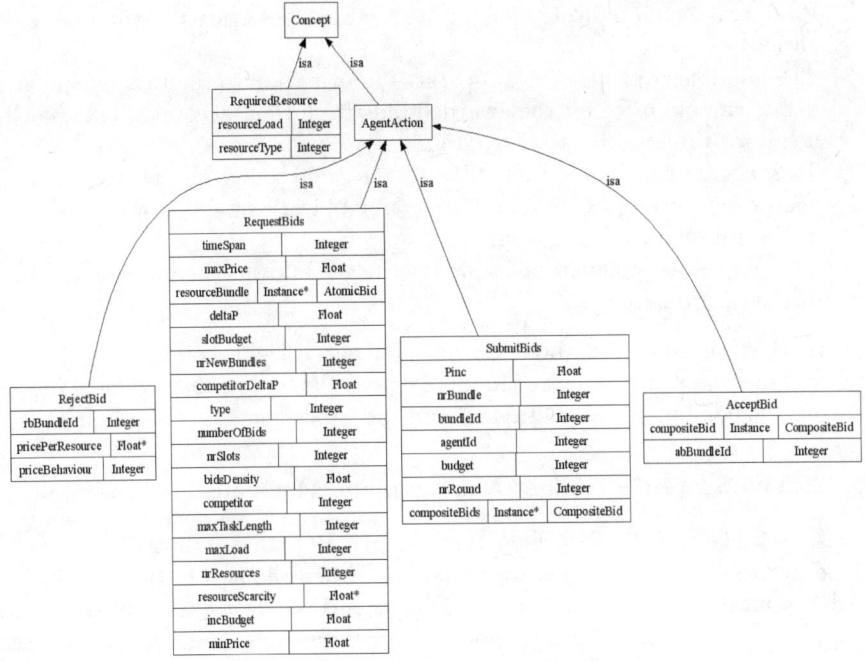

Fig. 6.5. Class diagram of the agents' ontology used in the combinatorial scheduling auction

6.2.1 The Combinatorial Scheduling Auction

In the following paragraph the course of the combinatorial Grid auction is described in the light of a UML sequence diagram based on the FIPA definition for the English auction[14] (steps are denoted in ○):

1. The auctioneer requests the resource agents to evaluate the available resource capacities and informs the bidders about the bidding terms. Then he announces the start of the auction. Additionally, the auctioneer awards an initial budget to the task agents.
2. Following the auctioneer's call-for-proposal, the task agents create their bids according to the desired resource combination. Bidders compute the associated bid price, dependent on their current pricing policy, their budget level, and the latest resource prices, if applicable.
3. The auctioneer receives the bids and calculates the return-maximizing combinatorial allocation. He informs the task agents about bid acceptance or rejection and requests the resource agents to reserve the resources awarded.

---

[14] FIPA Interaction Protocol Library Specification (Document number: XC00031D)

4. Resource agents inform the auctioneer about the status of the task execution.
5. The auctioneer propagates task status information to the task agents and debits the bid price for the awarded bids from their accounts, followed by a call-for-proposal for the next round.
6. Task agents can renew their bids in the next round in the case of non-acceptance or non-execution. The agents' bid pricing follows rules defined in the subsequent paragraph.
7. The process is repeated until the auctioneer informs them about the end of the auction.

In the following the three crucial elements of the combinatorial Grid scheduling system are described in more detail: the budget management mechanism, the combinatorial auctioneer and the task agents' bidding behavior.

6.2.2 The System's Budget Management Mechanism

Each task agent a_i holds a monetary budget BG_i that is initialized with a fixed amount BG^{ini} of *monetary units (MU)* when the negotiations begin. At the beginning of each round k the task agents' budgets are refreshed (see Figure 6.6, ○1) with an amount of MUs enabling the agents to acquire the resource bundles $b_{i,j}$ required for their ISIP provision task. The agents' budget refill can be done in two ways in our Grid economy:

- A fixed amount BG^{inc} that is defined by the system user is added to the agents' budgets independently of the production capacity they provide to the Grid system. This case, where task agents only behave as consumers, is defined as an *open Grid economy*. The *resource agents* which *own the resources* act independently of the task agents providing only resource availability and resource use information to the auctioneer. The resource agents are compensated by the auctioneer for the capacity provided proportionately to the auctioneer's income Inc^{acc}.
- Task and resource agents act as a unit of consumer and producer, both owning the resources of their peer system. This means a task and a resource agent reside simultaneously on each peer computer in the Grid. The resource agent does the reporting of resource use and provisioning for the task agent owning the peer computer resources (see Figure 6.6, ○1,4)[15]. The agents on the peer computer are compensated for the resources provided to the system. Starting with the initial budgets BG^{ini} the amount of MUs circulating in the system is kept constant for the *closed Grid economy*.

The accounting of the agents' budgets in the Grid system is performed by the combinatorial auctioneer (see Figure 6.6, ○1, and ○5).

---

[15] In Figure 1.2 this implies that resource agent 1 and task agent 1 reside on the same peer computer.

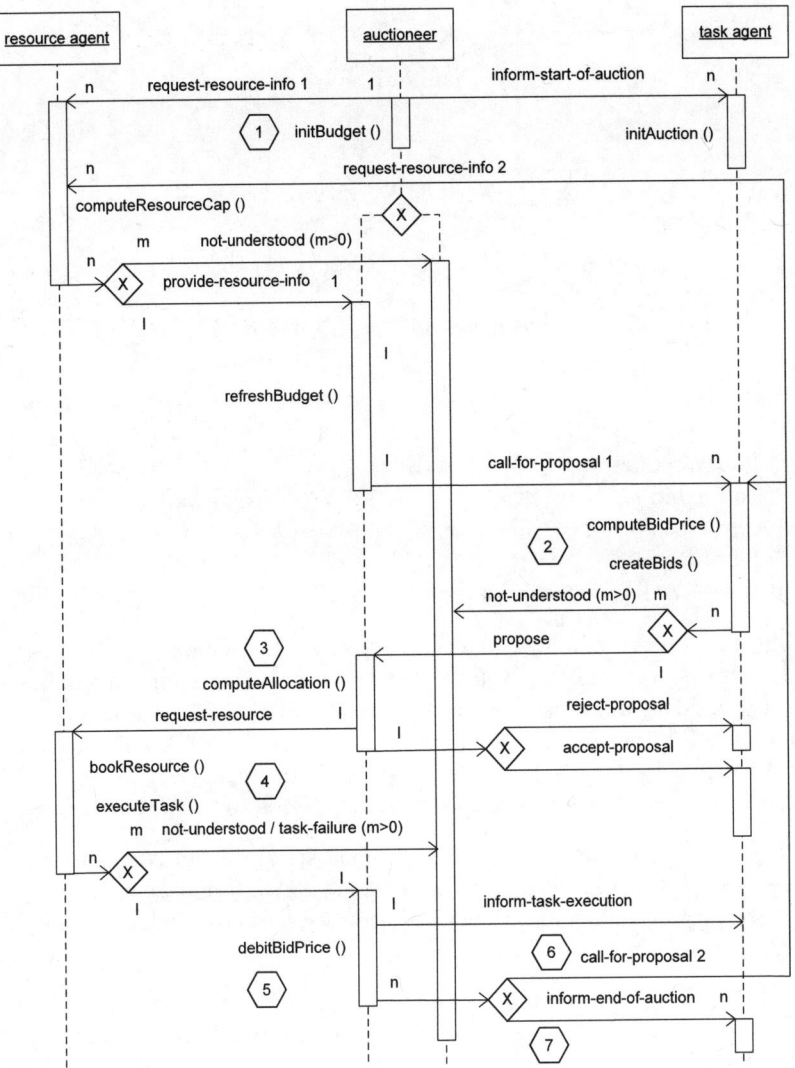

Fig. 6.6. FIPA AUML diagram for the iterative combinatorial scheduling auction

6.2.3 The Combinatorial Auctioneer

The combinatorial auctioneer controls the iterative allocation process of the Grid system. For this purpose the auctioneer awaits the bids that have been submitted by the task agents for the current round. The allocation is done according to the formal formulation of the CAP presented in section 6.1:

$$\max \sum_{i=1}^{I} \sum_{j=1}^{J_i} p_{i,j} \, x_{i,j}$$

subject to

$$\sum_{i=1}^{I} \sum_{j=1}^{J} q_{i,j}(o,t) \, x_{i,j} \leq q_{max}(o,t), \tag{6.7}$$

where $o \in \{1, \ldots, O\}, t \in \{1, \ldots, T\}$ and

$$\sum_{j=1}^{J} x_{i,j} \leq 1, \text{ where } i \in \{1, \ldots, I\} \text{ and } j \in \{1, \ldots, J\}.$$

Resources:	o	$\in \mathbb{R}$
Time slots:	t	$\in \mathbb{N}$
Resource requests:	$q_{i,j}(o,t)$	$\in \mathbb{N}$
Price for bid $b_{i,j}$:	$p_{i,j}$	$\in \mathbb{R}^+$
Acceptance variable:	$x_{i,j}$	$\in \{0,1\}$
Bid j of agent i:	$b_{i,j}$	$\in B$
Income for all accepted bids: Inc^{acc}		$\in \mathbb{R}^+$

The auctioneer's primary goal is to maximize the income Inc^{acc}. A second function of the auctioneer is to provide feedback about resource availability to allow the bidders to adjust their bid price $p_{i,j}$ for the rejected bids $p_{i,j} \in B^-$ in the following rounds. As shown in section 5.2.3 and 5.3.1 this is not trivial for CA settings because it is not always possible to calculate anonymous prices for the resources. The thesis proposes two methods to solve this problem:

- The use of *resource scarcity (RS)* as a substitute for pricing information.
- The application of interpolated *shadow prices (SP)* that guarantee approximate pricing information even in situations when regular shadow price calculation is not possible.

The following two sections illustrate the use of both methods as a basis for the agents' pricing behavior according to Schwind & Gujo (2006).

Resource Scarcity Information

Resource scarcity information can be used for valuation purposes instead of resource prices calculated directly by using the dual CAP formulation. This avoids the time-consuming SP calculation based on the nonlinear valuations of the bids (Xia et al. 2004). For this purpose a relative resource value v_o^{rel} is defined based on the auctioneer's information about resource scarcity. The relative value v_o^{rel} of resource o is defined as its scarcity $SR_o \in [0, 1]$ incremented by a constant 0.5 to enforce strongly increasing valuations:[16]

---

[16] The work of Schwind, Gujo & Stockheim (2006) uses a more sophisticated method for resource valuation based on resource scarcity information.

$$v_o^{rel} = SR_o + 0.5 \tag{6.8}$$

v_o^{rel} is defined via scarcity S_o of resource o:

$$SR_o = Q_o^{acc}/Q_o^{max} \tag{6.9}$$

The scarcity of the resources is computed as the ratio of the maximum allocable resource load Q_o^{max} given in the CM and the resource use Q_o^{acc} achieved by the auctioneer's bid acceptance in the previous round as indicated by the AM:

$$Q_o^{max} = \sum_{t=1}^{T} q_o^{max} \tag{6.10}$$

$$Q_o^{acc} = \sum_{t=1}^{T}\sum_{i=1}^{I}\sum_{j=1}^{J} q_{i,j}(o,t)\, x_{i,j} \tag{6.11}$$

Shadow Price Calculation

In many cases it is not possible to calculate *anonymous prices* but only *individual prices* for the goods (or resources) that are requested in the bids. Kwasnica et al. (2005) describes a pricing scheme for all individual goods in a CA, – descibed as RAD –, while approximating the prices in a divisible case based on the RSB approach first proposed by Rassenti et al. (1982). Both equilibrium methods have been intensively discussed in section 5.2.3. As in a similar approach by Bjørndal & Jørnsten (2001) they employ the *dual solution* of the relaxed WDP to calculate the shadow prices SP. In the simulation model presented here the dual LP approach of Kwasnica et al. (2005) is adopted. It includes accepted bids as well as rejected bids in the calculation:

$$min\ z = \sum_{o=1}^{O}\sum_{t=1}^{T} q_{max}(o,t) \cdot sp_{o,t}$$

subject to

$$\sum_{o=1}^{O}\sum_{t=1}^{T} q_{i,j}(o,t) \cdot sp_{o,t} = p_{i,j} \quad \forall\, b_{i,j} \in B^{+} \tag{6.12}$$

$$\sum_{o=1}^{O}\sum_{t=1}^{T} q_{i,j}(o,t) \cdot sp_{o,t} + \delta_{i,j} \geq p_{i,j} \quad \forall\, b_{i,j} \in B^{-}$$

Accepted bid set: $\qquad B^{+} \subseteq B$
Rejected bid set: $\qquad B^{-} \subseteq B$
Reduced cost: $\qquad \delta_{i,j} \in \mathbb{R}_0^{+}$
Shadow price variable: $sp_{o,t} \in \mathbb{R}_0^{+}$

The result of this formula is defined as *reduced shadow prices*.[17] The proposed SP calculation uses the *primal solution* of the LP problem delivered by the LP solver *LP_SOLVE 5.5*[18] to determine the sets B^+ and B^-:

$$SP_o(k) = \sum_{t=1}^{T} sp_{o,t} \quad \forall o \in O \qquad (6.13)$$

Shadow price: $SP_o(k) \in \mathbb{R}_0^+$

In general bid prices are not linear. This means that shadow prices SP_o cannot be calculated by the auctioneer in each round, i.e. there is no solution of the LP problem, or reduced costs equal zero for a number of bids (Bjørndal & Jørnsten 2001). In such cases the auctioneer relies on an approximate shadow price calculation based on pricing history H_{sp} in the last n rounds:[19]

$$\widehat{SP}_o(k) = \frac{\sum_{k-n}^{|H_{sp}|} h_{sp}}{|H_{sp}|} \quad \forall h_{sp} \in H_{sp} \wedge n \in \{1, \ldots, k-1\} \qquad (6.14)$$

$$h_{sp} = \begin{cases} \widehat{SP}_o(k) & \text{if} \quad SP_o(k) \neq 0 \wedge k \neq 1 \\ 0 & \text{if} \quad k = 1 \end{cases} \qquad (6.15)$$

Approximate shadow price: $\widehat{SP}_o(k) \in \mathbb{R}_0^+$

Now the *market value of a resource unit* v_o^{shad} can be determined based on the *shadow prices* SP_o. For this purpose the resource load in the final allocation is calculated by using the sums of $q_{i,j}(o,t)$ for each resource $o \in O$ and each time slot $t \in T$ for all accepted bids B^+:

$$v_o^{shad}(k) = \begin{cases} \dfrac{SP_o(k)}{\sum_t^T \sum_i^I \sum_j^J q_{i,j}(o,t)} & \forall b_{i,j} \in B^+ \quad \text{if} \ B^+ \neq 0 \wedge SP_o(k) \neq 0, \\[4mm] \dfrac{h_{sp}}{\sum_t^T \sum_i^I \sum_j^J q_{i,j}(o,t)} & \forall b_{i,j} \in B^+ \quad \text{if} \ B^+ \neq 0, \\[4mm] 0 & \text{otherwise} \end{cases} \qquad (6.16)$$

Market value of a resource unit: $v_o^{shad}(k) \in \mathbb{R}_0^+$

---

[17] SP_o are higher if rejected bids are omitted in (6.12) (Bjørndal & Jørnsten 2001).
[18] http://www.geocities.com/lpsolve/
[19] Xia et al. (2004) propose an iterative price adaptation process for those cases to achieve an approximate solution. However, due to the time criticality of our system the pricing history solution is used.

6.2.4 The Task Agents' Bidding Model

Based on the two measures of resource availability, scarcity SR_o and shadow prices SP_o, the task agents in the combinatorial simulation model try to acquire the resources needed for ISIP provision. Besides the resource availability, the task agents' bidding behavior is determined by their budget and an associated *bidding strategy*. The major goal of the task agents is to receive as many resource units as possible that are required for the performance of their ISIP processes at the lowest possible amount of MUs. Different *bidding strategies* and a related *utility function* will be tested and introduced in sections 6.3.5 and 6.3.6. The general bidding behavior of the task agents is similar for both pricing methods, used while employing both resource scarcity information and shadow prices:

- In each round k the task agents generate M new bids. The formation of the *new generated BMs* can be defined in the simulation setting according to the three matrix types presented in section 6.1.1: structured, substructured, and unstructured. The task agents can have the possibility to submit bids $b_{i,j}$ as exclusively eligible bundles. The eligibility is defined such that J BMs are treated as XOR bids in a bundle (OR-of-XOR).[20]
- Task agents repeat bidding for rejected bundles in the following round with a modified W2P while regarding the current resource availability.

$$
p_{i,j}(k) = \begin{cases} \dfrac{BG^{ini}}{L \cdot M \cdot J} & \text{if } k = 1 \\[2ex] \sum\limits_{o=1}^{S} \sum\limits_{t=1}^{N} \hat{v}_o(k-1) \cdot q_{i,j}(o,t) \cdot p^{inc} & \text{if } k > 1 \end{cases}
\tag{6.17}
$$

Task agents formulate their W2P for the bids required by the ISIP provision process based on the resource occupancy $q_{i,j}(o,t)$ in the BMs and a market value estimator $\hat{v}_o$ based either on resource scarcity information with $\hat{v}_o = v_o^{rel}$ or by using shadow prices $\hat{v}_o = v_o^{shad}$. To calculate the bundle prices, two cases must be considered:

- In the first round, a market value of the resources is not provided to the bidders. Therefore, bidder agents have to formulate the W2P for their *initial bids* with respect to the start-up budget BG^{ini} and their bidding strategy. This is done by calculating a mean bid price that guarantees the task agents' budget's lasting for the next L rounds if $M \cdot J$ new sets of XOR-bids are added (cp. line 1 of equation 6.17).
- In the following rounds, the task agents employ the estimated market values $\hat{v}_o$ of the resources to determine their W2P. The requested amounts of capacity are multiplied by the corresponding market value and, if the bid is initialized, a factor p^{ini} is included in the calculation so that initial bids may start below or above the current price level of the resource market.

---

[20] For the definition of OR-of-XOR and other variants of bids see section 5.1.7.

If a bid is rejected, the corresponding W2P is adapted by

$$p^{inc} = p^{ini} + (l \cdot \Delta p),\qquad(6.18)$$

Round of bid $b_{i,j}$: $l \in \mathbb{N}$
Multiplier increment: $\Delta p \in \mathbb{R}^+$
Initial price multiplier: $p^{ini} \in \mathbb{R}^+$

resulting in the above mentioned value of p^{ini} in round 1 (cp. line 2 of formula 6.17). To control the price adaption process, an additional price acceleration factor p^{inc} is introduced. At each round, p^{inc} is incremented by a constant Δp. Recalculating the price based on the current market value v_o results in a faster adoption process.

Rejected bids are *repeatedly* submitted with an updated W2P until the bid is accepted, but only for a limited number of rounds. Bids are *discarded* if they are not accepted after L rounds. Furthermore, the task agents' bidding behavior is limited by their budgets. When an agent's budget is exhausted, it formulates no new bids until the budget is refreshed in the next round $k + 1$. The bidding behavior of the task agents can be modified by varying parameters like J, L, M, and Δp. Based on these parameters, different bidding strategies will be evaluated in section 6.3.5.

6.3 Experimental Settings and Results

This section is dedicated to the evaluation of the combinatorial resource allocation system that has been specified in the previous sections of this chapter. In a first experimental series the system's auctioneer is benchmarked in the context of the ISIP scenario defined in section 6.1. A second experimental setting illustrates the system's behavior in the open and closed economy case, especially with respect to the reaching of stable price equilibria. A comparison of pricing behavior based on resource scarcity and shadow price calculation follows. This evaluation will be made in the context of resource failure situations. The testing of different bidding strategies with respect to efficiency measured using an utility function defined for the agents closes this experimental section.

6.3.1 Benchmarking the Combinatorial Auctioneer

The entire combinatorial simulation environment is now tested with respect to the allocation quality and calculation time that is achieved by the auctioneer's WDP algorithms. This is especially interesting with respect to real world environments where a trade-off has to be made between allocation quality and real time performance of the mechanisms. Unlike section 6.1.2 that is concerned with the development and comparison of the GR-SAA, SA-CAA and GA-CAA mechanisms, this paragraph deals with the agent-based simulation of five alternative approaches to solving the WDP: in addition to the three heuristics, the IG-CAA mentioned in section 6.1.2 and an *integer programming (IP-CAA)* method are tested. The IP-CAA employed here is implemented using the Open-Source IP module *LP_SOLVE 5.5*[21]. It provides the possibility of benchmarking the simulation results using the optimal CA allocation result.[22]

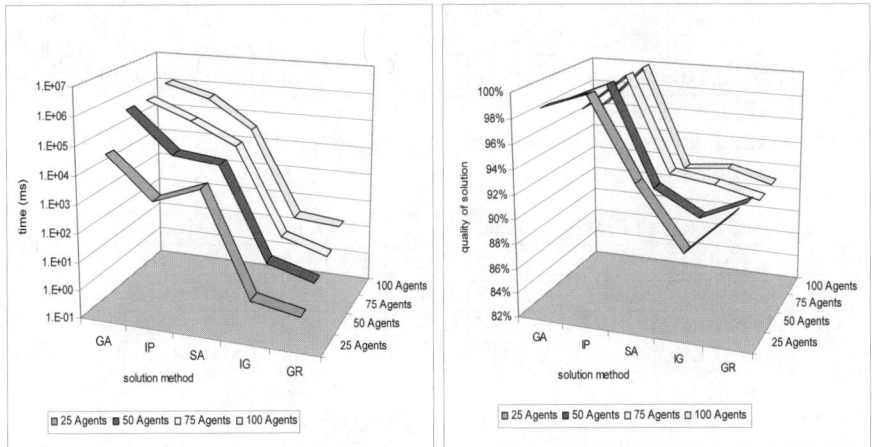

Fig. 6.7. Solution time in milliseconds (left) and allocation performance in percent of optimal allocation (right) for a GA-CAA, IP-CAA, SA-CAA, IG-CAA, and GR-CAA algorithm for structured bids with 25, 50, 75 and 100 agents

The experimental setting used in this benchmark is identical to the experimental environment employed in section 6.1.3. The AM $(q_{max} = 8)$ holds the four basic resources $O = 4$ over $T = 10$ time slots. The maximum resource request allowed in the bids was limited to $q_{bmax} = 3$. Each agent was allowed to submit three bids $J = 3$ of which only one could be allocated (XOR-condition) and the bids were generated with probability $p_{tso} = 0.33$. For the calculation of the bid price p_{min} was set to 1 and p_{max} set to 3. The number of agents varied between $I = 25$ and $I = 100$ in steps of 25 for each experiment.

The outcome is shown in Tables 6.11, 6.12, 6.13 and in Figure 6.7. The results were achieved using the benchmark mode of the simulator. The efficiency of the auctioneer's allocation was measured as a percentage of the optimal WDP solution achieved by the IP-CAA. Calculation time associated with the particular methods is denoted on the left part of Table 6.11. Considering all heuristics, GA performs best followed by the SA in terms of allocation efficiency. Computation time, however, is very extensive for the GA. In contrast to the GA, calculations using the greedy methods only take some milliseconds

---

[21] http://lpsolve.sourceforge.net/5.5/
[22] By contrast, in the experiments of section 6.1.3 the GR-CAA was used as a lower bound for performance measurement.

Table 6.11. Performance of the GA, IP, SA, IG, GR algorithm for 25 to 100 agents submitting three structured bids measured in CPU time (msec) and relative income for 100 simulations

agents	solution time in milliseconds					income relative to IP solution				
	GA	IP	SA	IG	GR	GA	IP	SA	IG	GR
25	49,319.2	1,871.0	7,480.4	2.0	0.9	98.75	100.00	93.49	88.57	92.06
50	646,113.7	23,994.5	16,100.5	9.2	2.9	98.39	100.00	92.06	90.06	91.76
75	612,491.7	144,568.6	25,757.2	19.5	5.9	96.99	100.00	92.18	91.67	90.80
100	1,167,030.0	423,363.6	38,053.6	28.7	19.5	96.41	100.00	91.56	91.92	90.96

Table 6.12. Performance of the CAP algorithms for 25 to 75 agents submitting unstructured bids

agents	solution time in milliseconds					income relative to IP solution				
	GA	IP	SA	IG	GR	GA	IP	SA	IG	GR
25	56,359.1	13,097.4	7,035.5	0.78	0.16	97.32	100.00	95.53	84.44	89.31
50	249,822.2	742,617.6	15,767.5	12.3	5.46	95.53	100.00	92.18	91.67	90.80
75	681,647.6	9,751,549.9	29,339.2	22.2	19.3	93.60	100.00	88.67	82.51	87.03

Table 6.13. CA algorithm performance for 25 agents submitting substructured bids

agents	solution time in milliseconds					income relative to IP solution				
	GA	IP	SA	IG	GR	GA	IP	SA	IG	GR
25	51,546.73	321,006.55	7,270.79	4.21	4.40	96.20	100.00	86.69	78.82	85.05

and lead to the lowest allocation quality. CAP calculation using the IP method takes a huge amount of CPU time for unstructured bids which are more complex than structured BMs (see Table 6.12). Computational effort strongly increases with the number of agents in this case. The most challenging bid structure for the IP-CAA is the substructured BM type. The exploding calculation time using the IP method is the reason for omitting the results for 100 agents in Table 6.12 and for 50, 75 and 100 agents in Table 6.13.[23] As a result, the benchmarking function of the CA simulator shows clearly that a greedy algorithm can be used in time critical environments if the quality of allocation is not very important. If a better quality of allocation is required, the SA-CAA and GA-CAA are a good choice whereas the GA has an advantage with respect to the solution quality. The IP-CAA can only be used for small problems and for bid structures that are less problematic than the substructured BM. However, one should keep in mind that heuristics are sensitive to parameter optimization and it is sensible to customize SA and GA variants for specific BM types.

---

[23] A single iteration of the benchmark would last several days on the Athlon X2 4200^+ system used for the simulations.

6.3.2 Comparing Open and Closed Economy Behavior

The next point in the empirical investigation of the combinatorial resource
scheduling system is the stability of allocation for the *open* and the *closed*
economy case. The evaluation of both systems is performed by running an
iterative allocation auction for the basic resources (NET,DSK,MEM,CPU)
with four task agents ($I = 4$) in a 50 round period. The agents were allowed
to formulate their bids according to the BM settings used in the benchmark
scenario of the previous section ($T = 10, q_{bmax} = 3, M = 5, J = 3$ with XOR-
condition, and $p_{tso} = 0.33$). Within this iterative auction the presence of
resource one is reduced from $q_{max} = 8$ to 4 units, thus 50% of the initial level,
at round 25 to test the stability of the system's budget and pricing behavior.

Open Economy

For the simulation of the open economy all agents were initially endowed with
a budget BG^{ini} of 200 MUs. In the following rounds an increment of $BG^{inc} =$
50 MUs was added for each agent in every round. The WDP calculation was
performed using the IP-CAA to enable the calculation and monitoring of
the shadow prices. For this purpose, the agents use the bidding mechanism
based on the SP-formulation described in equation (6.17) together with a price
increment factor of $\Delta p = 0.1$. The result of the simulations is depicted in the
upper part of Figure 6.8. The figure has the following subdivisions according
to the GUI representation of the simulator (see Figure 6.4). In the upper left
part the level of the agents' budgets BG_i is depicted in the course of the
auction rounds k, whereas the upper right part shows the scarcity SR_o of the
single resources. Beneath, the income of the auctioneer Inc^{acc} (left side) and
the estimated market values of the resources $\hat{v}_o$ (right side) are presented in
dependency on the auction round k.

As a result one can make the following observations for the open economy:

- After a few rounds of tâtonnement the auctioneer's income stabilizes at a
  level of approximate 220 MUs together with the discovery of the appropriate market prices for the resources. The entire system works in a state
  comparable to the 'Neo-Walrasian' equilibrium described in section 1.5.4.
- Together with the discovery of an equilibrium state for the prices, resource
  scarcity reaches a stable state on a high level of approximate $SR_o = 0.85$
  for all resources. It is important to mention that the resulting scarcity level
  depends partially on the allocation performance of the WDP algorithm
  employed and the number of bids submitted as shown in section 6.3.1.
  Lower performance of the WDP algorithm leads to lower scarcity levels.
  After reducing the availability of resource one by 50%, the scarcity of
  this resource, which is measured by the auctioneer, increases drastically to
  levels near 1.0, whereas the scarcity of the remaining resources diminishes
  to levels below 0.8. This is a result of the complementarities in the bids

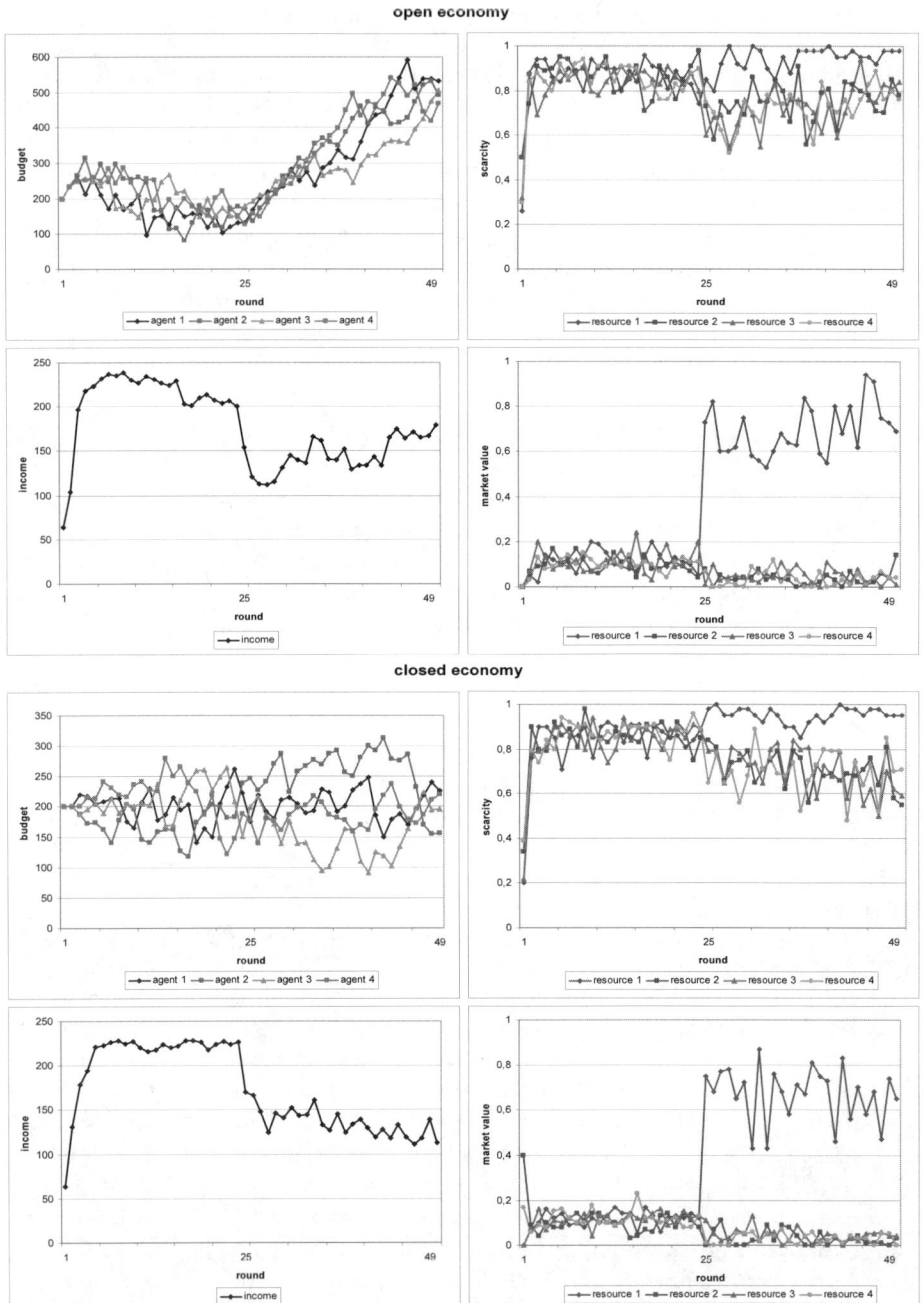

Fig. 6.8. System behavior of an open (upper part) and a closed economy (lower part) in a resource failure situation (shortening by 50% in round 25) in terms of resource scarcity, market value, agents' budget auctioneer's income

of the combinatorial allocation economy presented. Due to the fact that the demand for resource one can not be substituted by the agents using the remaining resources o_2 to o_4, availability of this resource type is the limitational factor in the allocation process.

- After shortening the availability of resource one by 50% the market value that is calculated based on the shadow prices inclines drastically, indicating the effective operation of the pricing mechanism.[24] However, despite the strong incline of the resource prices in the failure situation, the agents' monetary budgets BG_i steadily increase in each round k. This derives from the fact that the failure of resource one is rigorously limiting the expenditures of the agents, preventing them from totally reinvesting their refreshed budgets BG^{inc}. The SP mechanism is only partially capable of compensating for the failure of resource one by a further incline in its price level. The example shows clearly the shortcoming of the open economy. A stable state for agents' budgets is only achievable if BG^{inc} is adapted to SR_o in the allocation process. The pricing mechanism itself is not able to compensate for the heavy discrepancies in BG^{inc} and SR_o. In the case of a higher resource supply SR_o and a lower increase in budget BG^{inc}, agents show a tendency to diminish their budgets until depletion, as can be seen from Figure 6.8 (on top) for the first 25 rounds of the auction process.

Closed Economy

For the simulations in the closed economy, the same parameterization was chosen as for the open economy, except for the constant budget refresh BG^{inc} which is replaced with a mechanism that distributes a quarter of the auctioneer's income at the previous round to each agents' budget in the following round (see section 6.2.2). As a result, the closed system's behavior turns out to be closely related to that of the open system in terms of price and scarcity levels. Budget evolution, however, is different to the open case. As one could expect, agents were not able to accumulate high budget levels or end up with chronically exhausted budgets in this case. The system is forced to balance the agents' budgets with the market values and the availability of the resources. Let us have a short look at the course of the auctioneer's income Inc^{acc} in the iterative process with respect to the differences between the open and the closed economy. The characteristics of both time series are similar because resource availability is dominating both. Only if multiple bids can be submitted by the task agents due to higher budgets does the auctioneer's income increase slightly for the open economy case (see Figure 6.8 upper left for the final rounds). In analogy, Inc^{acc} declines for depleted budgets in the open economy, whereas such situations do not occur in the closed economy, which indicates a higher system stability.[25]

---

[24] An elaborate evaluation of pricing behavior will be presented in section 6.3.4.

[25] Various situations were tested for open vs. closed economy comparison. Results, however, underpin the representative character of the behavior described here.

6.3.3 Comparing Bidding on Scarcity and Shadow Prices

A main topic of this experimental section about dynamic pricing in a combinatorial resource allocation system is the evaluation of the system's differences between the two bidding mechanisms proposed in section 6.2.4. For this reason, comparative simulations between shadow price-based bidding and scarcity-oriented bid formulation will be performed for each factor treated in the following sections. To provide a first overview, the following experiments were executed: Four task agents were bidding for the resources using the dynamic pricing strategy, either based on shadow prices as formulated in equation or based on resource scarcity information as described in equations (6.8) and (6.17). Beginning with an initial budget of $BG^{ini} = 200$, agents submitted one bid bundle containing three alternative bids (XOR-bids, $J = 3$) in the first round. Then the task agents generated three additional bids in each following round k. The pattern and parameterization of the submitted bids was identical to the structured bid type used in the benchmark section 6.3.1 and will remain the same for all further experiments described in this experimental section. This means that the resulting BM structure of the submitted bids follows the type depicted in Table 6.3 with a parametrization of $q_{bmax} = 3, p_{tso} = 0.333$, and $t_{max} = 4$. To gain a broader overview of the system's behavior in both cases, the price acceleration factor Δp was varied between 0.1 and 1.0 by steps of 0.1. While performing 50 simulations with 50 auction rounds for each problem instance, estimated resource market value $\hat{v}_o$, volatility of $\hat{v}_o$, resource scarcity SR_o, and the volatility of SR_o were recorded for both pricing methods:

- In the case of *resource scarcity* used for an appropriate bid price formation, the average market value $\hat{v}_o$ of the resources was at a level of 0.14 MUs for all instances of Δp.[26] Volatility of $\hat{v}_o$ was 0.055 together with a measured average resource scarcity $SR_o = 0.83$ and an associated volatility of 0.1.
- For the *shadow price* based bidding process resource market values $\hat{v}_o$ are at a slightly lower level of 0.11 MUs associated with a corresponding volatility of 0.04. For the resource scarcity a higher value of average $SR_o = 0.86$ could be measured together with a volatility of 0.1.

As a result of this experiment it turns out that bidding on shadow price-based market information enables a better use of resources according to the difference of the measured scarcity values of 0.03 which indicates an improvement of allocation quality by 3%. This observation coincides with slightly lower market values for the resources due to the better use of the resources in the case of shadow pricing. The results confirm the conjecture that shadow price-based market information is a better foundation for the agents' bid formulation. It is a more precise source of information by construction.

---

[26] Interestingly the level Δp had no significant impact on the simulation results described in this paragraph. Therefore the values of the variables observed here, are considered to be representative for all instances of Δp in this model.

6.3.4 Evaluating Resource Pricing Behavior

An important goal of the combinatorial Grid simulation system is to investigate the allocation and pricing behavior in case of resource failure e.g. a drop in CPU capacity. Due to the complementarity in combinatorial bids such a failure has a strong impact on the use of the other resources as already discussed in section 6.3.2. The agents' strategy, that has been defined in the bidding model of section 6.2.4), is to invest more monetary units of their budget BG_i into the procurement of the scarce resource which represents the bottleneck. This means that the price of the bids has to be increased proportionally to the share of scarce resource capacity requested in the current bids. To test this behavior, a drop of capacity of resource one (o_1) is simulated and the resulting market price of the resource is recorded in the *closed economy*. The bidding strategy of the proxy-agents either uses a pricing strategy based on the shadow prices (see equation 6.13) or based on resource scarcity information (equation 6.9). The agents' bidding starts at a level of $p^{ini} = 0.8$ and continues with a price increment by $\Delta p = 0.1$ for the rejected bids. The task agents submit three new XOR-bundled bids according to the BM structure described in the previous section ($M = 3, J = 3, L = 5, q_{bmax} = 3, p_{tso} = 0.333, T = 10$, and $t_{max} = 4$). Similar experiments have been conducted by Schwind, Gujo & Stockheim (2006). However, their work compares a modified version of calculating $\hat{v}_o$ based on scarcity information with the bidding behavior based on shadow prices.

Figure 6.11 depicts an example of a simulation run with 50 auction rounds in a resource shortening situation where the capacity of resource one is cut from eight to three units at round 25. The plots indicate scarcity SR_o for the four resources that is measured by the auctioneer (upper part), the market value v_o of the resources that is based on the calculation of shadow prices SP_o (middle part), and the auctioneer's income Inc^{acc} received from the accepted bids per round k. The increase in scarcity SR_{o_1} measured after the restriction of capacity for resource one in round $k = 25$ is evident. It is coupled with a decline in the use of the other resource types (o_2 to o_4) and a drop in the auctioneer's income Inc^{acc}. The interesting question is now how strong the impact of a resource failure on the system's pricing behavior will be. For this purpose, different levels of remaining resource capacity were simulated and the resulting market values per resource unit were recorded.

Tables 6.14 and 6.15 show the results of this experiment (mean values for 20 simulation runs) while using the IP-CAA to solve the CAP. In the experimental setting used here, the capacity of resource one was stepwise diminished by one unit, beginning with seven units of remaining capacity for the first 20 simulations and ending up with one unit residual capacity for the last series of simulations to emulate a system failure situation. The simulations were performed in two ways: in the first simulation task agents used resource *scarcity information* to make their pricing decision; in the following simulation runs, agents employed *shadow prices* to calculate their W2P. As can be seen

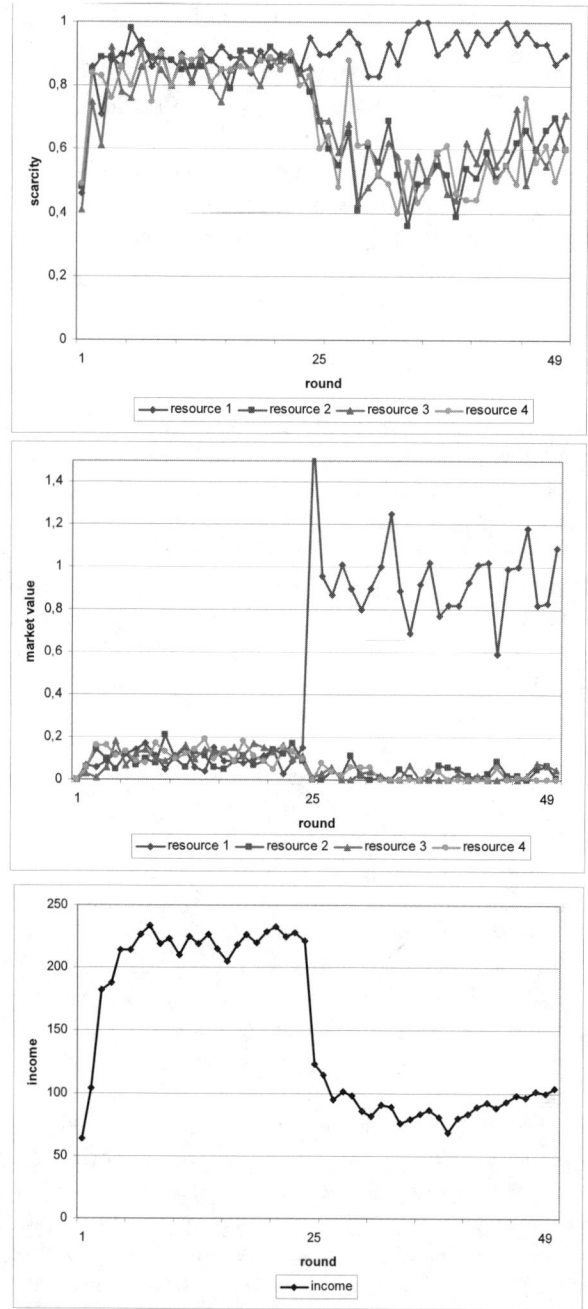

Fig. 6.9. Allocation and pricing behavior for a resource failure situation (shortening from 8 to 3 resource units in round 25) indicated by resource scarcity, market value of a resource unit, and auctioneer's income

Table 6.14. Market prices for resources o_1 to o_4 based on scarcity information and their volatility in dependency of enabled resource units from 1 to 7

market value after shortening						
enabled resource units of o_1						
resource	7	6	5	4	3	2	1
o_1	0.22633	0.34820	0.54608	0.83458	1.32743	2.47199	4.75169
o_2	0.12773	0.10415	0.08263	0.06026	0.04009	0.01023	0.00019
o_3	0.12568	0.10588	0.08053	0.06690	0.03977	0.00895	0.00044
o_4	0.12396	0.10880	0.08795	0.06138	0.04068	0.00993	0.00025

ratio of market value after/before shortening						
enabled resource units of o_1						
resource	7	6	5	4	3	2	1
o_1	1.65367	2.54275	3.79994	5.81576	9.64706	17.4475	34.0993
o_2	0.90263	0.74554	0.59740	0.42942	0.28867	0.07378	0.00123
o_3	0.88054	0.75526	0.58833	0.48812	0.28714	0.06373	0.00319
o_4	0.88033	0.77416	0.63831	0.44563	0.27884	0.07369	0.00190

market value volatility						
enabled resource units of o_1						
resource	7	6	5	4	3	2	1
o_1	0.07727	0.12884	0.22635	0.37710	0.64324	1.25214	2.46566
o_2	0.05625	0.05886	0.06108	0.06895	0.07551	0.07943	0.08412
o_3	0.05623	0.05800	0.06088	0.06598	0.07395	0.08083	0.08163
o_4	0.05713	0.05382	0.06329	0.06876	0.08229	0.07926	0.08022

Table 6.15. Market prices for resources o_1 to o_4 based on shadow prices and their volatility in dependency of enabled resource units from 1 to 7

market value after shortening						
enabled resource units of o_1						
resource	7	6	5	4	3	2	1
o_1	0.24760	0.41316	0.65852	0.94019	1.39200	2.30378	13.2626
o_2	0.12454	0.09561	0.05908	0.03803	0.02262	0.00662	0.00032
o_3	0.11470	0.09157	0.06563	0.04110	0.02074	0.00672	0.00039
o_4	0.12872	0.09361	0.06080	0.03870	0.01913	0.00640	0.00038

ratio of market value after/before shortening						
enabled resource units of o_1						
resource	7	6	5	4	3	2	1
o_1	1.80882	3.08386	4.79043	6.63305	10.1245	17.3053	90.8743
o_2	0.94854	0.68939	0.45242	0.27027	0.15931	0.04995	0.00242
o_3	0.85884	0.69836	0.46571	0.31345	0.14940	0.04847	0.00285
o_4	0.94172	0.69990	0.45506	0.28390	0.14489	0.04642	0.00261

market value volatility						
enabled resource units of o_1						
resource	7	6	5	4	3	2	1
o_1	0.09273	0.16454	0.28396	0.43047	0.66334	1.15314	11.8477
o_2	0.07124	0.06634	0.06620	0.07995	0.08369	0.08214	0.08048
o_3	0.06146	0.06335	0.07263	0.07243	0.07974	0.08470	0.08645
o_4	0.06446	0.06345	0.06968	0.07942	0.07586	0.08306	0.08218

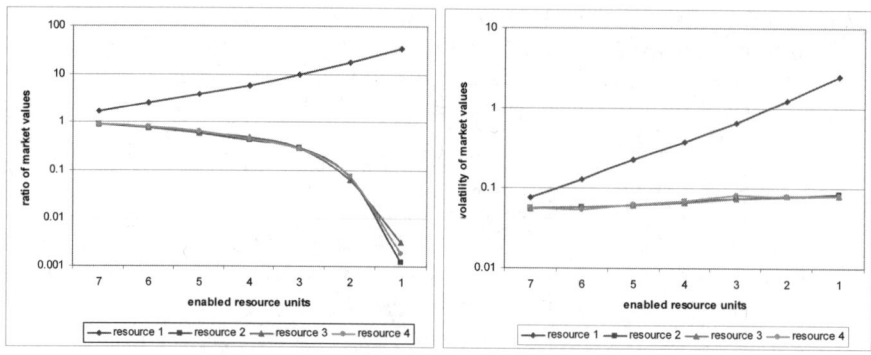

Fig. 6.10. Evaluation of resource pricing behavior in failure situations with bidding behavior based on scarcity information

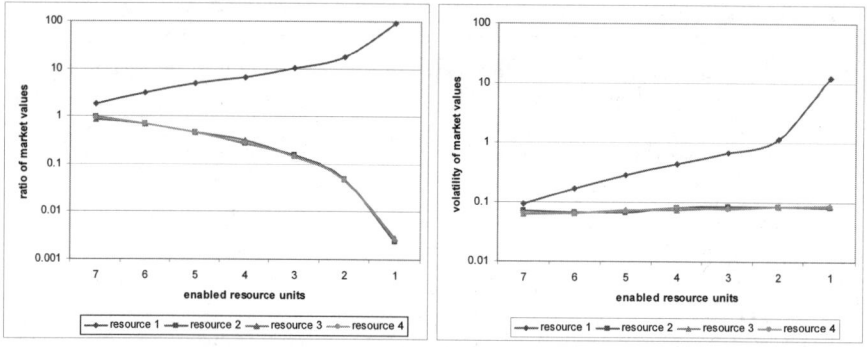

Fig. 6.11. Evaluation of resource pricing behavior in failure situations with bidding behavior based on shadow prices

from Figure 6.10 and Table 6.14, resource shortening had a strong impact on the market values $\hat{v}_o$ in the combinatorial economy while using *scarcity-based* bidding behavior. Figure 6.10 depicts the ratio of the mean market values for the resources after and before capacity cutting in a logarithmic scale. Market values measured by shadow prices incline from 0.23 MUs in the case of a resource reduction by one unit to a level of 4.75 MUs for a resource cut by 7 units. As a result of this behavior, to this behavior the volatility of the market resource prices increases from 0.1 to 2.5 as can be seen in Figure 6.10 (right side) and in Table 6.14 lower part.

In the case of bidding behavior that relies on *shadow prices*, the effect of resource cutting is even more dramatic. In analogy to Figure 6.10, Figure 6.11 exhibits the ratio of market values after/before resource shortening for different levels of remaining capacity together with their volatility. The absolute incline of the market price for a unit of resource one leads from 0.24 MUs for seven units of remaining capacity to 13.26 MUs in the extreme case of one unit of residual capacity (see Table 6.15). Additionally, the market values of the remaining resource types (o_2 to o_4) that are not subject to capacity reduction fall to a level of 0.003 MUs. This extreme pricing behavior is a result of the total absence of substitutionalities for the required resources in the ISIP provision process regarded in this work. In such borderline situations our pricing policy must fail, due to the fact that the system has no notable production capacities to perform its function. However, down to a level of two remaining resource units for resource one the pricing mechanism can be regarded as a useful instrument to reallocate resource load in the PCRA scenario. Despite the fact that the ISIP scenario employed in this thesis does not allow one to substitute a specific resource type (CPU,NET,DSK,MEM) by another resource at the same time, the bidding mechanism can effect an intertemporal shift of resource use in the failure situations considered here. The next section will discuss the effects of such an intertemporal shift of resource load in connection with the individual bidding behavior of the agents.

6.3.5 Testing Different Bidding Strategies

This section will look more closely at two different *bidding strategies* for the task agents and their underlying economic motivation. The results of these strategies will be tested in the light of two simple efficiency measures: the average acquisition price $\bar{p}$ per resource unit spent in the previous auction iterations and the average waiting time in fractions of rounds k until acceptance of a resource bundle required for ISIP provision. The agents' bidding strategies are distinguished by two properties: the price acceleration factor Δp used to dynamically adapt the price for rejected bids and the price multiplier p^{ini} that determines the level at which the bids are initially submitted (see equation 6.18). The economic rationale of these strategies can be described according to the argumentation in Schwind & Gujo (2006):

- An *aggressive bidding agent* that tries to achieve quick bid acceptance using a fast inclining pricing strategy. This behavior can be economically motivated for a task agent that bids for the execution of time critical tasks in an ISIP provisioning system. Such a time critical task can, for example be the hosting of a video conference that is scheduled for a narrow time window. The proxy agents must then prompt for a rapid fulfillment of their resource requests in the distributed computer system by quickly raising their bids to the market level.

- A *smooth bidding agent* that submits multiple XOR bid bundles (OR-of-XOR bids) to the auctioneer while patiently waiting for bid acceptance and slowly increasing bid prices. The economic rationale of this strategy can be motivated by the following example. The agent bids for resources which are required for the fulfillment of ISIP tasks that are not time-critical, like the computation of large time consuming database jobs on a Grid in a broad time window. It is then a plausible strategy for the task agent to acquire the necessary resource bundles at low market values if possible.

The bidding strategy experiments of this section are performed in the *closed loop* economy and assume a uniform production function for all the task agents. This leads to an equal distribution Inc^{acc}/I of the auctioneer's previous income Inc^{acc} ($BG^{ini} = 200MUs$) to the agents' budgets in each round as already described. The strategy applied was either to increase Δp for the rejected bids by a constant 0.2 (see equation 6.18) for the *smooth bidding* agents or to vary the bidding strategy in a range from $\Delta p = 0.1$ to 3.1 for the *aggressive bidding* agents (see Table 6.16 and 6.17). p^{ini} was left constant with 0.8 in this setting, but will also be altered in the experiments in the next section. The remaining parameterization of the bidding behavior was identical to that in the previous sections.[27] Two simulation types were performed in this setting: The first series of simulations employed a task agents' bidding behavior that is based on shadow prices SP_o, the second series used bids formulated on resource scarcity information SR_o.

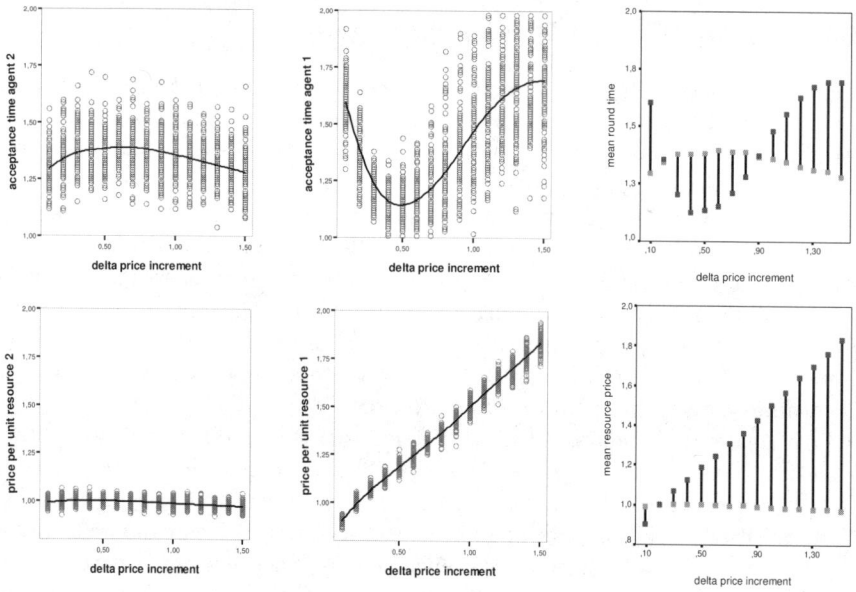

Fig. 6.12. Mean round time (up) and price per resource unit (below) in dependency on price increment Δp for smooth (agent 2) and aggressive bidder (agent 1) with price formation behavior based on shadow prices SP_o

---

[27] Starting with one bundle of three XOR bids ($J = 3$) in the first round, the task agent generated three additional bids ($M = 3$, OR-of-XOR) at each further round. The agents maintained and increased rejected bids over a maximum of $L = 5$ rounds. Newly generated BMs were of structured type ($q_{bmax} = 3, p_{tso} = 0.333, t_{max} = 4$). Maximum load in the CM was $q_{max} = 8$ per resource while T was ten time slots.

Figure 6.12 shows the results of the strategy simulations: 100 runs for each Δp in steps of 0.1 for the task agents in a bidding process based on shadow prices. The aggressive bidder acted competitively against three other smooth bidders. The upper part of Figure 6.12 shows the mean round time until bid acceptance whereas the lower part depicts the average budget spent per acquired resource unit. For an increasingly aggressive agent (higher Δp) the mean acceptance time $\bar{k}^{acc}_{aggr}$ (mean time elapsed until acceptance of a bid measured in fractions of rounds k) reduces by 0.25 for $\Delta p = 0.4$, 0.5, and 0.6 compared to the average acceptance time $\bar{k}^{acc}_{smoo}$ of the smooth bidder (see Table 6.16).

Table 6.16. Results for competing bidding strategies (*smooth, aggressive*) in terms of mean price per acquired resource unit $\bar{p}$ and average round time $\bar{k}^{acc}$ until bid acceptance for task agents with a pricing behavior based on shadow prices SP_o

	price increment Δp														
	0.1	0.2	0.3	0.4	0.5	0.6	0.7	0.8	0.9	1.0	1.1	1.2	1.3	1.4	1.5
$\bar{p}_{smoo}$	0.99	1.00	1.00	1.00	1.00	1.00	1.00	1.00	0.99	0.99	0.98	0.98	0.98	0.98	0.97
$\bar{p}_{aggr}$	0.90	1.00	1.07	1.13	1.19	1.25	1.31	1.36	1.43	1.51	1.57	1.64	1.70	1.76	1.83
$\bar{k}^{acc}_{smoo}$	1.30	1.34	1.38	1.38	1.38	1.40	1.39	1.39	1.37	1.36	1.35	1.33	1.31	1.30	1.28
$\bar{k}^{acc}_{aggr}$	1.60	1.36	1.20	1.13	1.13	1.15	1.21	1.28	1.37	1.48	1.56	1.63	1.67	1.69	1.69
$\Delta\bar{p}$	-0.09	0.0	0.07	0.12	0.19	0.25	0.31	0,36	0.43	0.52	0.58	0.66	0.72	0.79	0.86
Δk^{acc}	0.31	0.01	-0.18	-0.25	-0.25	-0.25	-0.18	-0.11	0.0	0.12	0.21	0.30	0.36	0.39	0.41

While raising Δp, the average acquisition price $\bar{p}$ per resource unit (over all resource types) increases linearly for the aggressive bidder (see Figure 6.12 lower part). As can be seen from Table 6.16 the optimal strategy for the aggressive bidder in terms of waiting time reduction is using a price multiplier increment of $\Delta p = 0.4$ that leads to an $\bar{k}^{acc}_{aggr}$ reduced by an average of 0.25 rounds. The round time reduction is coupled with an increase in mean price for resource acquisition by $\Delta\bar{p} = 0.12$ MUs. A further rise of Δp does not reduce the average acceptance time but leads only to higher $\bar{k}^{acc}_{aggr}$. This might be due to the fact that the aggressive agents' budget is exhausted by the increasingly expensive bidding strategy, temporarily preventing the submission of new bids in the following rounds. It should be noticed at this point that it is necessary to define a utility function for the task agents to determine the optimal bidding strategy while making a trade-off between acceptance time reduction and average acquisition price. This will be done for the entire strategy space (variation of $\Delta p, p^{ini}$) in section 6.3.6.

The results for the simulations using resource scarcity information in the agents' bidding process can be seen in Figure 6.13. Unlike the simulations based on shadow prices where mean acceptance time was approximately 1.35 for the smooth bidding agents, simulations using scarcity information started with $\bar{k}^{acc}_{smoo} = 0.74$, indicating that most initial bids were successful in the first round. With increasing Δp the aggressive bidder was able to reduce the average acceptance time to $\bar{k}^{acc}_{aggr} = 0.48$. This comes at additional acquisition costs of 0.3 MUs per resource unit compared with the smooth bidding agents'

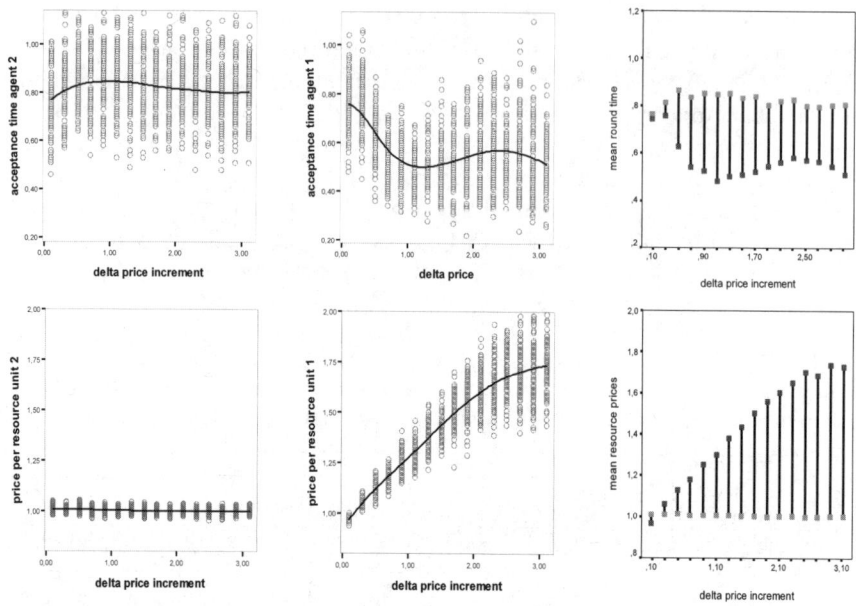

Fig. 6.13. Mean round time (up) and price per resource unit (below) in dependency of price increment Δp for smooth (agent 2) and aggressive bidder (agent 1) with price formation behavior based on resource scarcity information SR_o

average resource costs. Interestingly, a further increase in Δp did not lead to a steady incline of the average acceptance time from this point on as in the shadow price-based simulations. Rather, after reaching a local maximum of $\bar{k}_{aggr}^{acc} = 0.58$, the average acceptance time reduces slightly with an ascending Δp (see Table 6.17). This system behavior might be a little confusing at first sight, but in the light of a system stability reflection it seems to be explicable. A further increase in the aggressive bidders' Δp sustainable perturbs the economic equilibrium of the combinatorial allocation system. High bid prices for the rejected bids of the aggressive bidders prevent the smooth acting bidders from competing with their offers in the market. This is due to the scarcity information mechanism for price formation. The scarcity-based relative value v_o^{rel} in equation (6.17) does not provide sufficient information to adapt the smooth bidders' initial prices such that they are able to compensate for the strategic bidding gap Δp between smooth bidders and the aggressive bidder. However, this comes at a relatively high price, $\bar{p}_{aggr} = 0.73$ MUs, for the aggressive bidder. This is a severe drawback of the scarcity-based bidding mechanism. Fortunately this behavior only occurs in situations where the strategy gap between Δp_{aggr} and Δp_{smoo} is high.

Table 6.17. Results for competing bidding strategies (*smooth, aggressive*) in terms of mean price per acquired resource unit $\bar{p}$ and average round time $\bar{k}^{acc}$ until bid acceptance for task agents with a pricing behavior based on resource scarcity SR_o

	price increment Δp															
	0.1	0.3	0.5	0.7	0.9	1.1	1.3	1.5	1.7	1.9	2.1	2.3	2.5	2.7	2.9	3.1
$\bar{p}_{smoo}$	1.01	1.01	1.01	1.01	1.01	1.00	1.00	1.00	1.00	1.00	1.00	1.00	1.00	1.00	1.0	1.0
$\bar{p}_{aggr}$	0.97	1.06	1.13	1.18	1.25	1.30	1.38	1.43	1.50	1.56	1.60	1.65	1.70	1.69	1.74	1.73
$\bar{k}^{acc}_{smoo}$	0.76	0.81	0.86	0.84	0.85	0.85	0.85	0.83	0.84	0.80	0.82	0.82	0.80	0.80	0.80	0.81
$\bar{k}^{acc}_{aggr}$	0.74	0.75	0.63	0.54	0.53	0.48	0.50	0.51	0.52	0.54	0.56	0.58	0.57	0.56	0.54	0.51
$\Delta\bar{p}$	-0.05	0.05	0.12	0.18	0.24	0.30	0.38	0.43	0.50	0.56	0.60	0.65	0.70	0.69	0.74	0.73
$\Delta\bar{k}^{acc}$	-0.02	-0.06	-0.24	-0.30	-0.33	-0.37	-0.35	-0.33	-0.32	-0.26	-0.26	-0.24	-0.23	-0.23	-0.26	-0.29

6.3.6 Searching for Utility Optimizing Strategies

We will now look more closely at the agents' *bidding strategies* and their underlying economic motivation in terms of an utility function. The goal is to find the utility maximizing strategy for the agent types proposed in the *strategy space* spanned by Δp and p^{ini}.[28] Following Schwind, Stockheim & Gujo (2006) the utility function of an agent is defined by the following term:

$$U_a = \frac{\left(\sum_{(i,j)\in B_a} x_{i,j} \cdot \sum_{o=1}^{O} \sum_{t=1}^{T} q_{i,j}(o,t)\right)^{\gamma}}{(k_a^{acc})^{\beta}} \tag{6.19}$$

Utility function of agent a: $U_a \quad \in \mathbb{R}^+$
Bids of agent a: $B_a \quad \in \{(i,j)\mid i,j \in \mathbb{N}\}$
Bid acceptance time of agent a: $\bar{k}_a^{acc} \quad \in \mathbb{R}^+$
Parameters of the utility function: γ & $\beta \in \mathbb{R}^+$

While the amount of acquired ISIP resource units has a positive but diminishing marginal impact on the agent's utility, the number of periods an agent waits until his bids are accepted has a negative influence. To calculate the decreasing impact of the waiting time $\bar{k}_a^{acc}$[29] is used and β is introduced to adjust the force of the waiting time's impact. The force of the impact of the quantity is defined by γ with $0 < \gamma \leq 1$.[30] Two different types of agents are introduced based on this utility function: (i) the *quantity maximizer* with $\gamma = 0.5$ and $\beta = 0.02$ and (ii) the *impatient bidder* with $\gamma = 0.5$ and $\beta = 1.0$.[31]

(i) A *quantity maximizer* that tries to acquire a large amount of resource capacity. The hypothesis is that this agent follows a *smooth bidding strat-*

---

[28] Whereas the previous section only considered a variation of Δp, the strategy space defined in this section is extended to the dimensions $p^{ini} \times \Delta p$.

[29] $\bar{k}_a^{acc}$ denotes the average number of periods an agent waits until a bid is accepted.

[30] Within this experimental setting $\gamma = 0.5$ remains constant. β, however, varies to express the impact of waiting time depending on the task an agent has to fulfill.

[31] Unlike in the previous section, the agents of this stetting are primarily not intended to represent single bidding strategies, as the *smooth* or *aggressive* bidder do, but their characteristics are defined via their utility function.

egy, i.e. he increases the bid prices slowly. The economic rationale for this type of agent strategy follows the same argumentation as that provided for the smooth bidding behavior in the previous section 6.3.5.

(ii) An *impatient bidder* that suffers if he cannot use the resources instantaneously and will use an *aggressive bidding strategy*. This agent has to submit high initial prices, but overpaying will reduce the quantity he can acquire (see previous section). Moreover, we analyze whether a fast inclining pricing strategy can help to further increase the utility of this agent. The economic motivation of this behavior is analogous to the case of the proxy agent pursuing an *aggressive bidding strategy* presented in the section 6.3.5.

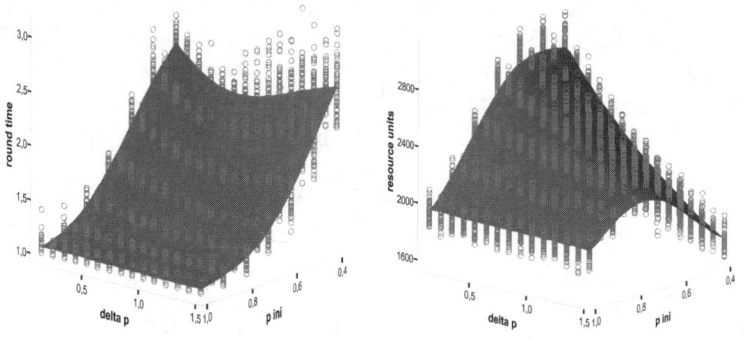

Fig. 6.14. Mean acceptance time and quantity of resource units for test agent bidding with varying price increment Δp and initial price p^{ini}

For the evaluation of the model, a four agent setting was used, with three agents representing a *smooth bidding* behavior using fixed $\Delta p = 0.2$ and $p^{ini} = 0.5$ and a single *test agent* with a varying bidding strategy between $\Delta p = 0.1 \ldots 1.5$ and $p^{ini} = 0.4 \ldots 1.0$. The applied strategy was either to increase Δp for the rejected bids (that were initially submitted with $p^{ini} = 0.5$) by a constant 0.2 for the *smooth bidding* agents or to vary the strategy (see Table 6.18 lower part) for the *test agents*.

The objective of the experiments was to discover the optimal bidding strategy for the test agent in competition with the remaining smooth bidding agents given the two types of utility function (quantity maximizer, impatient bidder) defined above. Except for Δp and p^{ini}, all agents used the same type of bid structure and bidding behavior. It is identical to that described in the previous sections (structured BMs OR-of-XOR with $J = 3$ and $M = 3, L = 5, q_{bmax} = 3, p_{tso} = 0.333, t_{max} = 4, q_{max} = 8, T = 10$). The pricing information provided by the auctioneer is based on the market values v^{shad} calculated by employing shadow prices SP_o as formulated in equation (6.13). A further discussion of this topic is provided in Stockheim, Schwind & Gujo (2006).

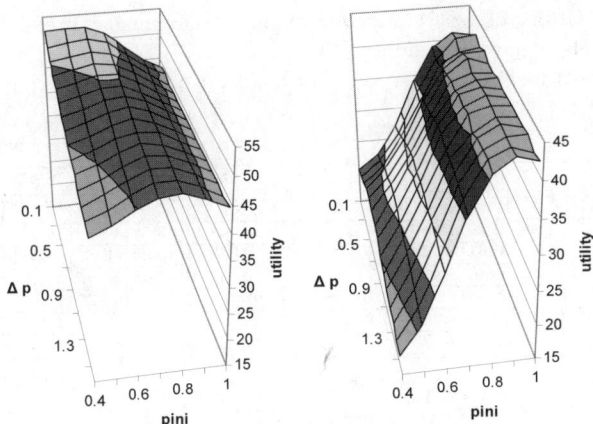

Fig. 6.15. Utility functions of the test bidder for quantity maximizing preference $\beta = 0.01$ (left) and impatient bidding behavior $\beta = 1.0$ (right) for determination of the optimal bidding strategy under varying price increment Δp and initial price p^{ini}

Figure 6.14 shows the results of the simulations, 50 runs for each Δp and p^{ini} in steps of 0.1, for the test agent in terms of resource units acquired per simulation run and the average bid acceptance time k_a. Due to the assumption of a closed loop economic system with equal production functions for all task agents within the setting used here, the same budget is available to all market participants for each simulation run. Figure 6.14 shows that for small Δp and p^{ini} most resource units could be acquired by the test agent. An aggressive strategy with high Δp and p^{ini}, in contrast, leads to a declining amount of acquired resources. While a significant shortening of acceptance time is mainly achieved by high p^{ini}, increasing Δp only impact on average acceptance time for low p^{ini}. The data depicted in Figure 6.14 serves as input for the calculation of the utility function according to equation (6.19). In Figure 6.15 the utility resulting from varying price increment Δp and initial pricing p^{ini} is depicted. It pays off for the quantity maximizing (patient) bidder to wait to see if its bids fit the current allocation by a relatively low price (low increment and initial price). In contrast, the aggressive bidder gains low utility from such a strategy (Figure 6.15 right side). The patient bidding agent receives the highest utilities by using an initial bid price close to the market value of the resources ($p^{ini} = 0.9$). Interestingly, the price increment in the following round does not have much impact on the acceptance time and thus on the utility of the aggressive test bidder. However, for bids exactly at market value utility declines sharply, signaling the peril of *'overbidding'* or simply paying too much for the ISIP resources required. This underlines the importance of accurate market value information to achieving good allocations. The shadow price controlled combinatorial grid enables agents to implement efficient bidding strategies according to the users' utility function.

Table 6.18. Utility of test bidder for quantity maximizing preference $\beta = 0.01$ (upper table) and impatient bidding behavior $\beta = 1.0$ (lower table) for determination of the optimal strategy under varying increment Δp and initial pricing behavior p^{ini}

p^{ini}	Δp														
	0.1	0.2	0.3	0.4	0.5	0.6	0.7	0.8	0.9	1.0	1.1	1.2	1.3	1.4	1.5
0.4	53.05	51.41	50.22	49.14	48.45	47.42	46.47	45.85	45.00	44.09	43.58	42.72	41.79	41.50	40.79
0.5	53.36	52.03	50.84	50.09	49.16	48.45	47.67	46.95	46.20	45.59	44.89	44.23	43.71	42.96	42.56
0.6	53.34	52.46	51.57	50.81	50.22	49.57	48.98	48.32	47.82	47.30	46.71	46.27	45.78	45.24	44.84
0.7	51.93	51.53	50.72	50.30	50.18	49.71	49.36	49.10	48.76	48.09	47.99	47.63	47.36	46.91	46.51
0.8	48.81	48.66	48.69	48.36	48.28	48.10	48.10	47.97	47.68	47.80	47.26	47.29	47.23	46.98	46.87
0.9	45.96	46.08	46.06	46.13	46.02	45.84	45.83	45.85	45.76	45.69	45.69	45.80	45.65	45.56	45.52
1.0	43.80	44.25	43.64	43.68	43.93	43.79	43.85	43.71	43.65	43.81	43.71	43.59	43.57	43.39	43.63

p^{ini}	Δp														
	0.1	0.2	0.3	0.4	0.5	0.6	0.7	0.8	0.9	1.0	1.1	1.2	1.3	1.4	1.5
0.4	21.08	21.48	21.96	22.73	22.85	22.42	21.60	21.38	20.30	19.30	18.84	17.84	17.45	16.86	16.64
0.5	23.34	23.91	24.57	25.65	25.72	26.25	26.33	25.67	25.03	24.59	24.04	23.17	22.33	21.49	20.98
0.6	27.73	28.01	28.88	29.80	30.38	30.81	31.43	30.78	30.92	30.74	30.02	30.23	29.36	28.43	28.04
0.7	34.74	34.30	34.10	35.56	36.06	36.06	36.98	37.36	37.56	36.72	36.89	36.56	36.72	36.14	35.57
0.8	39.69	38.95	39.44	40.13	40.56	40.81	40.78	41.51	41.01	41.43	40.24	40.99	40.97	41.15	40.78
0.9	41.77	41.95	41.12	42.13	42.65	41.70	41.96	42.31	42.43	42.34	42.20	42.66	42.39	42.35	42.64
1.0	41.15	42.28	41.50	42.14	41.50	41.49	42.10	41.71	41.97	42.05	41.93	41.93	42.33	41.44	41.83

In the course of this chapter an agent-based simulation system for combinatorial resource allocation has been developed and tested in various system states, like resource failure situations. Additionally, several WDP solving heuristics have been implemented and benchmarked using the simulation system. A major effort was made to enable the resource consuming task agents to submit suitable bids that express their W2P for resource bundles required for ISIP provision at a price that is close to a market clearing value identified by the combinatorial auctioneer. For this non-trivial task – due to non-linear valuations in the bundles the calculation of a unique resource price is not always possible from the mathematical point of view – two dynamic pricing methods have been proposed. One method relies on the scarcity of resources measured by the auctioneer, provided as relative price information SR_o to the bidding task agents. The other method uses the calculation of shadow prices SP_o in connection with an interpolation algorithm as part of a WDP solution employing IP methods. As a result of this chapter it turns out that shadow price-based resource allocation within the implemented combinatorial scheduler has several advantages in terms of stable pricing behavior in resource failure situations, allocation quality in general, and especially in connection with the use of different bidding strategies performed by the task agents. These bidding strategies have been assessed using a utility function that is capable of mapping differently weighted time and quantity preferences in the bidding process. These results will be turned into recommendations for the construction and use of combinatorial auctions in real world PCRA environments which will enable the provision of ISIP tasks in the concluding chapter.

Comparison of Reinforcement Learning and Combinatorial Auctions

The concluding chapter of this thesis will first summarize and compare the properties of the two main methods *reinforcement learning* and *combinatorial auctions* with respect to their application for PCRA in distributed computer systems. In a second step the main results of the empirical evaluation and simulations performed within this work will be discussed in terms of their application to the ISIP task allocation problem portrayed in the initial section 1.4. A catalog of recommendations derived from the results will be given for the practical application of the RL and CA methods investigated in real world IT environments. The chapter concludes with a survey of the possible development of this research area and provides a short survey of further questions that arise from the results of this thesis and are worth investigating in future research.

7.1 Properties of Reinforcement Learning and Combinatorial Auctions for ISIP Provision

This section discusses the properties of RL and CAs with respect to their application to ISIP provision processes. Two main facets are regarded within this comparison: *economic aspects* and *technological properties*.

7.1.1 Economic Properties of Dynamic Pricing Using RL and CA

Table 7.1 depicts the *economic properties* of dynamic pricing and resource allocation for RL-based yield management and combinatorial auctions. The properties investigated in this summary are deduced from the criteria that were introduced in the context of the dynamic pricing and automated resource allocation categorizations discussed in sections 2.1, 2.2, and 2.3.

- **Optimization objective:**
  The main optimization objective of the *market mediator* in the RL scenario as well as the in CA setting of this thesis was *income maximization*.

This optimization objective is necessary to guarantee the Pareto optimal allocation of the ISIP tasks according to the *market participants'* W2P. In the case of the RL mediator, an additional *yield maximizing* objective can be implied due to low marginal costs for capacity expansion in YM settings (see section 4.2.1).

- **Economic system:**
  Two types of economic systems have been investigated in the work presented. *Open-loop economies* were tested in both RL and CA settings and a *closed-loop economy* was evaluated for the CA allocation system. In the open-loop setting *task agents* only had the function of consumers with an externally defined income acquiring resources for ISIP provision, whereas in the closed-loop scenario *task agents* and *resource agents* are coupled together as prosumers in a "Neo-Walrasian" economy (see section 6.2.1 and 1.5.4).

- **Allocation problem:**
  The allocation problem type addressed in both the RL and the CA setting can be characterized a *combinatorial allocation* problem according to the classification given in section 2.2.3. This allows the consideration of complementarities between multiple resources. Since it is computationally easier to handle than the combinatorial allocation problem, a single resource variant of the *economic job-shop scheduling* problem was additionally used to evaluate the RL model (see section 4.4).

- **Allocation efficiency:**
  The quality of the resulting allocation achieved by employing RL and CA methods depends on the technological performance and the economic construction of the tested methods. In the case of the RL yield optimizing setting, the degree of price differentiation with respect to the task agents' W2P and the learning capability of the acceptance process plays the crucial role (see section 2.1.4 and 2.1.5). For the CA scenario the auctioneers' process guarantees an optimal allocation if bidders reveal their W2P truthfully, which can be assumed for the GVA and in the case of proxy-bidding agents (see section 5.1.8).

- **Response to market:**
  The time taken to respond to the market strongly depends on the iteration frequency of the auction in the CA setting (see section 2.1.2). Iteration frequency itself is coupled to the performance of the winner determination algorithms. Heuristics can help to increase the time taken to react to market fluctuations by reducing winner determination effort significantly for large problem scales (see section 6.1.2). For the RL yield optimizing system response to market is controlled by the learning rate of the RL algorithm (see section 2.1.5). RL rate can be adjusted within limits but also depends on the technological parameters of the RL system, like the existence and efficiency of state space compression methods (see section 4.4.6).

Table 7.1. Economic properties of dynamic pricing and resource allocation using reinforcement learning and combinatorial auctions

	Reinforcement Learning	**Combinatorial Auction**
Optimization Objective	Income Maximization, Revenue/Yield Maximization (see section 4.2)	Income Maximization (see section 5.1)
Economic System	Open-loop Economy (see section 6.2.2)	Open-loop, Closed-loop, "Neo-Walrasian" Economy (see section 6.2.2 and 1.5.4)
Allocation Problem	Economic Job-Shop Scheduling, Combinatorial Allocation (see section 2.2.3)	Combinatorial Allocation (see section 2.2.3)
Allocation Efficiency	High efficiency, if price differentiation classes and contingents are appropriate. (see section 2.1.4 and 2.1.5)	Pareto-optimal, if bidders are brought to express their W2P in the bid bundles truthfully. (see section 5.1.8)
Response to Market	Depends on the learning rate of the RL system, having a broad spectrum in the simulations. (see section 2.1.5)	Depends on the repetition frequency of the CA, which is high in the thesis's simulations. (see section 2.1.2)

7.1.2 Technology Properties of Dynamic Pricing Using RL and CA

As can be seen from the previous section, economic properties in the RL and CA scenarios considered here exhibit a direct relationship to the *technological properties* of the resource allocation methods. Therefore, the technological properties of dynamic pricing and resource allocation using RL and CA methods are discussed separately in this section. Table 7.2 depicts an overview of these technology properties that can be derived from the economic properties.

- **Optimization Problem:**
  For the RL yield maximizing scenario used within this thesis the underlying optimization problem can be regarded as a *multidimensional stochastic knapsack problem (MSKP)* which reduces to a single dimensional stochastic knapsack problem in a single resource allocation problem (see section 4.2.2). In the case of PCRA employing CAs, the underlying optimization problem has been identified as a *weighted set packing problem (SPP)* which is an instance of *multidimensional knapsack problems* (see section 5.1.5).
- **Optimization Algorithm:**
  In the case of the yield optimization scenario, acceptance decisions for ISIP tasks were primarily controlled using *reinforcement learning* techniques. Additionally, *artificial neural networks (ANN)*, *stochastic dynamic programming (SDP)*, and *genetic algorithms (GA)* were evaluated to solve the underlying MSKP (see section 4.5). For PCRA employing CAs to find an optimal allocation of ISIP tasks GAs have also been used to solve the weighted SPP. Furthermore, *integer programming (IP)*, *simulated annealing (SA)*, and *greedy heuristics* were employed to solve the *winner determination problem (WDP)* (see section 5.2).

Table 7.2. Technological properties of dynamic pricing and resource allocation using reinforcement learning and combinatorial auctions

	Reinforcement Learning	**Combinatorial Auction**
Optimization Problem	Multidimensional Stochastic Knapsack Problem (see section 4.2.2)	Multidimensional Knapsack Problem, Weighted Set Packing Problem (see section 5.1.5)
Optimization Algorithm	Reinforcement Learning, Genetic Algorithm, Artificial Neural Network, Stochastic Dynamic Programming (see section 4.5)	Integer Programming, Genetic Algorithm, Simulated Annealing, Greedy Heuristic (see section 5.2)
Computational Cost	High costs in the RL process esp. for NYM, low costs if the value function has been learned (see section 4.4 and 4.5)	High costs to determine the exact solution of the WDP for greater problem instances (see section 5.2.1)
Allocation Cost	Costs are a function of learning and evaluation effort, tendency to high aggregated costs, No resource price information (see section 4.4 and 4.5)	Costs solely depend on WDP calculation, strongly reducible by the use of heuristics, Resource prices are available (see section 6.3.1 and 6.2.3)

- **Computational Cost:**
  The calculation effort of PCRA for ISIP provision using a RL system as well as a CA scenario directly depends on the specific algorithm that is employed to solve the underlying optimization problem. Reinforcement learning systems, however, have to undergo two processing phases, the learning phase and the evaluation phase. Whereas the evaluation phase in general is performed in milliseconds, the learning phase tends to be very time consuming, especially if no adequate state space compression methods are employed, making the problem computationally intractable (see section 4.4 and 4.5). In the case of CA controlled ISIP resource allocation, winner determination is also intractable for larger problem sizes due to the $\mathcal{NP}$-hardness of the SPP. Complexity reduction measures and heuristics are able to keep the WDP within acceptable limits (see section 5.2.1).
- **Allocation Cost:**
  Allocation costs normally include all efforts to achieve a feasible resource allocation. This includes the effort of formulating the bids and the associated W2P along with the computational costs of winner determination. In this context, however, only computational costs that incur at the market mediator are considered as allocation effort for both the RL and CA scenario. This includes the auctioneer's effort to provide valuable pricing information to enable the task agents to formulate their bids. For the RL scenario, learning and evaluation costs are considered, as well as the effort for the provision of resource prices (see section 4.4 and 4.5). The allocation costs for the CA-based system do not include the proxy-agents' efforts to formulate the bids and their W2P (see section 6.3.1 and 6.2.3).

7.2 Main Results and Recommendations

The following section summarizes the main results of the thesis with respect to a real world application of the two resource allocation methodologies discussed: *reinforcement learning* and *combinatorial auctions*.

7.2.1 Reinforcement Learning

The primary research objective of deploying an RL system was to identify a yield maximizing allocation of production requests to resources in an ISIP provision environment. Resource requests, thus arrived in a stochastic process and the market mediator had to make an immediate decision whether to accept or to reject these tasks. After a fixed time interval task acceptance phase is closed in the RL-YM model of this thesis and the production period for the ISIP provision starts. The sum of the prices associated with accepted tasks is the final reward of the mediator in this model. The mediator's challenge was to find a suitable acceptance policy that yields a high total reward. This was attained by using RL methods two experimental scenarios were tested:

- **Single Resource Allocation**
  This scenario describes an *economic job-shop scheduling problem* where the tasks only require one resource type that has to be allocated by the market mediator. A slack time was associated with the desired production time and penalty costs were defined for job execution after due date. Compared to a simple allocation heuristic the income of the market mediator (*allocation heuristic* respectively) was about 10% to 30% higher if the RL system has been used (see section 4.4.6). However, the system's *allocation costs* were very high due to the long learning phase required to achieve a satisfying allocation result. Likewise, *response to market* time of the RL-YM system increased due to slow learning behavior, making the system only recommendable for PCRA environements with moderate fluctuations of demand and supply. Additionally, an extensive training phase had to be performed before deployment in real world applications. The single resource RL-YM system is constructed to be utilized in *open-loop economy* environments. For this reason, the system is more appropriate for deployment in PCRA settings where 'real money' is used to control the alllocation process than in *barter-oriented* systems with a virtual currency such as in the CA-controlled *closed-loop economy* considered in section 6.3.2. Naturally, resource complementarities can not be taken into account in this setting due to the single resource property of the scheduler. However, it is conceivable that the system could be applied in a *decentralized economy* where multiple RL schedulers compete for task execution and simultaneously learn a yield maximizing task acceptance function. An example of a real world application of the single resource RL-YM system could therefore be the leasing of *non-volatile memory* space (DSK) on different server

systems to database application providers. In such a case, demand fluctuations are expected to be low and the main bottleneck is DSK capacity making the consideration of resource complementarities obsolete.

- **Multiple Resource Allocation**

  The second setting for the RL yield maximizing system introduced complementarities and substitutionalities into the requested resource bundles. Bundles were expressed as bid matrices submitted to the market mediator by the task agents. Consequently the acceptance or rejection decision of the market mediator was then more complicated according to the underlying *multidimensional knapsack problem.* The resulting optimization problem is often called a *network yield management (NYM)* problem in the yield management context. RL was used to learn the correct valuation function for the yield maximizing acceptance or rejection decisions while including an ANN for state space compression. The results of the RL-YM system were compared to a solution employing a pure ANN, learning from the exact solution achieved by SDP, and a mixed method using an ANN and a GA. Compared with the exact solution calculated by SDP, the RL-YM system achieved the best performance, reaching about 95% of the optimal acceptance policy (see section 4.5.3). A drawback of the RL-YM system for multiple resources, however, is the fact that the system has been trained and evaluated only with a limited set of resource bundles. This kept *computational costs* in bounds and simultaneously reduced *allocation costs* to a sustainable degree. Like the single resource allocation system, the multi resource RL-YM is designed as an *open-loop economy* making the use of a 'real world currency' in connection with the task agents' budgets for resource acquisition a reasonable solution. Altogether, the multiple resource RL-YM method seems to be suitable for real world applications where the requested resource bundles do not have a complex structure and the requests follow a small number of predefined patterns. A suitable candidate for this dynamic allocation method is the assignment of ISIP requests for resource bundles in a distributed computer system, where the requests have predictable demand profiles. These tasks might be the scheduled calculation of transportation routes for a logistic services cooperation or the cyclical risk calculation for the assets of a financial institution. A major problem in all approaches performing a combinatorial allocation is the fact that the calculation of unambiquous resource prices for the single resources is not trivial. This problem was addressed in particular within the CA allocation setting. For the RL-YM system a satisfing solution for the pricing problem is even harder to achieve and was therefore omitted due to further increasing allocation costs. However, the absence of at least estimated prices in the RL-YM setting has direct impact on the *allocation efficiency* of the system because the bidders can not formulate their W2P associated with the bundles by using resouce market prices. This was a major reason for the introduction of CAs to solve combinatorial allocation problems, apart from the moderate *learning performance* and *market response* of RL.

7.2.2 Combinatorial Auctions

Motivated by the shortcomings of the RL-YM-based resource allocation solution especially for the interesting case of combinatorial allocation problems, a resource allocation system was built using a market mediator that performs a CA to schedule the ISIP task requests. The system can act as an *open-loop* as well as a *closed-loop* economy. In both cases a budget management system is used to control the monetary flow between task agents, resource agents and the market mediator in the resulting "Neo-Walrasian" economy. The bidding behavior of the individual agents can be configured in various ways to test the efficiency of bidding strategies (see section 6.2) giving the Grid-like system a high degree of flexibility. In order to reduce *computational cost* in the CA-controlled system, different *optimization algorithms* for the underlying weighted SPP *optimization problem* were developed and tested by integrating SA, GA and greedy methods together with an exact IP approach. Additionally, the combinatorial allocation system is able to provide market prices for the resource types to the task agents, enabling them to formulate an adequate W2P in their bid bundles. Two types of price determination methods are used in this context, a scarcity-based price information which is easy to calculate, and a computationally more demanding price calculation based on a dual LP (shadow prices) yielding more precise data.

In the following, a reasonable allocation system configuration will be illustrated in connection with four application examples:

- For a real world application demanding a fast allocation of ISIP tasks at low cost the connection of a winner determination heuristic such as, e.g., the SA method and a market price estimator based on resource scarcity information should be used within the system. Such an application could, e.g., be a combinatorial Grid that provides resources to a large scale Web-service in order to retrieve, replicate, and generate consumer customized portfolio data. These requests should have a high volatility in resource demand and a short access latency time making a quick response to market necessary.

- If a high efficiency is required in the combinatorial allocation process, the use of an IP winner determination algorithm in connection with shadow price information is advisable. This, however, is associated with high computational costs making the system unable to find a feasible allocation for bigger problem sizes. A precise allocation method is recommendable if bundle requests comprise a large amount of resource capacity making the packages a valuable asset. The booking of Web-based capacity for the TCP/IP-based transmission of a public event by a media content provider may serve as an example for such an application scenario.

- An alternative way to vary the CA system's behavior is to adjust the bidding strategies of the task agents in the Grid-like system. In some cases it is necessary to model aggressive bidding agents that submit combinatorial bids using a fast inclining pricing strategy. The execution of time critical tasks in an ISIP provisioning system might serve as motivation for such a behavior. Again the performance of a video conference in the distributed computer system seems to be an appropriate example if the conference has to be scheduled for a narrow time window. Task agents then are urged to bid for a prompt fulfillment of the tasks on the resources by using an aggressive pricing strategy (see section 6.3.5).

- If agents bid for the resources required to fulfill ISIP tasks that are not time-critical a moderate pricing strategy is indicated. An example of this may be the computation of large time consuming database jobs in the Grid like system where the execution is scheduled in a relaxed time window. A plausible strategy for the proxy bidding task agent is then to try to acquire the required resource capacity bundles at low market values by only slightly increasing their W2P. In an environment where different strategies were competing with each other, different bidding strategies for the task agents were identified (see section 6.3.5). These strategies were evaluated using a utility function that is capable of expressing a trade-off between waiting time for task acceptance and the quantity that can be be purchased with a given budget in an iterative CA (see section 6.3.6).

For both, CA and RL related systems empirical evidence was derived from a survey about *dynamic pricing preference* that was accomplished in the context of this thesis (see chapter 3). As the main result it turned out that dynamic pricing and allocation systems have to be easily integrable into existing IT infrastructure to gain economic importance for real world applications. In particular, ISIP-related businesses, like industries selling information on the Internet (e.g., media industry), are receptive to dynamic pricing procedures supporting the application of such techniques in Web-based systems. Though CAs are not widely known in industries in the context of dynamic pricing, bundle pricing is accepted as an important pricing mechanism in practice. This leads to the justified expectation that a further development of "easy-to-use" RL-YM and CA systems will result in a broader prevalence of dynamic pricing techniques, especially for the ISIP sector considered in this work. Notably, the CA-based allocation system developed within this thesis could help to achieve that goal, because it allows an individual configuration of the bidding agents to perform ISIP provision tasks for various industrial purposes. This intention is supported by the decision support framework for the design of CAs that was presented in section 5.3.2 and an evaluation procedure for the functionality of the designed CAs illustrated in section 5.3.3. Additionally, a configuration and communication language for CAs has been developed but not yet integrated into the framework of this thesis. This *combinatorial auction meta language (CAMeL)* (Schwind et al. 2004) is a further step towards reducing the effort for integration of CAs into existing IT structures because it allows the configuration, testing and performance of different CA designs that can afterwards be stored and retrieved for specific real world application situations. A highly standardized and preconfigured CA and YM software for PCRA should be the right direction to bring the proposed dynamic pricing technology in real world environments (Stockheim, Schwind & König 2003*a*). The next section will therefore discuss such future developments and will provide an outlook onto future research issues.

7.3 Outlook and Future Research

In this thesis a wide variety of *dynamic pricing* problems were investigated, including two main *automated allocation techniques* derived from the *artificial intelligence* and *operations research* domain: *reinforcement learning* and *combinatorial auctions*. For this purpose *multi-agent system (MAS)* settings were used to evaluate the technical and economic properties of the proposed resource allocation systems. This research method corresponds well with the still young tradition of *agent-based computational economics (ACE)* enabling *deeper empirical* and *normative understanding* on the one hand, and *qualitative insight* as well as *methodological advancement* on the other hand (see section 1.5.5). While thinking about future research many extensions of the existing models and experiments cross one's mind, especially for the combinatorial Grid allocation system. However, only some ideas should be mentioned here, organized in groups dealing with technical issues, economic questions and work for practical applicability.

Working areas for future research targeting to improve the technical performance of the RL-YM and CA systems presented could be:

- the introduction of improved learning and state-space compression techniques for RL-YM system to ameliorate the response-to-market behavior,
- the calculation of estimated resource prices in the combinatorial RL-YM setting in analogy to the pricing techniques used in the CA Grid simulation system (see section 4.5),
- the provision of more precise resource price estimators derived from scarcity information in the CA-based allocation system (see section 6.2.3).

Further research into open economic questions could be conducted along the following research directions:

- the use of parallel learning RL-YM agents competing in a distributed resource allocation economy from the perspective of evolutionary economics, meaning that agents will be withdrawn from the system if they fail to reach a minimum revenue goal,
- the introduction of incentive compatibility aspects into the agents' bidding behavior and the auctioneer's winner determination process in the CA setting (see section 5.1.8),
- the investigation of more than two differentiated bidding strategies within the same simulation setting to come closer to real world settings where users can individually define task agents' behavior for their application purpose (see section 6.3.5).

When coming to work on the practical application of the developed RL-YM and CA systems the following points seem of interest:

- the further development of the decision support system introduced in section 5.3.2, especially for the design task of ISIP allocation processes while integrating the experience gained from real world applications (see section 5.3.2),
- a deeper empirical evaluation of dynamic pricing preference, especially in the ISIP-related industry sector (see section 3.4),
- the validation of various CA designs for ISIP related resource allocation according to the two step review process sketched in section 5.3.3.

The practical implications of this work are about to be evaluated in a new research project called *financial business grid (FINGRID)*. It is intended to link networks of IT-infrastructure in enterprises of the financial industry by employing a Grid concept to provide a huge amount of IT-resource capacity for the users of complex financial software. FINGRID will allow the introduction of sophisticated financial engineering and analysis software without major investments in new specialized high-performance infrastructure. Besides this advantage of Grid technology, the question of accounting and billing for the use of the bundled resources arises in connection with the topic of an incentive compatible scheme that motivates the owners of IT infrastructure in different departments of an enterprises to offer their resources. The automated allocation system described in this book is able to provide the dynamic price information which is necessary to realize such a market-oriented Grid system. In this context the discussion about incentive compatible mechanisms in CAs (see section 5.1.8) should be intensified with respect to the pricing mechanism presented here. Furthermore the user acceptance of dynamic pricing mechanisms has to be evaluated more intensively with respect to the application domain of financial business and a direct comparison to fixed-price models has to be accomplished. This can be done by using the basic experience collected in the Internet survey on dynamic pricing methods that has been presented in chapter 3. Two types of markets are of interest within the upcoming FINGRID project:

- *intra-enterprise* resource markets
- *inter-enterprise* resource markets

Whereas in the case of an intra-enterprise resource market the incentive compatibility aspect retreats behind the pricing and accounting task, the inter-enterprise application case requires a highly elaborate incentive structure. Additionally, security aspects play a predominant role in inter-enterprise resource markets. All these aspects will be topics of research in the FINGRID project, while bringing the technical prerequisites that have been elaborated in this book into practice. Another aspect of extending the research of this thesis within the application oriented financial business grid would be not only to allow the bundling of hardware resources but also of software elements within a the complex engineering systems used in the financial industry. Software design and implementation is quite expensive in the financial domain and the number of potential users for this specialized software is limited. One way to organize a use-oriented amortization of such software components is to let bidders pay per use for bundles of software modules required in the intra-enterprise production process. This bundle pricing can be done in analogy to the resource bundle pricing presented in this work and combined with the accounting and pricing of the underlying IT-resources. A further facet that has not been discussed yet but may be of increased importance is the integration of quality of services factors into the service and resource provision models (Neumann et al. 2006). The quality level of the service provided could be taken into account within the CA model presented in section 6.2 by introducing a multi attributive auction where the required service level influences the bid formation of the users and acceptance decision of the auctioneer.

Another interesting aspect in the context of a market approach for the use of Grid services in the financial industry is the question of how to map the virtual prices and resulting debits for the consumption of resources and services to prices in a real currency. One simple way to do this might be to take the total cost of the entire Grid system and allocate it to the users of the Grid according to their proportion of the aggregated debit accounts of the users measured in the virtual currency. This should be a valid solution for the intra-enterprise approach, but is not very suitable for an inter-enterprise system, where Grid system partners are often interested in realizing some extra profit from the provision of their resources to other organizations. This question was recently addressed by a paper by Brunelle et al. (2006) which describes a scenario with markets for Grid resources in different organizations. These market places are allowed to issue different virtual currencies. The exchange rate between these currencies is to be determined by central bank-like institution by using equilibrium models. Additionally the system proposed by Brunelle et al. (2006) allows the free definition of a continuous utility function for the W2P determination processed by the proxy agents that bid for the system resources on behalf of the Grid users. Such a bidding behavior that is controlled by a continuous utility function seems to be a reasonable extension of the CA simulation model that has been presented in section 6.2. Finally, the formulation of a realistic production function for the resource agents in the closed-economy allocation model addressed in section 6.2 has to be tackled. Measuring the supply and demand function of real Grid users in the FINGRID environment will help us to evaluate an important aspect of economically inspired IT resource markets that has not been investigated profoundly in literature yet.

At the very end of this work we should risk a brief retrospective look at the paradigms in pricing and resource allocation in classic economic thought briefly discussed in section 1.5. Especially in the context of the combinatorial

Grid allocation framework presented in chapter 6 with its "Neo-Walrasian" character, one could imagine a development of an integrated theory of dynamic pricing in information services and information production based on findings gained through ACE simulations. F.A. Hayek's ideas about the catallactic character of the economic system and J.A. Schumpeter's thoughts concerning the creative destruction in an evolutionary economic process could provide the stimulation to extend the presented MAS-based CA framework with agents that are capable of learning optimal response and new strategies using RL technology. In the case of failure of such strategies, individual agents should disappear from the economic system and reappear as agents bearing new mated strategies in an evolutionary process controlled by a *genetic algorithm*. Questions of individual efficiency might be addressed in this extended dynamic pricing framework as well as the issue of global efficiency. Whereas individual efficiency leads more in the direction of marketing-oriented models like e.g. YM, global efficiency plays an important role for system stability issues. With such a general simulation tool for automated resource allocation at hand it should be possible to demonstrate the advantages of dynamic pricing, like increased allocation efficiency and higher robustness in system failure situations, for many application areas in the domain of information services and information production. At the end, there remains the expectation, that the technology developed and presented in the context of this this thesis will find its way into all kinds of real world application areas, and will revolutionize the domain of dynamic pricing, as is the case for the currently discussed economic inspired computational Grid systems.

A

Appendix

Internet Survey Questionnaire

In the following an abridged English version of the German 'Internettechnologie Fragebogen' is given, comprising only questions relevant to this book.[1]

GENERAL INFORMATION

1.1 Which industrial sector can your enterprise be assigned to?

○ agriculture and forestry, fishery, mining	**Services:**
○ manufacturing industry *(e.g. textile, engineering)*	○ hotel, restaurant industry
○ energy- and water supply	○ transportation sector
○ building industry	○ telecommunication
Trade:	○ banking and insurance
○ automobile trade, maintenance, repair	○ research and development
○ intermediation, wholesale (without automobile sector)	○ service provision for enterprises
○ retail (without automobile sector and gas stations); repair of consumer durables	○ property and housing sector, rental of mobiles
	○ data processing, databases *(includes software development, data acquisition etc.)*
○ other sector: _____	○ culture, sports and entertainment, provision of other public and privat services *(includes media industry)*

1.2 How many staff members are permanently employed in your enterprise?

1-9	10-19	20-49	50-99	100-199	200-249	250-499	≥ 500	don't know
○	○	○	○	○	○	○	○	○

1.3 How much is the total annual return of your enterprise (balance sheet total for banks and insurance companies) in millions of Euro?

by 2	by 5	by 10	by 50	by 100	by 250	by 500	≥ 500	don't know
○	○	○	○	○	○	○	○	○

1.4 What is the main customer segment of your enterprise?

○ end users, consumers (business-to-consumer B2C)	○ public sector (business-to-administration B2A)
○ business customers (business-to-business B2B)	

1.5 Which of the following categories is close to your hierachical level in your enterprise?

○ 1. level (e.g. CEO, director)	○ staff position
○ 2. level (e.g. division director)	○ project management
○ 3. level (e.g. department director)	○ other

1.6 How much percent _of return_ has been spent in your company for information and communication technology in 2003?

ca. _____ %	○ don't know

---

[1] The questionnaire can be downloaded at `www.combinatorial-auction.de`.

ACTIVITIES IN THE INTERNET AND „OFFLINE"

	no	yes	planned (< 2 years)	don't know
2.7 Does your enterprise create digital products or provide digital services (e.g. *software, news ticker, ringtones*)?	O	O	O	O
2.8 Is the direct distribution of digital products accomplishable via your Internet presence *(e.g. software download)*?	O	O	O	O

2.11 How much percent of your total turnover is resulting from Internet orders?

Internet share: _____ %   O don't know

2.13 Which part of return contributes to the total return of the following sources?	in the Internet		Conventional	
	share	don't know	share	don't know
sales of products / services	____ %	O	____ %	O
sales of contacts (e.g. *advertising space*)	+ ____ %	O	+ ____ %	O
sales of information (e.g. *user profiles, panel data*)	+ ____ %	O	+ ____ %	O
	= 1 0 0 %		= 1 0 0 %	

MARKETING AND SALES

5.9 If your company provides price information in the Internet, do you vary Internet pricing ...	no	yes	planned (< 2 years)	don't know
more often than your competitors in the „offline" world?	O	O	O	O
more individually than your competitors in the „offline" world *(e.g. price differentiation between customers)* ?	O	O	O	O

5.10 If your enterprise has other distribution channels besides the Internet, ...	no	yes	planned (< 2 years)	don't know
do you vary prices on the Internet **more often** than on other distribution channels?	O	O	O	O
do you vary prices on the Internet in **smaller steps** than on other distribution channels?	O	O	O	O
do you grant **specific Internet** rebates, premiums, etc. that would not be given otherwise?	O	O	O	O
do you grant rebates, premiums, etc. **more often** than on other distribution channels?	O	O	O	O
do you grant rebates, premiums, etc. on the Internet that are **higher** than on other channels?	O	O	O	O
do you form prices on the Internet **more individually** than on other distribution channels *(e.g. price differentiation between customers)*?	O	O	O	O

5.11 How much importance do you assign to the following barriers to flexible / individual pricing information in the Internet from your companies point of view?	high	rather high	rather low	low	none	don't know
no turnover increase attainable	O	O	O	O	O	O
costly determination of prices	O	O	O	O	O	O
cost of price variation *(menue costs)* too high	O	O	O	O	O	O
costly processing, maintaining and administration of the information base *(e.g. in ERP systems)*	O	O	O	O	O	O
costly integration into existing work flow / organization	O	O	O	O	O	O
low buyer or customer acceptance	O	O	O	O	O	O
opportunity to easily resell goods *(arbitrage)*	O	O	O	O	O	O
expected negative customer reaction due to perceived "inequality"	O	O	O	O	O	O
expected negative customer reaction due to perceived "price uncertainty"	O	O	O	O	O	O
legal aspects	O	O	O	O	O	O
peril of price wars with competitors	O	O	O	O	O	O
other barriers	O	O	O	O	O	O

5.14 How much importance do you assign to the following barriers to the <u>automated</u> communication of price and product information from your company's view?

	high	rather high	rather low	low	none	don't know
low buyer or customer acceptance	O	O	O	O	O	O
costly integration into existing IT infrastructure	O	O	O	O	O	O
costly integration into existing work flow / organization	O	O	O	O	O	O
costly processing, maintaining and administration of the information base (*e.g. in ERP systems*)	O	O	O	O	O	O
opportunity to easily reuse the provided information (*e.g. price comparison*)	O	O	O	O	O	O
costs exceed profit	O	O	O	O	O	O
peril of price war with competitors	O	O	O	O	O	O
other barriers	O	O	O	O	O	O

5.25 How do you assess the possible application of the following pricing models in the Internet for the sales department of your enterprise?

	high	rather high	rather low	low	none	don't know
fixed pricing (*e.g. catalog*)	O	O	O	O	O	O
bilateral price negotiation	O	O	O	O	O	O
English auctions (*highest bid wins*)	O	O	O	O	O	O
Dutch auctions (*price declines, first bid wins*)	O	O	O	O	O	O
combinatorial auctions (*auctioning of bundles of goods*)	O	O	O	O	O	O
bundle pricing (*price discount for a bundle of goods*)	O	O	O	O	O	O
reverse pricing (*customer sets a price, bid of the first customer that accepts a price below this hidden price threshold*)	O	O	O	O	O	O

5.26 How do you assess the acceptance of the following price models by your customers or clients?

	high	rather high	rather low	low	none	don't know
combinatorial auctions (*auctioning of bundles of goods*)	O	O	O	O	O	O
reverse pricing (*first bid below the hidden threshold is accepted*)	O	O	O	O	O	O

List of Symbols

Introduction and Chapter 1

$\mathbf{e}_i$	initial endowment for consumer i
$\mathbf{p_g}$	price vector of goods g
$\mathbf{y}_j$	production vector for producer a_j
a_i	consumer i $(1, \ldots, f)$
a_j	producer j $(m - f, \ldots, m)$
D_z	matrix of the first partial derivatives of z
Y_j	technologies available to producer a_j
$y_{j,g}$	producer j output of good g
z	excess demand function
T	time period divided into allocation intervals $t \in \{1, \ldots, T\}$
O	number of resources $o \in \{1, \ldots, O\}$
a	agent a $(1, \ldots, m)$
$q(o, t)$	quantity of a (requested) individual resource o at time t
BM	bid matrix for a resource bundle
AM	allocation matrix for the resources o within the period T
CM	constraint matrix denoting the maximum quantity q_{max}
p_g	price for good g
$\theta_{i,j}$	fraction of producer j owned by consumer i
u_i	utility function for consumer i
$\mathbf{x}_i$	consumption vector for consumer i
$x_{i,g}$	amount of good g consumed by agent i

Chapter 2

p_o	price for resource o
q_o	allocation for resource o

Chapter 4

A	action space
a_t	action at time t
α	learning rate
γ	discount rate parameter
λ	decay rate for TD learning
O	number of resources $o \in \{1, \ldots, O\}$
π	policy
$\mathcal{P}^a_{ss'}$	probability of transition from s to s' under action a
$Q^*(s, a)$	value of taking action a in state s under the optimal policy π
$Q^\pi(s, a)$	value of taking action a in state s under policy π
$\mathcal{R}^a_{ss'}$	expected reward for state transition from s to s' under action a
R_t	reward following t
r_t	reward at time t for action a_t
S	state space S
s_t	state at time t
$V^\pi(s)$	value of state s under policy π
i	index $i \in \{1 \ldots N\}$ of residual capacity for SDP
k	state index for SDP
π_a	acceptance policy for SDP
$\mathcal{P}^{\pi_a}_{ii'}$	probability of transition from i to i' under acceptance policy π_a
$V_k(i)$	value of the remaining residual capacity i at stage k for SDP
b_j	bid-date of the request for single resource RL-YM
d_j	job due-date for single resource RL-YM
g_f	schedule queue filling for single resource RL-YM
$q(o_j)$	request for job j with capacity o_j for single resource RL-YM
$q(o_j, t)$	request for job j with capacity o_j at time t for multiple resource RL-YM
l_j	job length for single resource RL-YM
p_j	contract penalty class of job j for single resource RL-YM
v_j	value class of job j for single resource RL-YM
$\delta(s_t, a_t)$	state transitions for the RL-YM system
w	neuron weights for the ANN used in multiple resource RL-YM

Chapter 5

$\mathcal{A}$	collection of elements (bids)
α	first element in a collection of bids $\mathcal{A}$
$\mathcal{B}$	branching rules in B&B

a^j	bidding agent j
$b^j(S)$	bid of bidder j for the requested subset S of goods i
C	constraint subset of permitted bid combinations $C \subseteq M$
x_j	binary acceptance variable for set j in SPP
c_i	allocation constraint vector
δ	depth of a search tree
ΔE	energy potential in SA
d_j	weight vector for set $j \in V$ in SPP
E	incidence matrix $E = (e_{ij})$
e_{ij}	element in incidence matrix indicating that j-th set of collection V includes element $i \in M$ in SPP
g	dummy good for CAP formulation with $M \cup g$
$\mathcal{E}_t$	set of bids eligible in round t in RAD
$f(\nu)$	evaluation function in search tree
$f(n)$	objective function in local search
f	transformation $f : M \to N$ of a set M onto a set N
Γ	graph representing remaining bids
γ	gradient in subgradient algorithm
$g(\nu)$	cost/return of a search path from starting to current node
$h(\nu)$	estimated cost/return of a search path from current to target node
ι	minimum bid increment in RAD
$\mathcal{L}$	lower bounding rules in B&B
l	number of items in a benchmark bid
L	number of winning bids in AkBA
λ	Lagrange multiplier in LAR
$G(w)$	dual problem in LP
$Q(x)$	primal problem in LP
$\mathcal{M}(x)$	series of solutions $\{\beta_0, \beta_1, \ldots, \beta_N\}$ in local search
M	set of goods (items) $i \in \{1 \ldots m\}$
$\mathcal{N}(n)$	neighboring solutions of n in local search
N	set of bidders $j \in \{1 \ldots n\}$
ν	node in search tree
ω	last element in a collection of bids $\mathcal{A}$
$\mathcal{P}$	family of permitted bid combinations C
p_{acc}	acceptance probability in SA
$p_{ask}(S)$	ask price for bundle S in iBundle
$p_{bid}(S)$	bid price for bundle S in iBundle
π	ranking rule in ordered list
Π	population of individuals in GA

p_i	price for an item i (good) in a bundle
p^j	price for a bundle j
$\mathcal{R}$	set of bids in the recent allocation
r^j	agent's a^j reported utility function in GVA
ρ	branching factor of a search tree
$\mathcal{S}$	search strategy in B&B
S	arbitrary combination of goods i (bundle set) with $S \subseteq M$
σ	step width in subgradient algorithm
ς	slack variable (reduced price) in RAD
$\mathcal{T}$	termination rules in B&B
Θ	temperature in SA
$\mathcal{U}$	upper bounding rules in B&B
u^j	agent's a^j utility function in GVA
V	collection of subsets $j \in V$ of elements $i \in M$ in SPP
v^j	agent's a^j bundle valuation
w_i	shadow prices of items i
$\wp \cdot b^j(S)$	partition of bid $b^j(S)$ in LP
$\mathcal{W}^-$	set of losing bids in RAD
$\mathcal{W}^+$	set of winning bids in RAD
$\mathcal{X}$	problem set in B&B
x_s	binary acceptance variable for bids $b^j(S)$
$\xi^j(S^*)$	feasible allocations for bidder j in AUSM
$y(S, j)$	binary acceptance variable for bids $b^j(S)$ with dummy goods g
Z	value of objective function in IP
$Z(\lambda)$	value of objective function in LAR
Z_{LP}	value of objective function in LP

Chapter 6

α	cooling factor for the SA-CAA
B	set of all bids $b_{i,j}$
B^+	set of accepted bids $b_{i,j}$
B^-	set of rejected bids $b_{i,j}$
β	Acceptance time preference in utility function of agent a
BG_i	monetary budget of agent i
BG^{inc}	monetary budget increment for agent i
BG^{ini}	initial monetary budget of agent i
$b_{i,j}$	bid j of agent i
$\delta_{i,j}$	reduced cost for bid $b_{i,j}$ within SP calculation

ΔE	energy potential of the SA-CAA
Δp	increment for bid price multiplier
γ	Resource quantity preference in utility function of agent a
H_{sp}	history of calculated SPs
I	number of bid bundles
J	number of bids in bundle
Inc^{acc}	auctioneers income derived from accepted bids B^+
k	round number of the iterative auction
$\bar{k}^{acc}_{aggr}$	mean round time elapsed until bid acceptance for aggressive bidder
$\bar{k}^{acc}_{smoo}$	mean round time elapsed until bid acceptance for smooth bidder
l	round of bid $b_{i,j}$
M	number of newly generated bids in each round k
T	time period divided into allocation intervals $t \in \{1, \ldots, T\}$
O	number of resources $o \in \{1, \ldots, O\}$
p_{acc}	acceptance probability for the SA-CAA
$\bar{p}$	average acquisition price per resource unit
$p_{i,j}$	price of a (requested) resource bundle j of agent i
p^{inc}	acceleration factor for bid price
p^{ini}	price multiplier for initial bid
p_{tso}	probability for the occupancy of a resource o at time t
Q^{acc}_o	accepted load for resource o
$q_{i,j}(o, t)$	quantity of a (requested) individual resource o at time t
$q^{ovl}_{i,j}$	overall load of resources o in a bid $b_{i,j}$
Q^{max}_o	maximum allocable load for resource o
q_{bmax}	maximum resource requestable by a bidder for $q_{i,j}(o, t)$
$q_{max}(o, t)$	maximum capacity of resource o at time t
r^g	sorted bid list for the GR-CAA
r^s	permutation list of key sequences for the GA-CAA
SP_o	shadow price of resource o
n	number of rounds included into the calculation of historical SPs
$sp_{o,t}$	shadow price variable for resource o at time t
SR_o	scarcity of resource o
Θ	temperature for SA-CAA
U_a	utility function of agent a
$\hat{v}_o$	market value estimator for resource o
v^{rel}_o	relative price of resource o
v^{shad}_o	market value of a resource o
$x_{i,j}$	acceptance variable for bid $b_{i,j}$

List of Figures

List of Tables

Glossary

ACE	agent-based computational economics	23, 239
ACL	agent communication language	62
AI	artificial intelligence	58
AkBA	ascending k-bundle auction	180
AM	allocation matrix	10
AMC	average marginal contribution	126
ANN	artificial neural network	113, 233
ANS	automated negotiation systems	31
ARA	automated resource allocation function	64
AUML	agent unified modeling language	62
AUSM	adaptive user selection mechanism	178
AUSMQ	adaptive user selection mechanism with queue	178
B2B	business-to-business	1
B2C	business-to-consumer	1
B&B	branch-and-bound	156, 162
BDI	believe desire intention	59
BFS	best-first search	163
BM	bid matrix	10
BrFS	breadth-first search	163
CA	combinatorial auction	66, 137
CABOB	combinatorial auction branch on bids	165
CAMel	combinatorial auction meta language	238
CAP	combinatorial auction problem	66, 139
CAS	complex adaptive system	23
CASANOVA	combinatorial auction search algorithm	171

IP	integer programming	22, 153, 233
IP-CAA	integer programming CA algorithm	213
IR	iterative repair	113
ISIP	information services and information products	6
IT	information technology	1
JADE	Java agent development framework	62
JPI	job providing interface	118
KIF	knowledge interchange format	62
LAR	Lagrange relaxation	175
LP	linear programming	153
LR	linear relaxation	156
MAJIC	multi-parameter auction for JINI components	55
MARIPOSA	contract net-based database management tool	48
MAS	multi-agent system	58, 239
MC	Monte-Carlo	96
MEM	volatile memory capacity	7
MIRAGE	microeconomic resource allocation system	57
MSKP	multidimensional stochastic knapsack problem	109, 233
MU	monetary unit	206
NBC	nested booking class	107
NBP	negotiation-based pricing	30
NET	network bandwidth	7
NGA	neural gas algorithm	135
NGS	neuro genetic scheduling	115
NIMROD-G	market-based Grid computing system	48
NSA	negotiation software agents	30
NSS	normalized shadow surplus	166
NYM	network yield management	109, 236
P2P	peer-to-peer	1
PCA	progressive combinatorial auction	180
PCRA	price controlled resource allocation	7
PREMIUM	Preis und Erlösmodelle im Internet	1
PRISE	pricing of distributed information services	1

References

Abramson, D., Buyya, R. & Giddy, J. (2002), 'A computational economy for grid computing and its implementation in the nimrod-g resource broker', *Future Generation Computer Systems* **18**(8).

Alstrup, J., Boas, S., Madsen, O. & Vidal, R. (1986), 'Booking policy for flights with two types of passengers', *European Journal of Operational Research* **27**, 274–288.

Altman, E., Boulogne, T., El-Azouzi, R., Jimenez, T. & Wynter, L. (2004), 'A survey on networking games in telecommunications', *Computers and Operations Research* .

Andersen, E. S. (1991), The core of schumpeter's work, Technical Report IKE Working Paper no. 68, Institute of Production, Aalborg University.

Andersen, E. S. (2004), From schumpeter's failed econometrics to modern evometric analysis: Creative destruction as a tale of two effects, *in* 'Proceedings of the Conference of the International Schumpeter Society', Milan.

Andersson, A., Tenhunen, M. & Ygge, F. (2000), Integer programming for combinatorial auction winner determination, *in* 'Proceedings of the Fourth International Conference on Multi-Agent Systems (ICMAS '00), Boston, MA', pp. 39–46.

Andersson, M. R. & Sandholm, T. (1998), Leveled commitment contracting among myopic individually rational agents, *in* 'Proceedings of the 3rd International Conference on Multi-Agent Systems', pp. 26–33.

Arrow, K. J. & Debreu, G. (1954), 'Existence for a competitive equilibrium for a competitive economy', *Econometrica* **22**(3), 265–290.

Arthur, B. W. (1994), 'Inductive reasoning and bounded rationality: (the el farol problem)', *American Economic Review* **84**(2), 406–411.

Ausubel, L. M., Cramton, P. & Milgrom, P. (2005), The clock-proxy auction: A practical combinatorial auction design, *in* Y. Shoham, P. Cramton & R. Steinberg, eds, 'Combinatorial Auctions', MIT Press.

Ausubel, L. M. & Milgrom, P. (2002), 'Ascending auctions with package bidding', *Frontiers of Theoretical Economics* **1**(1), 1–42.

Ausubel, L. M. & Milgrom, P. (2005), The lovely but lonely vickrey auction, *in* Y. Shoham, P. Cramton & R. Steinberg, eds, 'Combinatorial Auctions', MIT Press.

AuYoung, A., Chun, B. N., Snoeren, A. C. & Vahdat, A. (2004), Resource allocation in federated distributed computing infrastructures, *in* 'Proceedings of the 1st

Workshop on Operating System and Architectural Support for the On-demand IT InfraStructure, San Francisco, USA'.

Axelrod, R. (1987), The evolution of strategies in the iterated prisoner's dilemma, *in* L. Davis, ed., 'Genetic Algorithms and Simulated Annealing', Morgan Kaufman, pp. 32–41.

Axelrod, R. & Hamilton, W. D. (1981), 'The evolution of cooperation', *Science* **211**.

Axtell, R. L. (1999), The complexity of exchange, Technical report, The Center on Social and Economic Dynamics, Brookings Institution.

Ball, M., Donohue, G. & Hoffman, K. (2006), Auctions for the save, efficient and equitable allocation of airspace system resources, *in* R. Steinberg, Y. Shoham & P. Cramton, eds, 'Combinatorial Auctions', MIT-Press.

Banks, J. S., Ledyard, J. & Porter, D. (1989), 'Allocating uncertain and unresponsive resources: An experimental approach', *The RAND Journal of Economics* **20**(1), 1–25.

Bao, S. & Wurman, P. R. (2003), A comparison of two algorithms for multi-unit k-double auctions, *in* 'Proceedings of the Fifth International Conference on Electronic Commerce (ICEC-03), Pittsburgh'.

Barr, R., Golden, B. L., Kelly, J. P., Resende, M. & Stewart, W. (1995), 'Designing and reporting on computational experiments with heuristic methods', *Journal of Heuristics* **1**(1), 9–32.

Bean, J. C. (1992), Genetics and random keys for sequencing and optimization, Technical Report 92-43, Department of Industrial and Operations Engineering, University of Michigan, Ann Arbor, MI.

Bean, J. C. (1994), 'Genetic algorithms and random keys for sequencing and optimization', *ORSA Journal on Computing* **6**(2), 154–160.

Bellman, R. E. (1957), *Dynamic Programming*, Princeton University Press, Princeton, NJ.

Belobaba, P. P. (1987), Air Travel Demand and Airline Seat Inventory Management, PhD thesis, Massachusetts Institute of Technology, Cambridge, MA.

Belobaba, P. P. (1989), 'Application of a probabilistic decision model to airline seat inventory control', *Operations Research* **37**(2), 183–197.

Benoist, T., Bourreau, E., Caseau, Y. & Rottembourg, B. (2001), Towards stochastic constraint programming: A study of online multi-choice knapsack with deadlines, *in* 'Proceedings of the Seventh Conference on Principles and Practice of Constraint Programming - CP 2001', Vol. 2239 of *Lecture Notes in Computer Science*, Springer, Paphos, Cyprus, pp. 61–76.

Bergen, M., Ritson, M., Dutta, S., Levy, D. & Zbaracki, M. (2003), 'Shattering the myth of costless price changes: A framework for dynamic pricing', *European Management Journal* **21**(6), 663–669.

Bergenti, F. & Poggi, A. (2000), Exploiting UML in the design of multi-agent systems, *in* A. Omicini, R. Tolksdorf & F. Zambonelli, eds, 'Engineering Societies in the Agent World, First International Workshop (ESAW 2000), Berlin, Germany, Revised Papers', Vol. 1972 of *Lecture Notes in Computer Science*, Springer Verlag, Berlin, Germany, pp. 106–113.

Bernhardt, M. & Hinz, O. (2005), Creating value with interactive pricing mechanisms a web service-oriented architecture, *in* '7th International IEEE Conference on E-Commerce Technology 2005, München, Germany'.

Bertsch, L. (1990), *Expertengestützte Dienstleistungskostenrechnung*, Poeschel, Stuttgart, Germany.

Bertsekas, D. & Tsitsiklis, J. (1996), *Neuro-Dynamic Programming*, Athena Scientific, Belmont, MA.

Bertsimas, D. & Popescu, I. (2000), 'Revenue management in a dynamic network environment', *submitted to Transportation Science* .

Bichler, M., Kalagnanam, J., Katircioglu, K., King, A. J., Lawrence, R. D., Lee, H. S., Lin, G. Y. & Lu, Y. (2002), 'Applications of flexible pricing in business-to-business electronic commerce', *IBM Systems Journal* **2**.

Bichler, M., Kersten, G. E. & Strecker, S. (2003), 'Towards a structured design of electronic negotiations', *Group Decision and Negotiation* **12**(4), 311–335.

Bichler, M., Pikovsky, A. & Setzer, T. (2005), 'Kombinatorische Auktionen in der betrieblichen Beschaffung - Eine Analyse grundlegender Entwurfsprobleme', *Wirtschaftsinformatik* **2**.

Bichler, M. & Werthner, H. (2000), e-commerce: from fixed to dynamic pricing, Technical report, Department of Information Systems, Vienna University of Economics and Business Administration.

Bikhchandani, S. & Mamer, J. (1997), 'Competitive equilibrium in an exchange economy with indivisibilities', *Journal of Economic Theory* **74**(2), 385–413.

Bikhchandani, S. & Ostroy, J. M. (2002), 'The package assignment model', *Journal of Economic Theory* **107**, 377–406.

Bikhchandani, S. & Ostroy, J. M. (2005), From the assignment model to combinatorial auctions, *in* 'Combinatorial Auctions', MIT Press.

Bishop, C. M. (1995), *Neural Networks for Pattern Recognition*, Oxford University Press, Oxford, England.

Bitran, G. R. & Gilbert, S. M. (1996), 'Managing hotel reservations with uncertain arrivals', *Operations Research* **44**(1), 35–49.

Bitran, G. R. & Mondschein, S. V. (1995), 'An application of yield management to the hotel industry considering multiple day stays', *Operations Research* **43**(3), 427–443.

Bjørndal, M. & Jørnsten, K. (2001), An analysis of a combinatorial auction, Technical Report 2001-11, Department of Finance and Management Science, Norwegian School of Economics and Business Administration, Bergen, Norway.

Bossaerts, P., Fine, L. & Ledyard, J. (2002), 'Inducing liquidity in thin financial markets through combined-value trading mechanisms', *European Economics Review* **46**(9), 1671–1695.

Boutilier, C. & Goldszmidt, M. (1999), Continuous value functions approximation for sequential bidding policies, *in* 'Proceedings of the Fifteenth Annual Conference on Uncertainty in Artificial Intelligence (UAI-99), Stockholm', pp. 81–90.

Boyan, J. & Moore, A. (1998), Learning evaluation functions for global optimization and boolean satisfiability, *in* 'Proceedings of the Fifteenth National Conference on Artificial Intelligence', pp. 3–10.

Boyan, J. & Moore, A. (2000), 'Learning evaluation functions to improve optimization by local search', *Journal of Machine Learning Research* **10**, 77–112.

Brauer, W. & Weiss, G. (1998), Multi-machine scheduling - a multi-agent learning approach, *in* 'In Proceedings of the 3rd International Conference on Multi-Agent Systems', pp. 42–48.

Bredin, J., Maheswaran, R. T., Imer, C., Basar, T., Kotz, D. & Rus, D. (2000), A game-theoretic formulation of multi-agent resource allocation, *in* 'Proceedings of the 4th International Conference on Autonomous Agents', Barcelona, Spain, pp. 349–356.

Bronstein, I. N., Semendjajew, K. A., Musiol, G. & Mühlig, H. (2001), *Taschenbuch der Mathematik*, 5 edn, Verlag Harri Deutsch, Thun und Frankfurt/Main.

Brouwer, L. E. J. (1912), 'Über abbildung von mannigfaltigkeiten', *Mathematische Annalen* **71**, 97–115.

Brunelle, J., Hurst, P., Huth, J., Kang, L., Ng, C., Parkes, D. C., Seltzer, M., Shank, J. & Youssef, S. (2006), Egg: An extensible and economics-inspired open grid computing platform, *in* 'Proceedings of the 3rd International Workshop on Grid Economics and Business Models (GECON 2006), Singapore'.

Bry, F. (1989), Logic programming as constructivism: A formalization and its application to databases, *in* 'Proceedings of the Eighth ACM SIGACT-SIGMOD-SIGART Symposium on Principles of Database Systems, March 29-31, 1989, Philadelphia, Pennsylvania', ACM Press, pp. 34–50.

Brynjolfsson, E. (1993), 'The productivity paradox of information technology', *Communications of the ACM* **36**(12), 66–77.

Brynjolfsson, E. & Smith, M. (2000), 'Frictionless commerce? a comparison of internet and conventional retailers', *Management Science* **46**(4), 563–585.

Bubendorfer, K. & Hine, J. (2005), Auction based resource negotiation in nomad, *in* 'Twenty-Eighth Australasian Computer Science Conference (ACSC2005), Newcastle, Australia. CRPIT, 38', pp. 297–306.

Buhmann, J. M. & Hofmann, T. (1996), An annealed neural gas network for robust vector quantization, *in* 'Proceedings of the International Conference on Artificial Neural Networks (ICANN96)', pp. 151–156.

Buyya, R., Giddy, J. & Abramson, D. (2000), An evaluation of economy-based resource trading and scheduling on computational power grids for parameter sweep applications, *in* 'Proceedings of the 2nd International Workshop on Active Middleware', Kluwer Academic Press, Pittsburgh.

Buyya, R., Stockinger, H., Giddy, J. & Abramson, D. (2001), Economic models for management of resources in peer-to-peer and grid computing, *in* 'Proceedings of the SPIE International Conference on Commercial Applications for High-Performance Computing', Denver, USA.

Buyya, R. & Vazhkudai, S. (2001), Compute power market: Towards a market-oriented grid, *in* 'Proceedings of 1st IEEE International Conference on Cluster Computing and the Grid'.

Bykowsky, M. M., Cull, R. J. & Ledyard, J. (1995), 'Mutually destructive bidding: The fcc auction design problem', *Journal of Regulatory Economics* **17**(3), 205–228.

Cantillon, E. & Pesendorfer, M. (2006), Auctioning bus routes, *in* P. Cramton, Y. Shoham & R. Steinberg, eds, 'Combinatorial Auctions', MIT-Press.

Cao, X.-R., Shen, H.-X., Milito, R. & Wirth, P. (2002), 'Internet pricing with a game theoretical approach: concepts and examples', *IEEE ACM Transactions Networks* **10**(2), 208–216.

Caplice, C. (1996), An Optimization Based Bidding Process: A New Framework for Shipper-Carrier Relationships, PhD thesis, Massachusetts Institute of Technology.

Carvalho, A. X. & Puterman, M. L. (2005), 'Learning and pricing in an internet environment with binomial demands', *Journal of Revenue and Pricing Management* **3**(4), 320–336.

Casavant, T. & Kuhl, J. (1988), 'A taxonomy of scheduling in general-purpose distributed computing systems', *IEEE Transactions on Software Engineering* **14**(2), 141–154.

Catoni, O. (1998), 'Solving scheduling problems by simulated annealing', *SIAM Journal on Control and Optimization* **36**(5), 1539–1575.

Challet, D. & Zhang, Y.-C. (1997), 'Emergence of cooperation and organization in an evolutionary game', *Physica A* **246**(407).

Chandru, V. & Rao, M. R. (1998), Integer programming, Technical Report T.R.No. IISc-98-04, Department of Computer Science and Automotion, Indian Institute of Science, Bangalore, India.

Chang, S. S. (1998), A survey on reinforcement learning in global optimization, Technical report, Computer Science Department of the University of Berkeley, Berkeley, CA.

Chang, X. & Subramanian, K. R. (2000), A cooperative game theory approach to resource allocation in wireless atm networls, *in* 'Proceedings of Broadband Communications, High Performance Networking, and Performance of Communication Networks, IFIP-TC6, Networking 2000', Vol. 1815 of *Lecture Notes in Computer Science*, Springer, Berlin, pp. 969–978.

Chavez, A. & Maes, P. (1996), Kasbah: An agent marketplace for buying and selling goods, *in* 'Proceedings of the first International Conference on the Practical Application of Intelligent Agents and Multi-Agent Technology (PAAM)', Practical Application Company, London, UK, pp. 75–90.

Chen, V. C., Günther, D. & Johnson, E. L. (1999), 'Airline yield management: Optimal bid prices for single hub problems without cancellations', *submitted to Journal of Transportation Science* .

Cheng, J. & Wellman, M. (1996), 'The WALRAS algorithm: A convergent distributed implementation of general-equilibrium outcomes', *Computational Economics* **12**.

Chernev, A. (2003), 'Reverse pricing and online price elicitation strategies in consumer choice', *Journal of Consumer Psychology* **13**(1&2), 51–62.

Chevaleyre, Y., Endriss, U., Lang, J. & Maudet, N. (2005), Negotiating over small bundles of resources, *in* 'Proceedings of the 4th International Joint Conference on Autonomous Agents and Multiagent Systems (AAMAS-2005)', ACM Press.

Chu, P. C. & Beasley, J. E. (1995), A genetic algorithm for the set partitioning problem, Technical report, The Management School Imperial College London.

Chun, B. N., Buonadonna, P., AuYoung, A., Ng, C., Parkes, D. C., Shneiderman, J., Snoeren, A. C. & Vahdat, A. (2004), Mirage: A microeconomic resource allocation system for sensornet testbeds, *in* 'Proceedings of the 2nd IEEE Workshop on Embedded Networked Sensors (EmNetS-II); Sidney, Australia'.

Chun, B. N. & Culler, D. E. (2000), Market-based proportional resource sharing for clusters, Technical Report CSD-1092, University of California at Berkeley, Computer Science Division.

Chun, B. N. & Culler, D. E. (2002), User-centric performance analysis of market-based cluster batch schedulers, *in* 'Proceedings of the 2nd IEEE International Symposium on Cluster Computing and the Grid, Berlin, Germany', pp. 30–38.

Clarke, E. (1971), 'Multipart pricing of public goods', *Public Choice Journal* **11**, 17–33.

Clearwater, S. H. (1996), *Market-Based Control: A Paradigm for Distributed Resource Allocation*, World Scientific, Singapore.

Cole, R., Dodis, Y. & Roughgarden, T. (2003), Pricing networks with selfish routing, *in* 'Workshop on Economics of Peer-to-Peer Systems'.

Collins, J., Gini, M. & Mobasher, B. (2002), Multi-agent negotiation using combinatorial auctions with precedence constraints, Technical Report T.R.No. 02-009, University of Minnesota, Department of Computer Science and Engineering, Minneapolis, MN.

Collins, J., Tsvetovat, M., Mobasher, B. & Gini, M. (1998), MAGNET: A multi-agent contracting system for plan execution, *in* 'Proceedings of Workshop on Artificial Intelligence and Manufacturing: State of the Art and State of Practice', AAAI Press, pp. 63–68.

Conen, W. (2003), Economically-augmented job shop scheduling, *in* E. Giunchiglia, N. Muscettola & D. Nau, eds, 'International Conference on Automated Planning & Scheduling (ICAPS)', AAAI Press, Trento, Italy.

Conen, W. & Sandholm, T. (2001), Minimal preference elicitation in combinatorial auctions, *in* 'IJCAI Workshop on Economic Agents, Models, and Mechanisms, Seattle, August 2001'.

Conen, W. & Sandholm, T. (2002), Coherent pricing of efficient allocations in combinatorial economies, *in* 'AAAI Workshop Technical Report WA-02-06 of the AAAI 2002 Workshop on Game Theoretic and Decision Theoretic Agents (GTDT'02), Edmonton, Canada'.

Correa, J., Schulz, A. S. & Moses, N. S. (2004), 'Selfish routing in capacitated networks', *Mathematics of Operations Research* **29**(4), 961–976.

Costa, N. D. & Doria, F. A. (2005), Computing the future, *in* V. Velupillai, ed., 'Computability, Compexity and Constuctivity in Economic Analysis', Blackwell, Malden, pp. 15–50.

Cramton, P. (2005), Simulaneous ascending auctions, *in* P. Cramton, Y. Shoham & R. Steinberg, eds, 'Combinatorial Auctions', MIT Press, chapter 4.

Cramton, P., Shoham, Y. & Steinberg, R. (2005), Combinatorial auctions: Introduction, *in* R. Steinberg, Y. Shoham & P. Cramton, eds, 'Combinatorial Auctions', MIT Press.

Dantzig, G. B. & Thapa, M. N. (1997), *Linear Programming: Introduction*, Springer Series in Operations Research and Financial Engineering, Springer Verlag, Berlin.

Davenport, A. J. & Kalagnanam, J. (2001), Price negotiations for procurement of direct inputs, Technical Report IBM Technical Report RC 22078, IBM T. J. Watson Research Center, Yorktown Heights, New York.

Davis, R. & Smith, R. G. (1983), 'Negotiation as a metaphor for distributed problem solving', *Artificial Intelligence* **20**, 63–109.

Debreu, G. (1959), *Theory of Value*, John Wiley and Sons, New York, NY.

Deng, X., Papadimitriou, C. H. & Safra, S. (2002), On the complexity of equilibria, *in* 'Proceedings on 34th Annual ACM Symposium on Theory of Computing, 2002, Montréal, Québec, Canada', pp. 67–71.

Doerninger, W. (1984), 'Approximating general markovian decision problems by clustering their state- and action-spaces', *Math. Operationsforschung und Statistik: Ser. Optimization* **14**, 135–144.

Dooley, K. (2005), 'Complex adaptive systems : A nominal definition', http://www.eas.asu.edu/ kdooley/casopdef.html. accessed 1.7.2005.

Drexler, K. E. & Miller, M. S. (1988), Incentive engineering: for computational resource management, *in* B. A. Huberman, ed., 'The Ecology of Computation', Elsevier Science Publishers, North-Holland, pp. 231–266.

Durfee, E. H. (1999), Distributed problem solving and planning, *in* 'Multiagent Systems', The MIT Press, Cambrige, MA.

Dziong, Z. & Mason, L. G. (1996), 'Fair-efficient call admission control policies for broadband networks: A game theoretic framework', *IEEE ACM Transactions Networks* **4**(1), 123–136.

Easwaran, A. & Pitt, J. (2000), A brokering algorithm for cost & QoS-based winner determination in combinatorial auctions, *in* R. Loganantharaj & G. Palm, eds, 'Intelligent Problem Solving, Methodologies and Approaches, 13th International Conference on Industrial and Engineering Applications of Artificial Intelligence and Expert Systems (IEA/AIE 2000), New Orleans, LU', Vol. 1821 of *Lecture Notes in Computer Science*, Springer Verlag, Berlin, Germany.

Elendner, T. (2003), Scheduling and combinatorial auctions: Lagrangean relaxation-based bounds for the wjisp, Technical Report 570, Manuskripte aus den Instituten für Betriebswirtschaftslehre, Christian-Albrechts-Universität zu Kiel.

Elendner, T. (2004), *Winner Determination in Combinatorial Auctions: Market-based Scheduling*, Logos, Berlin.

Elmaghraby, W. & Keskinocak, P. (2003), 'Dynamic pricing: Research overview, current practices and future directions', *Management Science* **49**(10), 1288.

Epstein, R., Lysette, H., Catalan, J., Weintraub, G. & Martinez, C. (2002), 'A combinatorial auction improves school meals in chile', *Interfaces* **32**(6), 1–14.

Ernemann, C., Hamscher, V. & YahYapour, R. (2002), Economic scheduling in grid computing, *in* 'Job Scheduling Strategies for Parallel Processing, 8th International Workshop, JSSPP 2002, Edinburgh, Scotland', Lecture Notes in Computer Science, Springer Verlag.

Eymann, T. (2001), Co-evolution of bargaining strategies in a decentralized multi-agent system, *in* 'Proceedings of the AAAI Fall 2001 Symposium on Negotiation Methods for Autonomous Cooperative Systems, North Falmouth, MA, November 03-04'.

Eymann, T., Reinicke, M., Ardaiz, O., Artigas, P., Cerio, Freitag, F., Messeguer, R., Navarro, L. & Royo, D. (2003), Decentralized vs. centralized economic coordination of resource allocation in grids, *in* 'Proceedings of the 1. European Across Grids Conference'.

Fagerberg, J. (2003), 'Schumpeter and the revival of evolutionary economics: an appraisal of the literature', *Journal of Evolutionary Economics* **13**, 125–159.

Fan, M., Stallaert, J. & Whinston, A. (2001), 'Decentralized mechanism design for supply chain organizations using an auction market', *Information Systems Research* .

Fang, Z. & Bensaou, B. (2004), Fair bandwidth sharing algorithms based on game theory frameworks, *in* 'Proceedings of the 23rd Annual Joint Conference of the IEEE Computer and Communications Societies (INFOCOM 2004), Hong Kong, China', IEEE.

Feldman, M., Lai, K., Stoica, I. & Chuang, J. C.-I. (2004), Robust incentive techniques for peer-to-peer networks, *in* 'Proceedings 5th ACM Conference on Electronic Commerce (EC-2004), New York, NY, USA', ACM, pp. 102–111.

Feldman, M., Lai, K. & Zhang, L. (2005), A price-anticipating resource allocation mechanism for distributed shared clusters, *in* 'Proceedings of the ACM Conference on Electronic Commerce'.

Ferguson, D. (1989), The Application of Microeconomics to the Design of Resource Allocation and Control Algorithms, PhD thesis, Columbia University.

Ferguson, D., Nikolaou, C., Sairamesh, J. & Yemini, Y. (1996), Economic models for allocating resources in computer systems, *in* S. H. Clearwater, ed., 'Market-

Based Control: A Paradigm for Distributed Resource Allocation', World Scientific, Singapore.

FIPA (2000a), 'FIPA ACL message structure specification', http://www.fipa.org/specifications/fipa00061. accessed 15.6.2005.

FIPA (2000b), 'FIPA contract net interaction protocol specification', http://www.fipa.org/ specifications/fipa00029. accessed 15.6.2005.

FIPA (2000c), 'FIPA KIF content language specification', http://www.fipa.org/specifications/fipa00010. accessed 15.6.2005.

FIPA (2000d), 'FIPA ontology service specification', http://www.fipa.org/ specifications/fipa00086. accessed 15.6.2005.

Fisher, M. L. (1981), 'The lagrangian relaxation method for solving integer programming problems', *Management Science* **27**(1), 1–18.

Fisher, M. & Wolsey, L. (1982), 'On the greedy heuristic for continuous covering and packing problems', *SIAM Journal on Algebraic and Discrete Methods* **3**, 584591.

Foster, I., Jennings, N. R. & Kesselman, C. (2004), Brain meets brawn: Why grid and agents need each other, *in* 'Proceedings of the 3rd International Conference on Autonomous Agents and Multi-Agent Systems (AAMAS04)', New York, NY, pp. 8–15.

Foster, I., Kesselman, C. & Tuecke, S. (2001), 'The anatomy of the grid: Enabling scalable virtual organizations', *Lecture Notes in Computer Science* **2150**.

Fritzke, B. (1995), A growing neural gas network learns topologies, *in* G. Tesauro, D. Touretzky & T. Leen, eds, 'Advances in Neural Information Processing Systems', Vol. 7, MIT Press.

Fujishima, Y., Leyton-Brown, K. & Shoham, Y. (1999), Taming the computational complexity of combinatorial auctions: Optimal and approximate approaches, *in* 'Proceedings of the 16th International Joint Conference on Artificial Intelligence 1999 (IJCAI-99), Stockholm, Sweden', pp. 548 – 553.

Galstyan, A., Shashikiran, K. & Lerman, K. (2003), Resource allocation games with changing resource capacities, *in* 'AAMAS '03: Proceedings of the second international joint conference on Autonomous agents and multiagent systems, Melbourne, Australia', ACM Press, pp. 145–152.

Ganesh, A., Laevens, K. & Steinberg, R. (2001), Congestion pricing and user adaptation, *in* 'Proceedings of the IEEE Infocom 2001, the Annual Joint Conference of the IEEE Computer and Communications Societies (20th), 22-26 April 2001, Anchorage, Alaska', pp. 959–965.

Garey, M. R., Johnson, D. S. & Sethi, R. (1976), 'The complexity of flow shop and job shop scheduling', *Math. Operations Research* **1**, 117–129.

Garfinkel, R. & Nemhauser, G. (1969), 'The set partitioning problem: Set covering problem with equality constraints', *Operations Research* **17**(5), 848–856.

Genesereth, M. R. (1998), 'Knowledge interchange format: Draft proposed american national standard'. NCITS.T2/98-004.

Ghose, A., Choudhary, V., Mukhopadhyay, T. & Rajan, U. (2002), 'Electronic commerce and competitive first-degree price discriminators', *Review of Marketing Science Working Papers* **2**(1).

Giovanucci, A., Rodriguez-Aguilar, J., Cerquides, J., Reyes-Moro, A. & Noria, F. (2004), ibundler: an agent-based decision support service for combinatorial negotiations, *in* 'roceedings of the Nineteenth National Conference on Artificial Intelligence, Sixteenth Conference on Innovative Applications of Artificial Intelligence, 2004, San Jose, California, USA', pp. 1012–1013.

Goldberg, D. E. (1989a), *Genetic Algorithms in Search, Optimization and Machine Learning*, Addison Wesley, Reading, MA.

Goldberg, D. E. (1989b), *Genetic algorithms in search, optimization, and machine learning*, Addison-Wesley, Reading, MA.

Gomory, R. E. (1958), 'Outline of an algorithm for integer solutions to linear programs', *Bulletin of the American Mathematical Society* **64**, 275–278.

Gonen, R. & Lehmann, D. (2001), Linear programming helps solving large multi-unit combinatorial auctions, *in* 'Proccedings of the INFORMS Electronic Market Design Workshop 2001'.

Gosavi, A. (2004), 'A reinforcement learning algorithm based on policy iteration for average reward: Empirical results with yield management and convergence analysis', *Machine Learning* **55**(1), 5–29.

Gosavi, A., Bandla, N. & Das, T. K. (2002), 'A reinforcement learning appoach to airline seat allocation for multiple fare classes with overbooking', *IIE Transactions, Special Issue on Advances on Large-Scale Optimization for Logistics, Production and Manufacturing systems* **34**(9), 729–742.

Groves, T. (1973), 'Incentive in teams', *Econometrica* **41**(4), 617–631.

Gruber, T. (1995), 'Toward principles for the design of ontologies used for knowledge sharing', *International Journal of Human Computer Studies* **43**(5/6), 907928.

Guarino, N. (1995), 'Formal ontology, conceptual analysis and knowledge representation', *International Journal of Human Computer Studies* .

Gupta, A., Stahl, D. O. & Whinston, A. B. (1997), Priority pricing of integrated services networks, *in* J. Bailey & L. McKnight, eds, 'Internet Economics', MIT Press, Cambridge, pp. 323–352.

Gupta, A., Stahl, D. O. & Whinston, A. B. (1999), 'The economics of network management', *Communications of the ACM* **42**(9), 57 – 63.

Halblau, V. (2002), 'Algorithmische Konzepte in kombinatorischen Auktionen'. Diploma thesis, Frankfurt University, Institute for Information Systems.

Hardin, G. (1968), 'The tragedy of the commons', *Science* **162**(3859), 1243–1248.

Hart, P. E., Nilsson, N. J. & Raphael, B. (1968), 'A formal basis for the heuristic determination of minimum cost paths', *IEEE Transactions on Systems Science and Cybernetics* **4**, 100–107.

Hartung, J., Elpelt, B. & Klösener, K.-H. (1999), *Statistik: Lehr- und Handbuch der angewandten Statistik*, 12 edn, Oldenbourg Verlag, München, Wien.

Hayek, F. A. (1976a), Die Verwertung des Wissens in der Gesellschaft, *in* F. A. Hayek, ed., 'Individualismus und Wirtschaftliche Ordnung', Wolfgang Neugebauer, Salzburg, Austria.

Hayek, F. A. (1976b), *Law Legislation and Liberty*, Chicago University Press, Chicago, Illinois.

Hayek, F. A. (1996a), Der Wettbewerb als Entdeckungsverfahren, *in* F. A. Hayek, ed., 'Von Hayek bis White', MANZ Verlag, Vienna, Austria. First published in: Leube, K. R.(Ed.): The International Library of Austrian Economics (1968).

Hayek, F. A. (1996b), *Die Verhängnisvolle Anmassung: Die Irrtümer des Sozialismus*, Mohr Siebeck, Tübingen, Germany.

Hein, O., Schwind, M. & König, W. (2006), 'Scale-free networks - the impact of fat tailed degree distribution on diffusion and communication networks', *Wirtschaftsinformatik* **48**(4), 267–275.

Hein, O., Schwind, M. & Spiwoks, M. (2005), A microscopic stock market model with heterogeneous interacting agents in a scale-free communication network, *in*

'10th Annual Workshop on Economic Heterogeneous Interacting Agents (WE-HIA 2005), Essex, UK'.

Hildenbrand, A. P. & Kirman, W. (1976), *Introduction to Equilibrium Analysis*, North Holland, Amsterdam, The Netherlands.

Hirsch, M. D. & Papadimitriou, C. H. (1989), 'Exponential lower bounds for finding brouwer fixed points', *Journal of Complexity* **5**(4), 379–416.

Hohner, G., Rich, J., Ng, E., Reid, G., Davenport, A. J., Kalagnanam, J., Lee, H. S. & Chae, A. (2003), 'Combinatorial and quantity-discount procurement auctions benefit mars, incorporated and its suppliers', *Interfaces* **33**(1), 23–35.

Holland, J. H. (1975), *Adaptation in Natural and Artificial Systems: An Introductory Analysis with Applications to Biology, Control and Artificial Intelligence*, University of Michigan Press, Ann Arbor, MI.

Holland, J. H. (1998), *Emergence: From Chaos to Order*, Addison-Wesley, Redwood, CA.

Holland, J. & Miller, J. H. (1991), 'Artificial adaptive agents in economic theory', *American Economic Review* **81**(2), 365–370.

Hoos, H. H. & Boutilier, C. (2000), Solving combinatorial auctions using stochastic local search, *in* 'Proceedings of the 17th National Conference on Artificial Intelligence (IAAI), Austin, TX', AAAI, pp. 22–29.

Hoppmann, E. (1993), *Unwissenheit, Wirtschaftsordnung und Staatsgewalt*, Haufe Verlag, Freiburg, Germany.

Hornick, S. (1991), Value based revenue management - a new paradigm for airline seat inventory control, *in* 'Advanced Software Technology in Air Transport', AIT-Verlag, Hallbergmoos, Germany.

Hu, J. (1999), Learning in Dynamic Noncooperative Multiagent Systems, PhD thesis, University of Michigan, Ann Arbor.

Hu, J. & Wellman, M. P. (1998), Multiagent reinforcement learning: Theoretical framework and an algorithm, *in* 'Proceedings of the Fifteenth International Conference on Machine Learning, (ICML 1998), Madison, Wisconsin', Morgan Kaufmann, pp. 24–27.

Hu, J. & Wellman, M. P. (2003), 'Nash q-learning for general-sum stochastic games', *Journal of Machine Learning Research* **4**, 1039–1069.

Hu, J. & Zhang, Y. (2002), Online reinforcement learning in multiagent systems, Technical report, William E. Simon Graduate School of Business Administration, University of Rochester, New York.

Huberman, B. A. & Wu, F. (2002), The dynamics of reputations, Technical report, Hewlett-Packard Laboratories and Stanford University.

Hudson, B. & Sandholm, T. (2002), Effectiveness of preference elicitation in combinatorial auctions, *in* 'Proceedings of the Agent-Mediated Electronic Commerce (AMEC) workshop at AAMAS-02, Bologna, Italy'.

Hudson, B. & Sandholm, T. (2004), Effectiveness of query types and policies for preference elicitation in combinatorial auctions, *in* 'In Proceedings of the International Joint Conference on Autonomous Agents and Multiagent Systems (AAMAS), New York', pp. 386–393.

Huhns, M. N. & Stephens, L. M. (1999), Multiagent systems and societies of agents, *in* G. Weiss, ed., 'Multiagent Systems', MIT Press, Massachusetts, chapter 2, pp. 79–120.

Hultgren, G. & Eriksson, O. (2003), The notion of it-services from a social interaction perspective, *in* 'Proceedings of the Conference for the Promotion of

Research in IT at New Universities and University Colleges in Sweden, Promote IT, Gotland'.

Humair, S. (2001), Yield Management for Telecommunication Networks: Defining a New Landscape, PhD thesis, Massachusetts Institute of Technology, Cambridge, MA.

Il-Horn, H. & Terwiesch, C. (2003), 'Measuring the frictional costs of online transactions: The case of name-your-own-price channel', *Management Science* **49**(11), 1563–1579.

Jennings, N. R., Faratin, P., Lomuscio, A. R., Parsons, S., Sierra, C. & Wooldridge, M. (2001), 'Automated negotiation: Prospects, methods and challenges', *Journal of Group Decision and Negotiation* **10**(2), 199–215.

Judd, K. L. (2002), Solving dynamic stochastic competitive general equilibrium models, Technical report, Hoover Institution, Stanford University.

Kaelbling, L., Brandt, F., Bauer, W. & Weiss, G. (2000), Task assignment in multiagent systems based on vickrey-type auctioning and leveled commitment contracting, *in* 'Proceedings of the Fourth International Workshop on Cooperative Information Systems (CIA)', Vol. 1860 of *Lecture Notes in Computer Science*, Springer Verlag, Berlin, pp. 95–106.

Kaelbling, L., Littman, M. & Moore, A. (1996), 'Reinforcement learning: A survey', *Journal of Artificial Intelligence Research* **4**, 237–285.

Kakutani, S. (1941), 'A generalization of brouwer's fixed point theorem', *Duke Mathematical Journal* **8**, 457–459.

Kalagnanam, J. & Parkes, D. C. (2003), Auctions, bidding and exchange design, *in* M. Z. Shen, S. D. Wu & D. Simchi-Levi, eds, 'Supply Chain Analysis in the eBusiness Area', Kluwer Academic Publishers.

Kauffman, R. J. & Lee, D. (2004), Price rigidity on the internet: New evidence from the online bookselling industry, *in* 'Proceedings of the 25th International Conference on Information Systems (ICIS 2004), Washington, DC, December 2004'.

Kauffman, R. J. & Wang, B. (2001), New buyers arrival under dynamic pricing market microstructure: The case of group-buying discounts on the internet, *in* 'Proceedings of the 34th Hawaii International Conference on System Sciences - 2001, Maui, Hawaii', pp. 7034–7044.

Kephart, J., Hanson, J. E. & Greenwald, A. R. (2000), 'Dynamic pricing by software agents', *Computer Networks* **32**(6), 731–752.

Kephart, J. & Tesauro, G. (2000), Pseudo-convergent q-learning by competitive pricebots, *in* P. Langley, ed., 'Proceedings of the Seventeenth International Conference on Machine Learning (ICML 2000), Stanford University, Stanford, CA', Morgan Kaufmann, pp. 463–470.

Kimes, S. (1989), 'Yield management: A tool for capacity constrained firms', *Journal of Operations Management* **4**, 348–363.

Kirkpatrick, S., Jr., C. G. & Vecchi, M. (1983), 'Optimization by simulated annealing', *Science* **220**, 671–680.

Klemperer, P. (2004), *Auctions: Theory and Practice*, Princeton University Press.

Kleywegt, A. (2001), An optimal control problem of dynamic pricing, Technical report, School of Industrial and Systems Engineering, Georgia Institute of Technology, Atlanta, Georgia.

Kohonen, T. (1995), *Self-Organizing Maps*, Springer Series in Information Sciences, Springer Verlag, Berlin.

König, W. & Schwind, M. (2005), Entwurf von kombinatorischen Auktionen für Allokations- und Beschaffungsprozesse, in B. Rieger & D. Karagiannis, eds, 'Herausforderungen der Wirtschaftsinformatik: Festschrift für Prof. Krallmann', Vol. 3, Springer Verlag, Berlin, pp. 29–45.

Koopmans, T. & Beckmann, M. (1957), 'Assignment problems and the location of economic activities', *Econometrica* **25**(1), 53–76.

Korf, R. E. (1985), 'Depth-first iterative-deepening: an optimal admissible tree search', *Artificial Intelligence* **27**(1), 97–109.

Korilis, Y. A., Lazar, A. A. & Orda, A. (1995), 'Architecting noncooperative networks', *IEEE Journal on Selected Areas in Communication* **13**(7), 1241–1251.

Korilis, Y. A., Lazar, A. A. & Orda, A. (1997), 'Capacity allocation under noncooperative routing', *IEEE Transactions on Automatic Control* **42**(3), 309–325.

Kubler, F. (2005), 'Notes on Arrow-Debreu economies', http://www.vwl.uni-mannheim.de/kuebler/prs/cge1.pdf.

Kuhlen, R. (1995), *Informationsmarkt - Chancen und Risiken der Kommerzialisierung von Wissen*, UVK: Universitätsverlag Konstanz, Konstanz.

Kurose, J. & Shima, R. (1989), 'A microeconomic approach to optimal resource allocation in distributed computer systems', *IEEE Transactions on Computers* **38**(5), 705 – 717.

Kwasnica, A. M., Ledyard, J., Porter, D. & DeMartini, C. (2005), 'A new and improved design for multi-objective iterative auctions', *Management Science* **51**(3), 419–434.

Kwok, Y.-K., Song, S. & Hwang, K. (2005), Selfish grid computing: Game-theoretic modeling and nas performance results, in 'Proceedings of the International Symposium on Cluster Computing and the Grid, Cardiff, UK'.

Lai, K. (2005), Markets are dead, long live markets, Technical Report arXiv:cs.OS/0502027, HP Labs, Palo Alto, CA, USA.

Lai, K., Huberman, B. A. & Fine, L. (2004), Tycoon: A distributed market-based resource allocation system, Technical Report arXiv:cs.DC/0404013, HP Labs, Palo Alto, CA, USA.

Lai, K., Rasmusson, L., Adar, E., Sorkin, S., Zhang, L. & Huberman, B. A. (2004), Tycoon: an implementation of a distributed market-based resource allocation system, Technical Report arXiv:cs.DC/0412038, HP Labs, Palo Alto, CA, USA.

Lawler, E. L. & Wood, E. D. (1966), 'Branch-and-bound methods: A survey', *Operations Research* **14**(4), 699–719.

Ledyard, J., Olson, M., Porter, D., Swanson, J. A. & Torma, D. P. (2002), 'The first use of a combined-value auction for transportation services', *Interfaces* **32**(5), 4–12.

Ledyard, J., Porter, D. & Rangel, A. (1997), 'Experiments testing multiobject allocation mechanisms', *Journal of Economics & Management Strategy* **6**(3), 639–675.

Lehmann, D., Müller, R. & Sandholm, T. (2005), The winner determination problem, in R. Steinberg, Y. Shoham & P. Cramton, eds, 'Combinatorial Auctions', MIT Press.

Lehmann, D., O'Callaghan, L. & Shoham, Y. (1999), 'Truth revelation in approximately efficient combinatorial auctions', *Journal of the ACM* **49**(4), 577–602.

Lesser, V. R. (1995), 'Multiagent systems: An emerging subdiscipline of ai', *ACM Computing Surveys* **27**(3), 340–342.

Levine, D. (1994), A Parallel Genetic Algorithm for the Set Partitioning Problem, PhD thesis, Argonne National Laboratory, Argonne, IL.

Levy, L., Blumrosen, L. & Nisan, N. (2001), On line markets for distributed object services: the majic system, *in* 'Proceedings of the 3rd USENIX Symposium on Internet Technologies and Systems USITS 2001, San Francisco, CA', USENIX, pp. 85–96.

Leyton-Brown, K., Pearson, M. & Shoham, Y. (2000), Towards a universal test suite for combinatorial auction algorithms, *in* 'ACM Conference on Electronic Commerce', pp. 66–76.

Likhodedov, A. & Sandholm, T. (2005), Approximating revenue-maximizing combinatorial auctions, *in* 'Proceedings of the National Conference on Artificial Intelligence (AAAI), Pittsburgh, PA'.

Littman, M. (1994), Markov games as a framework for multi-agent reinforcement learning, *in* 'Proceedings of the Eleventh International Conference on Machine Learning, San Francisco, CA', Morgan Kaufmann, pp. 157–163.

MacKie-Mason, J. K., Murphy, L. & Murphy, J. (1996), The role of responsive pricing in the internet, *in* J. Bailey & L. McKnight, eds, 'Internet Economics', MIT Press, pp. 279–304.

MacKie-Mason, J. K. & Varian, H. R. (1995), 'Pricing congestible network resources', *IEEE Journal on Selected Areas in Communications* **13**(7), 1141–1149.

Maes, P., Guttman, R. & Moukas, A. G. (1999), 'Agents that buy and sell', *Communications of the ACM* **42**(3), 81–91.

Maheswaran, R. T. & Basar, T. (2001), Decentralized network resource allocation as a repeated noncooperative market game, *in* 'Proceedings of the 40th IEEE Conference on Decision and Control', Orlando, Florida, pp. 4565–4570.

Malinvaud, E. (1974), *Lectures on Microeconomic Theory*, 4th edn, North Holland, Amsterdam, The Netherlands.

Malone, T., Fikes, R. E., Grant, K. R. & Howard, M. T. (1988), Enterprise computation, *in* B. Huberman, ed., 'The Ecology of Computation', Elsevier Science Publishers, North-Holland, pp. 177–205.

Martin, B. (2004), Combinatorial aspects of yield management, a reinforcement learning approach, *in* 'Proceedings of The 2004 European Simulation and Modelling Conference ESM2004', Magdeburg, Germany.

Mas-Colell, A., Whinston, M. & Green, J. (1995), *Microeconomic Theory*, Oxford University Press, Oxford.

McAfee, P. & McMillan, J. (1987), 'Auctions and bidding', *Journal of Economic Literature* **25**, 699–738.

McCarthy, J. (2004), 'What is artificial intelligence?', http://www-formal.stanford.edu/jmc/whatisai/whatisai.html. Accessed 11.5.2005.

McMillan, J. (1994), 'Selling spectrum rights', *Journal of Economic Perspectives* **8**(3), 145–162.

McMillan, J. (1995), 'Why auction the spectrum?', *Telecommunications Policy* **19**, 191–199.

Mehra, P. & Wah, B. W. (1993), Population-based learning of load balancing policies for a distributed computer system, *in* 'Proceedings of the Computing in Aerospace 9 Conference Oct. 1993, San Diego, CA', American Institute of Aeronautics and Astronautics, pp. 1120–1130.

Mehra, P. & Wah, B. W. (1997), 'Automated learning of load-balancing strategies in multiprogrammed distributed systems', *International Journal of System Sciences* **28**(11), 1077–1100.

Metropolis, N., Rosenbluth, A. W., Rosenbluth, M. N., Teller, A. H. & Teller, E. (1953), 'Equation of state calculations by fast computing machines', *Journal of Chem. Phys.* **21**(6), 1087–1092.

Milgrom, P. (1989), 'Auctions and bidding: A primer', *Journal of Economic Perspectives* **3**(3), 3–22.

Milgrom, P. (2000), 'Putting auction theory to work: The simultaneous ascending auction', *Journal of Political Economy* **108**(2), 245–272.

Milgrom, P. (2004), *Putting Auction Theory to Work*, Cambridge University Press.

Miller, M. S. & Drexler, K. E. (1988a), Comparative ecology: A computational perspective, *in* B. A. Huberman, ed., 'Ecology of Computation', Elsevier Science Publishers, Amsterdam, pp. 51–76.

Miller, M. S. & Drexler, K. E. (1988b), Markets and computation: Agoric open systems, *in* B. A. Huberman, ed., 'The Ecology of Computation', Elsevier Science Publishers, North-Holland, pp. 133–176.

Miller, M. S., Krieger, D., Hardy, N., Hibbert, C. & Tribble, E. D. (1996), An automated auction in ATM network bandwidth, *in* 'Market-based control: A Paradigm for Distributed Resource Allocation', World Scientific Publishing Co., Inc., River Edge, NJ, USA, pp. 96–125.

Morris, J., Ree, P. & Maes, P. (2000), Sardine: Dynamic seller strategies in an auction marketplace, *in* 'Proceedings of the 2nd ACM Conference on Electronic Commerce (EC-00)', Minneapolis, MN, USA, pp. 128–134.

Mu'alem, A. & Nisan, N. (2002), Truthful approximation mechanisms for restricted combinatorial auctions, *in* 'Proceedings of the Eighteenth National Conference on Artificial Intelligence and Fourteenth Conference on Innovative Applications of Artificial Intelligence, Edmonton, Alberta, Canada. AAAI Press, 2002', pp. 379–384.

Mui, L., Mohtashemi, M. & Halberstadt, A. (2002), A computational model of trust and reputation, *in* 'Proceedings of the 35th Hawaii International Conference on System Science (HICSS), Big Island'.

Mullen, T. & Wellman, M. P. (1995), A simple computational market for network information services, *in* 'Proceedings of the First International Conference on Multiagent Systems', AAAI Press / MIT Press, San Francisco, CA, pp. 283–189.

Mullen, T. & Wellman, M. P. (1996), Market-based negotiation for digital library services, *in* 'Proceedings of the Second USENIX Workshop on Electronic Commerce', Oakland, CA.

Nair, S. & Bapna, R. (2001), 'An application of yield management for internet providers', *Naval Research Logistics* **48**, 348–362.

Narahari, Y., Raju, V. L. P. & Ravikumar, K. (2005), 'Dynamic pricing models for electronic business', *Sadhana, Indian Academy of Sciences Proceedings in Engineering Sciences* . Special Issue on Electronic Commerce and Electronic Business.

Nash, J. F. (1950), 'The bargaining problem', *Econometrica* **18**, 155–162.

Neumann, D., Holtmann, C. & Orwat, C. (2006), 'Grid-economics', *Wirtschaftsinformatik* **48**(3), 206–209.

Ng, C., Parkes, D. C. & Seltzer, M. (2003a), Strategyproof computing: Systems infrastructures for self-interested parties, *in* 'Proceedings of the 1st Workshop on the Economics of Peer-to-Peer Systems, Berkeley, CA'.

Ng, C., Parkes, D. C. & Seltzer, M. (2003b), Virtual worlds: Fast and strategyproof auctions for dynamic resource allocation, *in* 'Proceedings of the third ACM

Conference on Electronic Commerce (EC-2003), San Diego, CA', ACM, pp. 238–239.

Nieschlag, R., Dichtel, E. & Hörschgen, H. (1994), *Marketing*, 17th edn, Duncker & Humboldt, Berlin, Germany.

Nisan, N. (2000), Bidding and allocation in combinatorial auctions, *in* 'Proceedings of the 2.nd ACM Conference on Electronic Commerce (ACM EC'00), Minneapolis, MN', ACM, pp. 1–12.

Nisan, N. (2005), Bidding languages, *in* R. Steinberg, Y. Shoham & P. Cramton, eds, 'Combinatorial Auctions', MIT-Press.

Nisan, N., Mu'alem, A. & Lavi, R. (2003), Towards a characterization of truthful combinatorial auctions, *in* 'Proceedings of the 44th Annual IEEE Symposium on Foundations of Computer Science 2003, Cambridge, MA, USA'.

Nollau, V. & Hahnewald-Busch, A. (1978), 'Approximating general markovian decision problems by clustering their state- and action-spaces', *Math. Operationsforschung und Statistik: Ser. Optimization* **9**, 109–117.

Norman, B. A. (1995), Scheduling Using the Random Keys Genetic Algorithm, unpublished PhD thesis, University of Michigan, Ann Arbor, Michigan.

Norman, B. A. & Bean, J. C. (1994), Random keys genetic algorithm for job shop scheduling, Tech. Rep. No. 94-5, The University of Michigan, Ann Arbor, MI.

Norman, B. A. & Bean, J. C. (1997), Operation sequencing and tool assignment for multiple spindle CNC machines, *in* 'Proceedings of the Forth International Conference on Evolutionary Computation', IEEE, Piscataway, NJ, pp. 425–430.

Norman, B. A. & Bean, J. C. (2000), 'Scheduling operations on parallel machines', *IIE Transactions* **32**(5), 449–459.

Norman, B. A., Smith, A. E. & Arapoglu, R. A. (1998), Integrated facility design using an evolutionary approach with a subordinate network algorithm, *in* A. E. Eiben, T. Bäck, M. Schoenauer & H.-P. Schwefel, eds, 'Parallel Problem Solving from Nature, PPSN V', Springer-Verlag, Berlin, pp. 937–946.

Odell, J. H., van Dyke Parunak, H. & Bauer, B. (2000), Extending uml for agents, *in* 'Proceedings of the Agent-Oriented Information Systems Workshop at the 17th National Conference on Artificial Intelligence', pp. 3–17.

Odell, J., Parunak, V. D. & Bauer, B. (2001), Representing agent interaction protocols in UML, *in* P. Ciancarini, & M. J. Wooldridge, eds, 'Agent-Oriented Software Engineering', Springer-Verlag, Berlin, Germany, pp. 121–140.

Oppenheimer, D., Albrecht, J., Patterson, D. & Vahdat, A. (2005), Design and implementation tradeoffs forwide-area resource discovery, *in* 'Proccedings of the 14th IEEE International Symposium, Research Triangle Park, North Carolina'.

Osborne, M. J. & Rubinstein, A. (1994), *A Course in Game Theory*, MIT Press, Cambridge, MA.

Ostwald, J. & Lesser, V. R. (2004), Combinatorial auctions for resource allocation in a distributed sensor network, Computer science technical report, University of Massachusetts.

Papadimitriou, C. H. (1981), 'On the complexity of integer programming', *Journal of the ACM* **28**(4), 765–768.

Papadimitriou, C. H. (1994), 'On the complexity of the parity argument and other inefficient proofs of existence', *Journal of Computer and Systems Sciences* **48**(3), 498–532.

Papadimitriou, C. H. (2001), Algorithms, games, and the internet, *in* 'Proceedings on 33rd Annual ACM Symposium on Theory of Computing', pp. 749–753.

Papadimitriou, C. H. & Steiglitz, K. (1998), *Combinatorial Optimization: Algorithms and Complexity*, Dover Publications, Mineola, NY.

Papastavrou, J., Rajagopalan, S. & Kleywegt, A. (1996), 'The dynamic and stochastic knapsack problem with deadlines', *Management Science* **42**, 1706–1718.

Pareto, V. (1906), *Manuale di economia politica con una introduzione alla scienza sociale*, Società Editrice Libraria, Milano.

Parkes, D. C. (1999), iBundle: An efficient ascending price bundle auction, *in* 'Proceedings of the First ACM Conference on Electronic Commerce (ACM EC'99), Denver, CO', ACM, pp. 148–157.

Parkes, D. C. (2001*a*), Iterative Combinatorial Auctions: Achieving Economic and Computational Efficiency, PhD thesis, Department of Computer and Information Science, University of Pennsylvania.

Parkes, D. C. (2001*b*), An Iterative Generalized Vickrey Auction: Strategy-proofness without complete revelation, *in* 'Proceedings of the AAAI Spring Symposium on Game Theoretic and Decision Theoretic Agents, Stanford, CA', AAAI.

Parkes, D. C. (2005), Iterative combinatorial auctions, *in* R. Steinberg, Y. Shoham & P. Cramton, eds, 'Combinatorial Auctions', MIT Press, pp. 41–77.

Parkes, D. C. & Kalagnanam, J. (2005), 'Models for iterative multiattribute vickrey auctions', *Management Science* **51**, 435–451. Special Issue on Electronic Markets.

Parkes, D. C. & Shneidman, J. (2004), Distributed implementations of vickrey-clarke-groves mechanisms, *in* 'Proceedings of the 3rd International Joint Conference on Autonomous Agents and Multi Agent Systems, Bologna, Italy', pp. 261–268.

Parkes, D. C. & Ungar, L. H. (2000), Iterative combinatorial auctions: Theory and practice, *in* 'Proceedings of the 17th National Conference on Artificial Intelligence (AAAI-00)', pp. 74–81.

Poggi, A. & Bergenti, F. (2001), A development toolkit to realize autonomous and inter-operable agents, *in* 'Proceedings of the the the 5th International Conference on Autonomous Agents (Agents 2001), Montreal, Canada', pp. 632–639.

Polk, C. & Schulman, E. (2000), Enhancing the liquidity of bond trading, *in* G. Russell & J. Rosen, eds, 'The Handbook of Fixed Income Technology', Summit Group Press.

Porter, D., Rassenti, S. J. & Smith, V. L. (2003), Combinatorial auction design, Technical report, Interdisciplinary Center for Economic Science.

Principe, J. C., Euliano, N. R. & Lefebvre, W. C. (1999), *Neural and Adaptive Systems: Fundamentals through Simulations*, Wiley & Sons, New York, NY.

Puterman, M. L. (1994), *Markov Decision Problems*, John Wiley and Sons, New York, NY.

Raju, V. L. P., Narahari, Y. & Kumar, R. (2005), 'Learning dynamic prices in multi-seller electronic markets with with price sensitive customers, stochastic demands, and inventory replenishments', *IEEE Transactions on Systems, Man, and Cybernetics, Part C*. Special Issue on Game-theoretic Analysis and Stochastic Simulation of Negotiation Agents.

Rassenti, J. S., Smith, V. L. & Bulfin, R. L. (1982), 'A combinatorial auction mechanism for airport time slot allocation', *The Bell Journal of Economics* **13**(2), 402–417.

Regev, O. & Nisan, N. (1998), The popcorn market: an online market for computational resources, *in* 'Proceedings of the first international conference on Information and computation economies', Charleston, SC, pp. 148 – 157.

Reinartz, W. (2001), 'Customising prices in online markets', *European Business Forum* **6**, 35–41.

Reinicke, M., Eymann, T., Ardaiz, O., Artigas, P., Freitag, F. & Navarro, L. (2003), Self-organizing resource allocation for autonomic networks, *in* '14th International Workshop on Database and Expert Systems Applications'.

Riedmiller, S. C. & Riedmiller, M. A. (1999), A neural reinforcement learning approach to learn local dispatching policies in production scheduling, *in* 'Proceedings of the International Joint Conference on Artificial Intelligence (IJCAI'99), Stockholm, Sweden', pp. 764–771.

Ronen, A. & Nisan, N. (2000), Computationally feasible vcg mechanisms, *in* 'Proceedings of the 2nd ACM Conference on Electronic Commerce (ACM EC-00), Minneapolis', pp. 242–252.

Rosenschein, J. & Zlotkin, G. (1994), *Rules of Encounter*, MIT Press, Cambridge, MA.

Rothkopf, M. H., Pekeč, A. & Harstad, R. M. (1998), 'Computationally manageable combinatorial auctions', *Management Science* **44**(8), 1131–1147.

Rothlauf, F., Goldberg, D. E. & Heinzl, A. (2002), 'Network random keys – A tree network representation scheme for genetic and evolutionary algorithms', *Evolutionary Computation* **10**(1), 75–97.

Roughgarden, T. (2002), How unfair is optimal routing?, *in* 'Proceedings of the 13. Annual ACM-SIAM Symposium on Discrete Algorithms', pp. 203–204.

Russel, S. & Norvig, P. (2003), *Artificial Intelligence: A Modern Approach*, 2 edn, Prentice Hall, New Jersey.

Sackmann, S. & Strüker, J. (2005), *Electronic Commerce Enquête 2005 - 10 Jahre Electronic Commerce: Eine stille Revolution in deutschen Unternehmen*, KIT Verlag, Institut für Informatik und Gesellschaft, Telematik, Universität Freiburg.

Sandholm, T. (1993), An implementation of the contract net protocol based on marginal cost calculations, *in* 'Proceedings of the Eleventh National Conference on Artificial Intelligence, Washington, DC', pp. 256–262.

Sandholm, T. (1995), Distributed rational decision making, *in* G. Weiss, ed., 'Multiagent Systems', MIT Press, Massachusetts, pp. 201–258.

Sandholm, T. (1996), Negotiation among Self-Interested Computationally Limited Agents, PhD thesis, University of Massachusetts, Amherst, MA.

Sandholm, T. (2002*a*), 'Algorithm for optimal winner determination in combinatorial auctions', *Artificial Intelligence* **135**(1-2), 1–54.

Sandholm, T. (2002*b*), 'eMediator: A next generation electronic commerce server', *Computational Intelligence* **18**(4), 656–676.

Sandholm, T. (2005), Optimal winner determination algorithms, *in* R. Steinberg, Y. Shoham & P. Cramton, eds, 'Combinatorial Auctions', MIT Press.

Sandholm, T. & Lesser, V. R. (1997), 'Coalitions among computationally bounded agents', *Artificial Intelligence, special issue on Economic Principles of Multiagent Systems* **94**(1-2), 99–137.

Sandholm, T., Suri, S., Gilpin, A. & Levine, D. (2005), 'CABOB: A fast optimal algorithm for combinatorial auctions', *Management Science* **51**(3), 374–390.

Satterthwaite, M. A. & Williams, S. R. (1989), 'Bilateral trade with the sealed bid k-double auction: Existence and efficiency', *Journal of Economic Theory* **48**, 107–133.

Savit, R., Brueckner, S. A., van Dyke Parunak, H. & Sauter, J. A. (2003), 'Phase structure of resource allocation games', *Physics Letters A* **311**(4-5), 359364,.

Scarf, H. (1982), The computation of equilibrium prices: An exposition, *in* M. Intriligator & K. J. Arrow, eds, 'Handbook of Mathematical Economics', Vol. 2, North-Holland Publishing, Amsterdam, chapter 21, pp. 1007–1061.

Scarf, H. (1994), 'The allocation of resources in the presence of indivisibilities', *Journal of Economic Perspectives* **8**(4), 111–128.

Scarf, H. F. & Hansen, T. (1973), *Computation of Economic Equilibrium*, Yale University Press, New Haven, Connecticut.

Schaerf, A., Shoham, Y. & Tennenholtz, M. (1995), 'Adaptive load balancing: A study in multi-agent learning', *Journal of Artificial Intelligence Research* **2**, 475–500.

Schillo, M., Bürckert, H.-J., Fischer, K. & Klusch, M. (2001), Towards a definition of robustness for market-style open multi-agent systems, *in* 'Proceedings of the fifth international conference on Autonomous agents (Agents '01)', ACM Press, Montreal, Quebec, Canada.

Schneider, J. G., Boyan, J. A. & Moore, A. W. (1998), Value function based production scheduling, *in* 'Proceedings 15th International Conference on Machine Learning', Morgan Kaufmann, San Francisco, CA, pp. 522–530.

Schumpeter, J. (1908), *Das Wesen und der Hauptinhalt der Theoretischen Nationalökonomie*, 3. Auflage, 1998 edn, Duncker & Humboldt Berlin.

Schumpeter, J. (1911), *Theorie der wirtschaftlichen Entwicklung*, 3. Auflage, 1997 edn, Duncker & Humboldt Berlin.

Schumpeter, J. (1950), *Kapitalismus, Sozialismus und Demokratie*, UTB für Wissenschaft, 7. Auflage, 1998 edn, A. Francke Verlag, Tübingen und Basel.

Schwefel, H.-P. (1995), *Evolution and Optimum Seeking*, Wiley-Interscience, New York, NY.

Schwind, M. (2003), Bekräftigungslernen, *in* W. König, H. Rommelfanger, D. Ohse, O. Wendt, M. Hoffmann, M. Schwind, K. Schäfer, H. Kuhnle & A. Pfeifer, eds, 'Taschenbuch der Wirtschaftsinformatik und Wirtschaftsmathematik', 2nd edn, Harri Deutsch, Frankfurt a. M., Thun, pp. 188–197.

Schwind, M. (2005a), Design of combinatorial auctions for allocation and procurement processes, *in* '7th International IEEE Conference on E-Commerce Technology 2005, München, Germany'.

Schwind, M. (2005b), Report on the NUI Galway Masterclass in computable and behavioural economics, COBERA 2005, Technical report, National University of Galway and Institute for Information Science Frankfurt.

Schwind, M. & Grolik, S. (2003), Softwareagenten, *in* W. König, H. Rommelfanger, D. Ohse, O. Wendt, M. Hoffmann, M. Schwind, K. Schäfer, H. Kuhnle & A. Pfeifer, eds, 'Taschenbuch der Wirtschaftsinformatik und Wirtschaftsmathematik', 2nd edn, Harri Deutsch, Frankfurt a. M., Thun, pp. 340–345.

Schwind, M. & Gujo, O. (2006), Using shadow prices for resource allocation in a grid with proxy-bidding agents, *in* 'Proceedings of the 8th International Conference on Enterprise Information Systems (ICEIS 2006), Paphos, Cyprus'.

Schwind, M., Gujo, O. & Stockheim, T. (2006), Dynamic resource prices in a combinatorial grid system, *in* 'Proceedings of the IEEE Joint Conference on E-Commerce Technology (CEC'06) and Enterprise Computing, E-Commerce and E-Services (EEE'06), San Francisco, CA'.

Schwind, M. & Meyer, S. (2001), Bekräftigungslernende Agenten als Basis für die gewinnmaximale dynamische Bepreisung von Informationsprodukten im Internet, Technical report, Institut für Wirtschaftsinformatik, Johann Wolfgang Goethe Universität, Frankfurt.

Schwind, M., Stockheim, T. & Gujo, O. (2006), Agents' bidding strategies in a combinatorial auction controlled grid environment, *in* 'Proceedings of the AAMAS 2006 Trading Agent Design and Analysis/Agent-Mediated Electronic Commerce Joint Workshop, Hakodate, Japan'.

Schwind, M., Stockheim, T. & Rothlauf, F. (2003), Optimization heuristics for the combinatorial auction problem, *in* 'Proceedings of the Congress on Evolutionary Computation CEC 2003', pp. 1588–1595.

Schwind, M., Stockheim, T. & Seibel, S. (2003), Price controlled resource allocation for the provision of information products and services employing combinatorial auctions, *in* 'Proceedings of the 11th European Conference on Information Systems (ECIS 2003), Naples Italy'.

Schwind, M., Weiss, K. & Stockheim, T. (2004), CAMeL - Eine Meta-Sprache für Kombinatorische Auktionen, Technical report, Institut für Wirtschaftsinformatik, Johann Wolfgang Goethe Universität, Frankfurt.

Schwind, M. & Wendt, O. (2002), Dynamic pricing of information products based on reinforcement learning: A yield-management approach, *in* M. Jarke, J. Koehler & G. Lakemeyer, eds, 'KI 2002: Advances in Artificial Intelligence, 25th Annual German Conference on AI (KI 2002), Aachen, Germany', Vol. 2479 of *Lecture Notes in Computer Science*, Springer Verlag, Berlin, Germany, pp. 51–66.

Simon, H. (1992), *Preismanagement*, 2nd edn, Gabler, Wiesbaden, Germany.

Simsek, B., Albayrak, S. & Korth, A. (2004), Reinforcement learning for procurement agents of the factory of the future, *in* 'Proceedings of the Conference of Evolutionary Computation CEC 2004, Portland, OR'.

Smale, S. (1975), Price adjustment and global newton method, *in* 'Contributions to Economic Analysis', North-Holland Publishing, pp. 191–205.

Smale, S. (1976), 'Dynamics in general equilibrium theory', *American Economic Review* **66**(2), 288–294.

Smith, B., Leimkuhler, J. & Darrow, R. (1992), 'Yield management at american airlines', *Interfaces* **1**, 8–31.

Smith, V. L. (1994), 'Economics in laboratory', *The Journal of Economic Perspectives* **8**(1), 113–131.

Snoek, M. (2000), Neuro-genetic order acceptance in a job shop setting, *in* 'Proceedings of the 7th International Conference on Neural Information Processing (ICONIP), Taejon, Korea', pp. 815–819.

Sowa, J. (2000), *Knowledge Representation: Logical, Philosophical and Computational Foundations*, Brookes/Cole, Pacific Groove, CA.

Spann, M., Skiera, B. & Schäfers, B. (2004), 'Measuring individual frictional costs and willingness-to-pay via name-your-own price mechanisms', *Journal of Interactive Marketing* **18**(4), 22–36.

Sridharan, M. & Tesauro, G. (2000), Multi-agent q-learning and regression trees for automated pricing decisions, *in* 'Proceedings of the 4th International Conference on Multi-Agent Systems (ICMAS 2000), Boston, MA', IEEE Computer Society, pp. 447–448.

Steinberg, R. (2000), A general combinatorial auction procedure, Technical Report WP 17/00, Judge Institute of Management, University of Cambridge.

Stockheim, T. & Schwind, M. (2004), Agent-based scheduling in supply chain management, *in* 'Proceedings of the Second European Workshop on Multi-Agent System EUMAS 2004, Barcelona, Spain'.

Stockheim, T., Schwind, M. & Gujo, O. (2006), Agent's bidding strategies in a combinatorial auction, *in* K. Fischer, I. Timm, E. Andre & N. Zhong, eds, 'Multiagent System Technologies', Vol. 4196 of *Lecture Notes in Artificial Intelligence*, Springer, Heidelberg, pp. 37–48.

Stockheim, T., Schwind, M. & König, W. (2003*a*), A model for emergence and diffusion of software standards, *in* 'Proceedings of the 36th Hawaii International Conference on System Sciences (HICSS-36), Big Island, Hawaii'.

Stockheim, T., Schwind, M. & König, W. (2003*b*), A reinforcement learning approach for supply chain management, *in* 'Proceedings of the First European Workshop on Multi-Agent Systems EUMAS 2003, Oxford, UK'.

Stockheim, T., Schwind, M., Korth, A. & Simsek, B. (2003), Supply chain yield management based on reinforcement learning, Technical Report 2003-77, Chair of Business Administration, esp. Information Systems - Goethe University of Frankfurt.

Stockheim, T., Schwind, M. & Weiss, K. (2006), 'Modeling diffusion of communication standards', *International Journal of IT Standards and Standardization Research* **4**(2), 24–42.

Stockheim, T., Wendt, O. & Schwind, M. (2005), A trust-based negotiation mechanism for decentralized economic scheduling, *in* 'Proceedings of the 38th Hawaiian International Conference on System Sciences (HICSS-38), Hilton Waikoloa Village, Big Island, Hawaii'.

Stoica, I., Abdel-Wahab, H. & Pothen, A. (1995), A microeconomic scheduler for parallel computers, *in* 'Proceedings of the Workshop on Job Scheduling Strategies for Parallel Processing, IPPS'95, Santa Barbara, CA', Vol. 949 of *Lecture Notes in Computer Science*, Springer, pp. 200–218.

Stonebraker, M., Devine, R., Kronacker, M., Litwin, W., Pfeffer, A., Sah, A. & Staelin, C. (1994), An economic paradigm for query processing and data migration in mariposa, Sequoia 2000 94/49, University of California, Berkeley.

Stützle, T. (1999), *Stochastic Local Search-Foundations and Applications*, Infix, Sankt Augustin.

Sundermann, E. & Lemahieu, I. (1995), PET image reconstruction using simulated annealing, *in* 'Proceedings of the SPIE Medical Imaging '95 (Image Processing)', SPIE, pp. 378–386.

Sutton, R. & Barto, A. (1998), *Reinforcement Learning: An Introduction*, MIT-Press, Cambrige, MA.

Sycara, K. (1998), 'Multiagent systems', *AI Magazine* **10**(2), 79–93.

Sycara, K., Paolucci, M., Giampapa, J. & van Velsen, M. (2003), 'The retsina mas infrastructure', *Autonomous Agents and Multi-Agent Systems* **7**(1,2), 29–48.

Talluri, K. & van Ryzin, G. (1998), 'An analysis of bid-price controls for network revenue management', *Management Science* **44**(11), 1577–1593.

Tanner, A. & Mühl, G. (2003), A combinatorial exchange for autonomous traders, *in* 'Proceedings of the 4th International Conference on Electronic Commerce and Web Technologies', LNCS, Springer Verlag.

Tardos, E. & Roughgarden, T. (2002), 'How bad is selfish routing?', *Journal of the ACM (JACM)* **49**, 236–259.

Tesauro, G. (2001), Pricing in agent economies using neural networks and multi-agent q-learning, *in* C. L. Giles & R. Sun, eds, 'Sequence Learning - Paradigms, Algorithms, and Applications', Vol. 1828 of *Lecture Notes in Computer Science*, Springer, pp. 288–307.

Tesauro, G. & Kephart, J. (2001), 'Pricing in agent economies using multi-agent q-learning', *Autonomous Agents and Multi-Agent Systems* **5**(3), 289–304.

Tesfatsion, L. S. (2001), 'Introduction to the computational economics (special issue on ACE)', *Computational Economics* **18**(1), 1–8.

Tesfatsion, L. S. (2006), A constructive approach to economic theory, *in* K. L. Judd & L. S. Tesfatsion, eds, 'Handbook of Computational Economics', Vol. Agent-Based Computational Economics of *Handbooks in Economic Series*, North-Holland.

Tsvetovat, M., Sycara, K., Chen, Y. & Ying, J. (2000), Customer coalistions in the electronic marketplace, *in* 'Proceedings of the Fourth International Conference on Autonomous Agents, Barcelona, Spain', ACM Press.

Tucker, P. & Berman, F. (1996), On market mechanisms as a software technique, Technical report, Technical report, University of California, San Diego TR CS96-513.

Ulph, D. & Vulkan, N. (2000), E-commerce, mass customisation and price discrimination, Technical Report 00/489, School of Economics, Finance and Management, University of Bristol, UK.

Uzawa, H. (1962), 'Walras' existence theorem and brower's fixed point theorem', *The Economic Studies Quarterly* **8**(1), 59–62.

van Ryzin, G. & McGill, J. (1999), 'Revenue management: Research overview and prospects', *Transportation Science* **33**(2), 233–256.

van Slyke, R. & Young, Y. (2000), 'Finite horizon stochastic knapsacks with applications to yield management', *Operations Research* **48**(1), 155–172.

Varian, H. R. (1995), Economic mechanism design for computerized agents, *in* 'Proceedings of the first USENIX Workshop on Electronic Commerce', USENIX, pp. 13–21.

Varian, H. R. (1996), 'Differential pricing and efficiency', http://www.firstmonday.org/issues/issue2/different/.

Varian, H. R. & MacKie-Mason, J. (1994), Generalized Vickrey Auctions, Technical report, University of Michigan, Ann Arbor, MI.

Velupillai, V. (2004*a*), The foundations of computable general equilibrium theory, Technical report, Deparment of Economics, National University of Ireland, Galway.

Velupillai, V. (2004*b*), The unreasonable ineffectiveness of mathematics in economics, Technical Report 6, University of Trento, Italy, Department of Economics.

Vengerov, D. (2005), A reinforcement learning framework for utility-based scheduling in resource-constrained systems, Technical Report TR-2005-141, Sun Microsystems Labs Institute, Menlo Park.

Vickrey, W. (1963), 'Counterspeculation, auctions, and competitive sealed tenders', *Journal of Finance* **16**(1), 8–37.

Vriend, N. J. (2002), 'Was hayek an ace?', *Southern Economic Journal* **68**(4), 811–840.

Vriend, N. J. (2004), 'Ace models of market organization', *Revue d'Economie Industrielle* **107**, 63–74.

Vries, S. D. & Vohra, R. (2001), 'Combinatorial auctions: A survey', *INFORMS Journal on Computing* **15**(3), 284–309.

Waldspurger, C. A. (1995), Lottery and Stride Scheduling: Flexible Proportional-Share Resource Management, PhD thesis, Massachusetts Institute of Technology.

Waldspurger, C. A., Hogg, T., Huberman, B. A., Kephart, J. & Stornetta, S. (1991), 'Spawn: A distributed computational economy', *Software Engineering* **18**(2), 103–117.

Waldspurger, C. A. & Weihl, W. (1994), Lottery scheduling: Flexible proportional-share resource mangement, *in* 'Proceedings of the First Symposium on Operating Systems Design and Implementation (OSDI '94), Monterey, CA', pp. 1–11.

Waldspurger, C. A. & Weihl, W. (1995), Stride scheduling: Deterministic proportional share resource managemen, Technical Report MIT/LCS/TM-528, Massachusetts Institute of Technology, Cambridge, MA.

Walras, L. (1896), *Élements d'économie politique pure ou Théorie de la richesse sociale*, Guillaimin et Cie, Paris.

Watkins, C. J. (1989), Learning from Delayed Rewards, PhD thesis, Cambridge University, Cambridge, MA.

Weatherford, L. & Bodily, S. (1992), 'A taxonomy and research overview of perishable-asset revenue management: Yield management, overbooking and pricing', *Operations Research* **40**, 831–844.

Weiss, G. & Sen, S. (1996), *Adaption and learning in multi-agent systems*, Lecture Notes in Artificial Intelligence, Springer-Verlag, Berlin.

Weiss, R. M. & Mehrotra, A. (2001), 'Online dynamic pricing: Efficiency, equity and the future of e-commerce', *Virginia Journal of Law and Technology* **6**(2), 1–6.

Wellman, M. P. (1993), 'A market-oriented programming environment and its application to distributed multicommodity flow problems', *Journal of Artificial Intelligence Research* **1**, 1–23.

Wellman, M. P. (1994), A computational market model for distributed configuration design, *in* 'In Proceedings of the 12th National Conference on Artificial Intelligence (AAAI-94), Seattle, Washington', pp. 401–407.

Wellman, M. P. (1995), Market-oriented programming: Some early lessons, *in* S. H. Clearwater, ed., 'Market-Based Control: A Paradigm for Distributed Resource Allocation', World Scientific Publishing, Singapore.

Wellman, M., Walsh, W., Wurman, P. & MacKie-Mason, J. (2001), 'Auction protocols for decentralized scheduling', *Games and Economic Behavior* **35**(1/2), 271–303.

Wendt, O. (1995), *Tourenplanung durch den Einsatz naturanaloger Verfahren: Integration von genetischen Algorithmen und Simulated Annealing*, Gabler Verlag, Wiesbaden, Germany.

Wendt, O., Stockheim, T., Grolik, S. & Schwind, M. (2002), Distributed ontology management: Prospects and pitfalls on our way towards a web of ontologies, *in* 'Dagstuhl Workshop (Event Nr. 02212) - DFG-SPP 1083 Intelligente Softwareagenten und betriebswirtschaftliche Anwendungsszenarien'.

Wirtz, J., Kimes, S. E., Theng, J. H. P. & Patterson, P. (2003), 'Revenue management: Resolving potential customer conflicts', *Journal of Revenue and Pricing Management* **2**, 216–226.

Wolfstetter, E. (1996), 'Auctions: An introduction', *Journal of Economic Surveys* **10**, 367–420.

Wolski, R., Plank, J. S., Brevik, J. & Bryan, T. (2001), 'Analyzing market-based resource allocation strategies for the computational grid', *The International Journal of High Performance Computing Applications* **15**(3), 258–281.

Wolski, R., Plank, J. S., Bryan, T. & Brevik, J. (2001), G-commerce: Market formulations controlling resource allocation on the computational grid, *in* 'Proceedings of the 15th International Parallel & Distributed Processing Symposium (IPDPS-01), San Francisco, CA', IEEE Computer Society, pp. 46–55.

Wooldrige, M. J. & Jennings, N. (1995), 'Intelligent agents: Theory and practice', *The Knowledge Engineering Review* **10**(2), 115–152.

Wurman, P. R. & Wellman, M. P. (2000), AkBA: A progressive, anonymous-price combinatorial auction, *in* 'Proceedings of the Second ACM Conference on Electronic Commerce, Minneapolis', pp. 21–29.

Wurman, P. R., Wellman, M. P. & Walsh, W. E. (1998), The Michigan Internet AuctionBot: A configurable auction server for human and software agents, *in* 'Proceedings of the Second International Conference on Autonomous Agents, Minneapolis'.

Xia, M., Koehler, G. J. & Whinston, A. B. (2004), 'Pricing combinatorial auctions', *European Journal of Operational Research* **154**(1), 251–270.

Xiong, L. & Liu, L. (2003), A reputation-based trust model for peer-to-peer ecommerce communities, *in* 'Proceedings of the IEEE Conference on E-Commerce (CEC'03)'.

Yaiche, H., Mazumdar, R. R. & Rosenberg, C. (2000), 'A game-theoretic framework for bandwidth allocation and pricing in broadband networks', *EEE ACM Transactions on Networking* **8**(5), 667–678.

Ygge, F. (1998), Market-Oriented Programming and its Application to Power Load Management, PhD thesis, Lund University.

Ygge, F. & Akkermans, H. (1999), 'Decentralized markets versus central control: A comparative study', *Journal of Artificial Intelligence Research* **11**, 301–333.

Ygge, F. & Akkermans, H. (2000), 'Resource-oriented multicommodity market algorithms', *Autonomous Agents and Multi-Agent Systems* **3**(1), 53–71.

Ygge, F., Wellman, M. P. & Walsh, W. E. (2000), Combinatorial auctions for supply chain formation, *in* 'Proc. ACM Conference on Electronic Commerce (ACM EC'00)', Minneapolis, Minnesota, pp. 260–269.

Zhang, W. (1996), Reinforcement Learning for Job-Shop Scheduling, PhD thesis, Oregon State University, Corvallis, OR.

Zhang, W. (1999), *State-Space Search: Algorithms, Complexity, Extensions, and Applications*, Springer, Berlin.

Zhang, W. & Dietterich, T. G. (1995), A reinforcement learning approach to jobshop scheduling, *in* 'Proceedings of the 14th International Joint Conference on Artificial Intelligence (IJCAI-95)', Morgan Kaufmann, Orlando, FL, pp. 1114–1120.

Zhang, W. & Dietterich, T. G. (2000), 'Solving combinatorial optimization tasks by reinforcement learning: A general methodology applied to resource-constrained scheduling', *Submitted to the Journal of Artificial Intelligence Research* .

Zlotkin, G. & Rosenschein, J. S. (1992), A domain theory for task oriented negotiation, Technical Report 92-13, Leibniz Center for Computer Science, Hebrew University, Jerusalem, Israel.

Zlotkin, G. & Rosenschein, J. S. (1993), Compromise in negotiation: Exploiting worth functions over states, Technical Report 93-3, Leibniz Center for Computer Science, Hebrew University.

Zlotkin, G. & Rosenschein, J. S. (1996), 'Mechanism for automated negotiation in state oriented domains', *Journal of Artificial Intelligence Research* **5**, 163–238.

Zurel, E. & Nisan, N. (2001), An efficient approximate allocation algorithm for combinatorial auctions, *in* 'Proceedings of the third annual ACM Conference on Electronic Commerce (ACM EC'01), Tampa, FL', pp. 125–136.

Zweben, M., Daun, B. & Deal, M. (1994), Scheduling and rescheduling with iterative repair, *in* M. Zweben & M. S. Fox, eds, 'Intelligent Scheduling', Morgan Kaufmann, Orlando, FL, chapter 8, pp. 121–140.

Zweben, M., Daun, B., Drascher, E., Deale, M., Eskey, M. & Davis, E. (1992), 'Learning to improve constraint-based scheduling', *Artificial Intelligence* **58**, 271–296.

Index

Lecture Notes in Economics and Mathematical Systems

For information about Vols. 1–496
please contact your bookseller or Springer-Verlag

Printing: Krips bv, Meppel
Binding: Stürtz, Würzburg